New Developments on Fundamental Problems in Quantum Physics

Fundamental Theories of Physics

An International Book Series on The Fundamental Theories of Physics: Their Clarification, Development and Application

Volume 81

New Developments on Fundamental Problems in Quantum Physics

edited by

Miguel Ferrero
University of Oviedo,
Oviedo, Spain

and

Alwyn van der Merwe
University of Denver,
Denver, Colorado, U.S.A.

KLUWER ACADEMIC PUBLISHERS
DORDRECHT / BOSTON / LONDON

A C.I.P. Catalogue record for this book is available from the Library of Congress.

ISBN 0-7923-4374-3

Published by Kluwer Academic Publishers,
P.O. Box 17, 3300 AA Dordrecht, The Netherlands.

Sold and distributed in the U.S.A. and Canada
by Kluwer Academic Publishers,
101 Philip Drive, Norwell, MA 02061, U.S.A.

In all other countries, sold and distributed
by Kluwer Academic Publishers,
P.O. Box 322, 3300 AH Dordrecht, The Netherlands.

Printed on acid-free paper

Printed in the Netherlands

CONTENTS

Photograph reproduced with kind permission from Foto Nebot, Oviedo, Spain.

1. C.González
2. J. Klackauer
3. J. Unturbe
4. P.J. Salas
5. T. Angelidis
6. E. Santos
7. A.L. Sanz
8. Z. Mate
9. J.L. Sánchez-Gómez
10. H.M. França
11. C. Cabrillo
12. M. Plenio
13. J.L. Cereceda
14. R. Risco
15. P. Fleming
16. I. Cirac
17. A. Sanpera
18. L. Hardy
19. B. Ischi
20. I. Gonzalo
21. J. Suárez Capitaine
22. T. Walther
23. G. García Alcaine
24. A. Peres
25. A. Fine
26. S. Fdez Huelga
27. T. Brun
28. J.L. Rosales
29. M. Lorente
30. O. Manko
31. P. Mittelstaedt
32. L. Diósi
33. J. Levitan
34. W. Strunz
35. N.J. Cerf
36. N. Gisin
37. R. Omnès
38. D.J. Miller
39. L.P. Horwitz
40. F. Hasselbach
41. O. Scully
42. M. Mathar
43. O. Cohen
44. A. Ekert
45. J. Ospino
46. P. Kwiat
47. W. lamb
48. J. Rarity
49. A. Cabello
50. A. Lombardi
51. A. Afriat
52. A. Bramon
53. D. Vitali
54. L. Bonci
55. A. Dobado
56. I. Percival
57. M. Nowakoski
58. Z. Onyskiewicz
59. T. Calarco
60. G. García de Polavieja
61. M.A. González
62. H.D. Zeh
63. R. Tarrach
64. F. Selleri
65. C. Ragancharyulu
66. M. Simonius
67. A. Fdez-Rañada.
68. P. Bogar
69. M. Revzen
70. M. Ferrero
71. A. Solana
72. L. Toffolatti
73. N. Corte
74. M. Laucelli
75. J. Pykacz
76. T. M. Marshall
77. J.L. Trueba
78. F. Junquera
79. F. Martín Maroto
80. A.M. Cetto
81. J.J Alvarez
82. F. Sols
83. L. de la Peña
84. J. Martínez-Linares

PREFACE

The *International Symposium on Fundamental Problems in Quantum Physics* was held from July 21 to 26, 1996, in Oviedo, Spain. During these six days, the participants, meeting in the magnificent surroundings of an old baroque building, were able to create the relaxed atmosphere conducive to thoughtful discussions of a spate of experimental techniques newly developed to test some of the stranger predictions of quantum physics, undoubtedly one of the most fascinating and successful constructs in the intellectual history of mankind. The advances under consideration included: recent experiments in quantum optics, electron and ion interferometry, photon down conversion in nonlinear crystals, single trapped ions interacting with laser beams, atom-field coupling in micromaser cavities, quantum computation, quantum cryptography, the quantum statediffusion model, decoherence and macroscopic quantum effects, quantum gravity, the quantum mechanics of cosmology and, finally, quantum nonlocality along with the continuing debate surrounding the interpretation of quantum mechanics.

In the latter context, we were specially honored by the presence of Prof. Willis Lamb, who gave a masterful survey of the history of this interpretation, interlaced with many of his own contributions to the subject.

We would like to express our gratitude to all participants in the *International Symposium* for making this meeting a truly memorable experience and for the excellent quality of the papers they contributed, all of which are now collected in the present volume.

The symposium could not have taken place without the financial support of the following institutions and organizations, to whom we are greatly indebted:

- European Commission, DG XII, Brussels.
- Ministerio de Educacion y Cultura, DGICYT, Madrid.
- Consejería de Cultura del Principado de Asturias, FICYT, Oviedo.
- Fundación Principe de Asturias, Oviedo.
- Caja de Asturias, Oviedo.
- Universidad de Oviedo.
- Universidad Autónoma de Madrid.
- Universidad de Cantabria, Santander.
- Universidad Complutense de Madrid.
- Sociedad Asturiana de Filosofía, Oviedo.

We are also thankful to Professors A. F. Rañada, J. L. Sánchez-Gómez, and E. Santos for their help in turning the organizational activities into an easy exercise. Very special thanks should finally go to Dr. Susana F. Huelga, Ana María Espino, José Manuel G. Sancho, and Catalina Loring for their smooth handling of all the work that daily needed to be done.

The Editors

THE HIDDEN-MEASUREMENT FORMALISM: QUANTUM MECHANICS AS A CONSEQUENCE OF FLUCTUATIONS ON THE MEASUREMENT

Diederik Aerts and Sven Aerts

TENA, Free University of Brussels
Pleinlaan 2, B-1050 Brussels, Belgium

We expose a formalism, that we have called the 'hidden measurement formalism', where the quantum structure is due to the presence of 'fluctuations' on the interaction between the system and the measurement apparatus. In this formalism the quantum mechanical probabilities are not ontological but arrise as a consequence of 'lack of knowledge' about this interaction. We study the quantum classical limit and the EPR problem in the light of this explanation.

Key words: hidden measurements, determinism, paradoxes.

1. HIDDEN MEASUREMENTS

To understand quantum mechanics many have sought to supplement the quantum state with additional variables that would allow a more complete description of the system. One problem associated with these so called hidden variables is that, once we supplement the state of the entity with additional variables, one restores the Kolmogorovian character of the probabilities. In this case a measurement is an observation of an entity and the probabilistic character arises because there is a lack of knowledge about the precise state of the entity. In a sense this explanation is 'too classical' as is demonstrated by the many no-go theorems [11]. Bohr stressed the importance of taking into account the whole experimental setup because the setup would determine in an essential way the possible set of expected outcomes. We want to push this idea into rigorous mathematics and develop a theoretical framework wherein the probabilities of quantum mechanics arise not because of the indefinite character of the state of the entity, but because of the indefinite state of the measurement equipment and the lack of knowledge about the interaction between system and measurement apparatus. Such an

M. Ferrero and A. van der Merwe (eds.),
New Developments on Fundamental Problems in Quantum Physics, 1–6.

approach has been elaborated [1,2,3] by the Brussels group under the following two hypotheses: (1) a measurement induces a 'real' state transition (2) the probability related to an outcome of an experiment is due to a lack of knowledge about the interaction between the measurement apparatus and the entity under observation. Let us first introduce a model that explicitly incorporates these features and constitutes a model of a two-dimensional Hilbert space quantum entity.

2. THE QUANTUM MACHINE

The machine (see also [1,2,3]) that we consider consists of a physical entity S that is a point particle P that can move on the surface of a unit sphere, denoted *surf*. The unit-vector v where the particle is located on *surf* represents the state p_v of the particle (see Fig.1(a)). For each point $u \in surf$, we introduce the following experiment e_u. We consider the diametrically opposite point $-u$, and install a piece of elastic of length 2, such that it is fixed with one of its end-points in u and the other end-point in $-u$. Once the elastic is installed, the particle P falls from its original place v orthogonally onto the elastic, and sticks on it (Fig. 1(b)). Then the elastic breaks and the particle P, attached to one of the two pieces of the elastic (Fig. 1(c)), moves to one of the two end-points u or $-u$ (Fig. 1(d)). Depending on whether the particle P arrives in u (as in Fig. 1) or in $-u$, we give the outcome o_1^u or o_2^u to e_u. We can easily calculate the probabilities corresponding to the two possible outcomes. Therefore we remark that the particle P arrives in u when the elastic breaks in a point of the interval L_1 (which is the length of the piece of the elastic between $-u$ and the point where the particle has arrived, or $1 + \cos\theta$) , and arrives in $-u$ when it breaks in a point of the interval L_2 ($L_2 = L - L_1 = 2 - L_1$). We make the hypothesis that the elastic breaks uniformly, which means that the probability that the particle, being in state p_v, arrives in u, is given by the length of L_1 divided by the length of the total elastic (which is 2). The probability that the particle in state p_v arrives in $-u$ is the length of L_2 (which is $1 - \cos\theta$) divided by the length of the total elastic. If we denote these probabilities respectively by $P(o_1^u, p_v)$ and $P(o_2^u, p_v)$ we have:

$$P(o_1^u, p_v) = \frac{1 + \cos\theta}{2} = \cos^2\frac{\theta}{2}, \tag{1}$$

$$P(o_2^u, p_v) = \frac{1 - \cos\theta}{2} = \sin^2\frac{\theta}{2}, \tag{2}$$

These transition probabilities are the same as the ones related to the outcomes of a Stern-Gerlach spin experiment on a spin -1/2 quantum particle, of which the quantum spin-state in direction $v = (\cos\phi\sin\theta,$

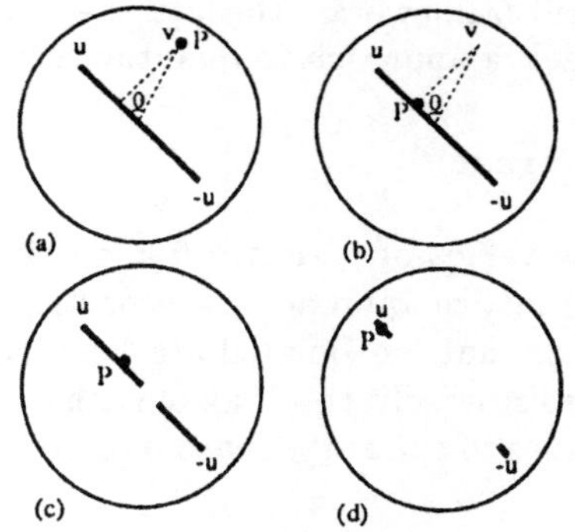

Fig. 1. A representation of the quantum machine. In (a) the physical entity P is in state p_v in the point v, and the elastic corresponding to the experiment e_u is installed between the two diametrically opposed points u and $-u$. In (b) the particle P falls orthogonally onto the elastic and sticks to it. In (c) the elastic breaks and the particle P is pulled towards the point u, such that (d) it arrives at the point u, and the experiment e_u gets the outcome o_1^u.

$\sin\phi\sin\theta, \cos\theta)$, denoted by ψ_v, and the experiment e_u corresponding to the spin experiment in direction $u = (\cos\beta\sin\alpha, \sin\beta\sin\alpha, \cos\alpha)$, is described respectively by the vector and the self adjoint operator

$$\psi_v = (e^{-i\phi/2}\cos\theta/2, e^{i\phi/2}\sin\theta/2), \tag{3.a}$$

$$H_u = \frac{1}{2}\begin{pmatrix} \cos\alpha & e^{-i\beta}\sin\alpha \\ e^{i\beta}\sin\alpha & \cos\alpha \end{pmatrix} \tag{3.b}$$

of a two-dimensional complex Hilbert space.

3. QUANTUM-CLASSICAL, NON LOCALITY AND EPR

A most interesting consequence of viewing quantum probabilities as being due to fluctuations of the interaction between the measurement and the entity that is being measured, is that one can easily parameterize these fluctuations. This was done [7] with a parameter called epsilon and the model was baptized the epsilon-model (see also [5]). One can study the behaviour of the epsilon model for varying epsilon values, and we have done this within three different frameworks: the lattice approach, the probabilistic approach and the *-algebra approach to quantum mechanics.

3.a. The Lattice Approach

For intermediate values of ϵ, that is $0 < \epsilon < 1$, we find that 2 of the 5 axioms needed in the lattice approach to reconstruct quantum mechanics are violated. The axioms that are violated are the weak-modularity and the covering law, and it are precisely those axioms that are needed to recover the vector space structure of the state space in quantum mechanics (see [7] for details).

3.b. The Probabilistic Approach

If we take the case of vanishing fluctuations, do we obtain the Kolmogorovian theory of probability? This would be most interesting, since then we would have constructed a macroscopic model with an understandable structure (i.e., we can see how the probabilities arise) and a quantum and a classical behaviour. One of the authors [3] proposed to test the polytopes for a family of conditional probabilities . The calculations can be found in [9] and the result was what we hoped for: a macroscopic model with a quantum and a Kolmogorovian limit (see also [12] for a detailed analysis of the quantum-classical relation in this framework).

3.c. The *-Algebra Approach

The *-algebra provides a natural mathematical language for quantum mechanical operators. We applied the concepts of this approach to the epsilon model to find that an operator corresponding to an ϵ-measurement is linear if and only if $\epsilon = 1$ [10].

Using the sphere model, we can illustrate the issue of non-locality. Picture the sphere model and a quantum entity located somewhere on the sphere. Metaphorically, we visualise the space context and the momentum context as orthogonal observables, thus as orthogonal directions for the piece of elastic. A measurement of the position of the entity will give a definite outcome, dragging the entity into one of the possible spatial outcomes, that is, dragging the entity into the space context. Similarly, a measurement of the momentum observable will drag that same entity along an orthogonal direction into one of the eigenstates of the momentum operator. This means that the locus and the momentum are partly created during the measurement. The state of the entity before the measurement has no definite momentum or locus (in the ontological sense), but is nevertheless definite itself, being a pure state. We arrive at a conclusion that is very difficult for us to visiualise: Space is not the theatre of reality, but a *context*. The same applies for momentum (see [4] for more details).

In [8] we concentrated on the question as to the nature of the classical limit procedure of a general quantum entity. The result gives one an easy way of picturing the classical limit. Given a probability distribution $\rho(x)$ in the case of zero fluctuations, we obtain the distribution $\rho_\epsilon(x)$, for the case of ϵ fluctuations, by the following procedure. By means of a constant function we cut off a piece of the original distribution $\rho(x)$, such that the surface contained in the cutoff piece equals ϵ. Next we move this cut function down to the x-axis and renormalise by dividing with ϵ. If we proceed this way for ever decreasing values of ϵ, we finally obtain a Dirac distribution for the classical limit in the case we have only one maximum, with the peak of the distribution located at the original maximum.

In 1996 [6] we focused on the question of the meaning of the Bell-inequality. A model [2] for singlet experiments like Aspect's based on the hidden measurements that reproduces the predictions of quantum mechanics for all possible angles of the polarizer was constructed in 1991. The "non-local" element was introduced explicitly by means of a rod that connected two sphere-models. We parametrized the correlation that the rod provided. The result of this construction is that one violates the Bell-inequalities even more for classical but non-locally connected systems, that is, $\epsilon = 0$, illustrating that the violation of the Bell-inequality is due to the non-locality rather then to the indeterministic character of quantum theory.

4. ACKNOWLEDGEMENTS

This work is supported by IUAP-III nr 9. D. Aerts is Senior Research Associate of the Belgian Fund for Scientific Research.

5. REFERENCES

1. D. Aerts, *J. Math. Phys.* **27**, 202 (1986).
2. D. Aerts, *Helv. Phys. Acta* **64**, 1 (1991).
3. D. Aerts, *Int. J. Theor. Phys.* **34**, 1165 (1995).
4. D. Aerts, *Int. J. Theor. Phys.* **31**, 1815 (1992).
5. D. Aerts, S. Aerts, B. Coecke, T. Durt, B. D'Hooghe, and F. Valckenborgh, "A Model with varying fluctuations on the measurement context," these proceedings.
6. D. Aerts, S. Aerts, B. Coecke, and F. Valckenborgh, "The meaning of the violation of Bell's inequality: Non-local correlation or quantum behaviour," preprint, Free University of Brussels (1996).

7. D. Aerts and T. Durt, *Found. Phys.* **24**, 1353 (1994).
8. D. Aerts, T. Durt, and B. Van Bogaert, *Symposium on the Foundations of Modern Physics 1992*, T. Hyvonen, ed. (World Scientific, Singapore, 1993), p. 53.
9. S. Aerts, "Conditional probabilities with a quantal and a Kolmogorovian limit," *Int. J. Theor. Phys.* **35**, 2201 (1996).
10. D. Aerts and B. D'Hooghe, "The operator structure of a non-quantum non-classical entity," *Int. J. Theor. Phys.* **35**, 2241 (1996).
11. B. Coecke, B. D'Hooghe, and F. Valckenborgh, "Classical physical entities with a quantum description," these proceedings.
12. T. Durt, *From Quantum to Classical, a Toy Model*, PhD thesis, Free University of Brussels, 1996

A MODEL WITH VARYING FLUCTUATIONS IN THE MEASUREMENT CONTEXT

Diederik Aerts, Sven Aerts, Bob Coecke, Bart D'Hooghe, Thomas Durt, and Frank Valckenborgh

TENA, Free University of Brussels
Pleinlaan 2, B-1050 Brussels, Belgium

In the "hidden measurement formalism," we try to explain the emergence of the quantum probabilities from the presence of fluctuations in the measurement context. We use a model that was constructed by Aerts et al. as a metaphorical model to extend these ideas.

Key words: hidden measurements.

1. A CLASSICAL SPIN-1/2 MODEL

In a paper that was published in 1986 [1], Aerts introduced a hidden variable model where the quantum probabilities arise because of a lack of knowledge on the experimental context during the measurement process. One version of this model consists of a particle, corresponding to the physical entity, moving on the surface of the two-dimensional unit sphere S^2; the experimental setup corresponds to the choice of two opposite points on the sphere. The interaction during the measurement consists in the following: the particle is orthogonally projected on the connecting line between these two points. To visualise the model, we assume the connecting line to be a sticky elastic that will break at a random place after contact with the particle, and pulls it to one of the endpoints (Fig. 1). If we know both the initial state and the breaking point of the elastic, we can predict the result of the measurement with absolute certainty. This model is isomorphic to the case of a spin-1/2 particle in a Stern-Gerlach experimental device. Since the hidden parameter belongs to the global experimental set-up, this idea was aptly dubbed the "hidden measurement" approach. One can generalise and prove that it is always possible to find such a hidden measurement explanation for general non-classical probability structures [2,6,7].

M. Ferrero and A. van der Merwe (eds.),
New Developments on Fundamental Problems in Quantum Physics, 7–9.

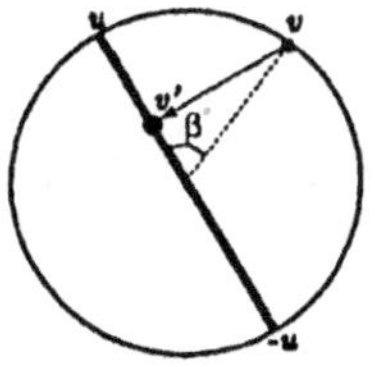

Fig. 1. Illustration of a measurement **u** on an Aerts' spin-$\frac{1}{2}$ entity when the initial state is **v**.

For this model we can easily prove our claim in [8]. If we consider a single measurement in the **u**-direction, we can introduce a parameter $\lambda, -1 \leq \lambda \leq 1$ that indicates the place of rupture on the measurement elastic. It is easy to see that we can use the same set $\Lambda = [-1, +1]$ for all measurements, in all possible directions. We shall consider Jauch and Piron's no-go theorem [9] and take a closer look at the dispersion-free components $(\mathbf{p}, \lambda)$ in our model. If we perform an experiment in the **u**-direction we find:

$$(\mathbf{p}, \lambda)(\mathbf{u}) = \begin{cases} 0 & \lambda > \cos(\beta), \\ 1 & \lambda \leq \cos(\beta), \end{cases}$$

with β the angle between **p** and **u**. If we take two experiments in different directions **u** and **v** and an initial state **p** and choose $\lambda < min\{\cos(\beta_1), \cos(\beta_2)\}$, we see that $(\mathbf{p}, \lambda)(\mathbf{u}) = (\mathbf{p}, \lambda)(\mathbf{v}) = 1$, while $(\mathbf{p}, \lambda)(\mathbf{u} \wedge \mathbf{v}) = (\mathbf{p}, \lambda)(0) = 0$. Therefore the dispersion-free components do not satisfy the Jauch-Piron condition, and we can not apply the theorem.

2. GENERALISATION OF THE MODEL

In a generalisation of this model, constructed by Aerts et al. [3,4,5], one considers the lack of knowledge on the measurement context to disappear in a continuous way. Using the metaphorical elastic of our earlier example, one can easily accomplish this by introducing a parameter $\epsilon, 0 \leq \epsilon \leq 1$, describing in a simple but quantitative way the magnitude of the fluctuations, with 2ϵ the length of the fragile domain on this string and **-u** as lower bound for an experiment in the **u**-direction. For fixed ϵ, the resulting model is a particular case of the so-called ϵ-model. It is easy to compute the transition probabilities if we perform such a measurement in the **u**-direction, for an entity in a state **v**:

$$\mathcal{P}[\mathbf{u}|\mathbf{v}] = 1 - \mathcal{P}[-\mathbf{u}|\mathbf{v}] = 1 \quad (2\epsilon - 1 < \mathbf{u} \cdot \mathbf{v}),$$

$$\mathcal{P}[\mathbf{u}|\mathbf{v}] = 1 - \mathcal{P}[-\mathbf{u}|\mathbf{v}] = \frac{1+\mathbf{u}\cdot\mathbf{v}}{2\epsilon} \quad (2\epsilon - 1 \geq \mathbf{u}\cdot\mathbf{v}).$$

If $\epsilon = 1$, we can represent the propositional structure of the model in the two-dimensional Hilbert space $\mathcal{C}^2$. Varying ϵ, the model yields a structural transition between quantum behaviour on the one hand ($\epsilon = 1$) and a classical structure on the other ($\epsilon = 0$), yielding purely deterministic experiments in the latter case. Indeed, only in these two limit cases the propositional lattice generated by the model satisfies all five of Piron's axioms necessary to apply his well-known representation theorem [10]. For intermediate ϵ, the lattice is not weakly modular and it doesn't satisfy the covering law [3]. Both these axioms are needed to impose a vector space structure on the representation space of the entity.

In a following step, we can slightly generalise this model by permitting fluctuations of varying magnitude in the measurement context, but always letting $\epsilon > 0$. In this case, the classical situation would only appear if we are able to accurately control the measurement environment, corresponding to the possibility of choosing an experiment with an arbitrary small value of ϵ among the set of all possible experiments. It can be shown that in this case one generates the standard topology on S^2, while the strictly classical case would correspond with the discrete topology [5].

3. ACKNOWLEDGEMENTS

This work is supported by IUAP-III n°9. D.A. is Senior Research Associate, and B.D. and F.V. are Research Assistants at the NFWO. T.D. is Research Assistant at the IIKW.

4. REFERENCES

1. D. Aerts, *J. Math. Phys.* **27**, 202 (1986).
2. D. Aerts, *Found. Phys.* **24**, 1227 (1994).
3. D. Aerts and T. Durt, *Found. Phys.* **24**, 1353 (1994).
4. D. Aerts, B. Coecke, T. Durt, and F. Valckenborgh, *Tatra Mountains Mathematical Publications* **10** (1996a), to appear.
5. D. Aerts, B. Coecke, T. Durt, and F. Valckenborgh, *Tatra Mountains Mathematical Publications* **10** (1996b), to appear.
6. B. Coecke, *Found. Phys. Lett.* **8**, 437 (1995).
7. B. Coecke, *Helv. Phys. Acta* **69** (1996), to appear.
8. B. Coecke, B. D'Hooghe, and F. Valckenborgh, these proceedings.
9. J.M. Jauch and C. Piron, *Helv. Phys. Acta.* **36**, 827 (1963).
10. C. Piron, *Foundations of Quantum Physics* (Benjamin, Reading, 1976).

SHORT TIME BEHAVIOR AND ZENO EFFECT IN RELATIVISTIC QUANTUM FIELD THEORY

Ramón F. Alvarez-Estrada

Departamento de Física Teórica, Universidad Complutense
28040-Madrid, Spain

José L. Sánchez-Gómez

Departamento de Física Teórica
Universidad Autónoma de Madrid
Cantoblanco, 28049-Madrid, Spain

We study the short-time evolution of decaying particles in the framework of relativistic quantum field theory. Significant differences between the relativistic and non-relativistic (that is, non-relativistic quantum mechanics) cases are found; in particular the so-called quantum Zeno effect is not present in relativistic quantum field theory, at least in the models studied herein.

Key words: relativistic unstable particles, quantum Zeno effect.

1. INTRODUCTION

The analysis of the finite-time evolution in relativistic quantum field theory (RQFT) and, in particular, the behavior at very short times of the survival amplitude of unstable particles in RQFT, has not attracted, in general, much interest. That is probably due to the fact that most of the interesting (from an experimental point of view) properties of elementary particles are studied in the framework of the S-matrix approach, which involves transition amplitudes from $t = -\infty$ to $t = \infty$, so that finite times do not play any relevant role in such questions.

Yet the short-time behavior of the survival amplitude of unstable particles is an important matter as far as the foundations of quantum mechanics are concerned, and, as we will try and show here, has some peculiarities in the RQFT case which make it to be worth studying. But before going to

M. Ferrero and A. van der Merwe (eds.),
New Developments on Fundamental Problems in Quantum Physics, 11–16.

RQFT, let us briefly recapitulate the main features of the survival amplitude (SA) at short times in non-relativistic quantum mechanics (just QM henceforth): see, for instance, [1] for a comprehensive review.

Let $P(t)$ $(t > 0)$ be the survival probability (SP) for an unstable system which is represented in QM by some normalized state, $|\Psi\rangle$, and let H be the Hamiltonian governing the evolution of the quantum system. One has $P(t) = |\langle\Psi | \exp(-itH) | \Psi\rangle|^2$ (with $\hbar = 1$). Then, assuming the finiteness of the energy dispersion in the initial (unstable) state, i.e.,

$$(\Delta E)^2 = \langle\Psi | H^2 | \Psi\rangle - (\langle\Psi | H | \Psi\rangle)^2 < \infty, \tag{1}$$

as indeed happens in QM, one can show that at very short times

$$P(t) \simeq 1 - (\Delta E)^2 t^2, \quad t \to 0^+. \tag{2}$$

Notice that the last equation shows the violation of the exponential decay law at very short times. The quadratic t-behavior gives rise to the so-called quantum Zeno effect, whose consequences and possible experimental detection have been profusely discussed: see [2] and [3] for recent work on this effect.

In the next section, we shall show that the above quadratic t-behavior does not hold in RQFT any longer. On the contrary, a linear behavior is to be expected for the time dependence —at very short times— of the SP of unstable particles in the framework of RQFT.

2. THE SHORT-TIME SURVIVAL AMPLITUDE FOR UNSTABLE PARTICLES IN RQFT

In this section we are going to analyze the short time decay of unstable particles in RQFT (in 3 space dimensions, of course). Now, one should be careful about both vacuum polarizing interactions and ultraviolet divergent renormalization. Let us, first, clearly state the problem.

We consider —with respect to a given reference frame— some "free" Hamiltonian H_0 which describes particles for $t \leq 0$ in RQFT in the Schrödinger picture. Also, let $\boldsymbol{P}$ be the total three-momentum operator, with $[\boldsymbol{P}, H_0] = 0$, and let $|i, \boldsymbol{p}_i\rangle$ the initial state representing the unstable particle at $t = 0$. By assumption, $|i, \boldsymbol{p}_i\rangle$ is a common eigenstate of both H_0 and $\boldsymbol{P}$, and we will, naturally, select a frame in which the unstable particle is at rest, i.e., $\boldsymbol{p}_i = 0$. If $|0\rangle$ represents the vacuum of H_0, then

$$|i, \boldsymbol{p}_i = 0\rangle \equiv |i\rangle = a^+(\boldsymbol{p}_i = 0)\,|0\rangle, \tag{3}$$

where $a^+(\boldsymbol{p}_i)$ is the associated creation operator (in Schrödinger's picture).

The interaction Hamiltonian H_I, which acts for $t > 0$ and is responsible for the decay, has, by assumption, the following properties:

1. It commutes with $\boldsymbol{P}$
2. It polarizes vacuum
3. It gives rise to either a superrenormalizable local RQFT or to a renormalizable one and, so, it includes the corresponding ultraviolet divergent renormalization counterterms.
4. It involves a small coupling constant, in the sense that one can reliably apply perturbation theory. In this regard, a good example would be the interaction Hamiltonian responsible for the conjectured proton decay in grand unified theories (GUT's): in such a case, the order of magnitude of the corresponding H_I is much smaller than that of the strong (QCD) Hamiltonian, which in turn would be the analogue of H_0. (See the last section of this work.)

Upon going over to the interaction picture (ip) and after some straightforward calculations, we get the basic formula enabling us to compute the SA in renormalized perturbation theory

$$A(t) = \sum_{n=0}^{\infty} \frac{(-1)^n}{n!} \int_0^t dt_1 \cdots \int_0^t dt_n \, \langle 0 \mid T[a(t)H_I(t_1)\ldots H_I(t_n)a^+(t=0)] \mid 0\rangle \langle i \mid i\rangle^{-1}, \tag{4}$$

where, for simplicity, we have omitted ip in the operators, as well as $\boldsymbol{p}_i = 0$ in a and a^+. Of course, T stands for time-ordered product. One can now evaluate $A(t)$ by using (4) and Wick's theorem in the standard way. Let m_i be the renormalized rest mass of the unstable particle. Then, it will be useful to introduce the following factorized form of the SA

$$A(t) = \exp(-im_i t)\tilde{A}(t), \tag{5}$$

where the "reduced" survival amplitude $\tilde{A}(t)$ tends to 1 for any $t > 0$ if $H_I \to 0$.

Now we have to deal with the divergences generically known as disconnected vacuum contributions (dvc). For the sake of brevity, we do not present here the details of the calculation (see [4] for a more elaborated treatment) which yields the following result

$$\tilde{A}(t) = A_{dvc}(t)A_{ph}(t). \tag{6}$$

$A_{ph}(t)$, to be called hereafter the physical survival amplitude, is the sum of all perturbative contributions to $\tilde{A}(t)$ in each of which $a(t)$ is not contracted with $a^+(t=0)$. Therefore all the contributions to A_{ph} are free of infinite-volume divergences; all the latter are included in the factor $A_{dvc}(t)$. We should notice that $A_{ph} \to 1$ as$H_I \to 0$.

Let us now briefly discuss the results obtained with the two models previously mentioned.

2.a. A Superrenormalizable Model

We shall consider an unstable neutral scalar boson (i), with mass m_i, which can decay into two different stable neutral bosons (a, b), with masses m_a, m_b ($m_i > m_a + m_b$). H_0 is the sum of the corresponding free Hamiltonians of the three bosonic particles, whereas the interaction Hamiltonian for $t > 0$ in the Schrödinger picture is, by assumption

$$H_I = \lambda \int d^3x\, \Phi_i(\boldsymbol{x})\Phi_a(\boldsymbol{x})\Phi_b(\boldsymbol{x}) - \delta m_i^2 \int d^3x\, N \mid \Phi_i(\boldsymbol{x})\mid^2 + \text{o.c.t.}, \tag{7}$$

where the Φ's represent the corresponding quantized bosonic fields, λ is a coupling constant (with dimensions of mass) and δm_i^2 is the ultraviolet mass renormalization counterterm for the unstable i boson (see below). The symbol N denotes normal ordering of field operators, while o.c.t. means other divergent renormalization counterterms: specifically, mass renormalization counterterms for the a and b bosons, which will not be explicitly written as they are not needed in this work.

The RQFT model described by Eq. (7) is superrenormalizable. The mass renormalization for the i boson is ($\epsilon \to 0^+$)

$$\delta m_i^2 = \frac{i\lambda^2}{(2\pi)^4} \int d^4k \, \frac{1}{(k^2 - m_a^2 + i\epsilon)(k^2 - m_b^2 + i\epsilon)}. \tag{8}$$

After a long calculation, whose details can be found in [4], we get the following expression for the physical survival probability (SP) up to order λ^2 ($\hbar = c = 1$):

$$P(t) \equiv \mid A_{ph}(t)\mid^2 = 1 - \frac{\lambda^2}{16\pi\, m_i} t + \frac{\lambda^2(m_a + m_b)}{16\pi^2 m_i} t^2 + \ldots. \tag{9}$$

We then see that, for $t \ll (m_a + m_b)^{-1}$,

$$P(t) \simeq 1 - \frac{\lambda^2}{16\pi m_i} t, \tag{10}$$

so that the SP decreases *linearly* with t for very small values of t and, hence, the Zeno effect "disappears".

For consistency reasons, it is worth considering the SP for large values of t. When $t \to \infty$, we get

$$P(t) \to 1 - \frac{\lambda^2 Q}{8\pi m_i^2} t \to \exp(-\Gamma t), \text{ with } \Gamma = \frac{\lambda^2 Q}{8\pi m_i^2}, \tag{11}$$

where Q is the CM momentum of the final particles. The last equation is just the expression of the Fermi golden rule in the present case.

2.b. A Renormalizable Model

We shall now consider a (unstable) relativistic neutral spin-1/2 fermion (i), with renormalized mass m_i, which can decay into two relativistic stable neutral particles (a, b) with renormalized masses $m_a, m_b > 0$, so that: i) $m_i > m_a + m_b$, and ii) b is a spin-1/2 fermion while a is a scalar spinless boson (our considering all particles as neutral is just for simplicity; the general case is physically equivalent).

Now H_0 is the sum of three free Hamiltonians: one, corresponding to particle a, being similar to those in the previous case and the other two, of particles i and b, being the well-known for relativistic free Dirac fermions. For $t > 0$ the interaction Hamiltonian in the Schrödinger picture is

$$H_I = \lambda \int d^3x \, [\bar{\Psi}_b(\boldsymbol{x})\Psi_i(\boldsymbol{x})]\Phi_a(\boldsymbol{x}) + \text{h.c.} - \delta m_i \int d^3x \, N[\bar{\Psi}_i(\boldsymbol{x})\Psi_i(\boldsymbol{x})] + \text{o.c.t.}, \tag{12}$$

where λ is a renormalized dimensionless coupling constant. The mass renormalization counter term for the unstable i-fermion is to order λ^2

$$\delta m_i = \frac{i\lambda^2}{(2\pi)^4} \int d^4k \, \frac{-k.\gamma + m_b}{(k^2 - m_a^2 + i\epsilon)(k^2 - m_b^2 + i\epsilon)}, \tag{13}$$

where $k.\gamma = k^0\gamma^0 - \boldsymbol{k}\gamma$, the γ's being, of course, the Dirac matrices. In the present case, o.c.t. should include the contribution corresponding to the wave-function renormalization, which is not given here for the sake of brevity. The calculation now turns out to be much more elaborated than in the previous case. We will simply present the final result (see [4] for further details).

For very small values of t, the survival probability is given by

$$P(t) = 1 - \frac{\lambda^2}{16\pi}(2m_b + m_i)t + \ldots. \tag{14}$$

In the present case, the quadratic term is much more complicated than in the previous one and is not given explicitely here as it will not be needed for subsequent discussion. In any case, we see that the behavior of $P(t)$ at very small t linear, so that, again, the Zeno effect does not appear. Eq. (14) gives the right expression of the SP in the renormalizable case, which was incorrectly given in [4].

3. DISCUSSION

The main result of this work is that, in RQFT (or at least in the RQFT models we have studied so far), the short-time behavior of the survival probability for the decay of an unstable particle is linear in t, instead of quadratic as it is the case in QM. The origin of the different behaviors in QM and RQFT seems to be due to the fact that in the latter there are vacuum polarization and ultraviolet divergences which made the energy dispersion in the initial (unstable) state to be infinite, then invalidating the main assumption of QM, as already commented in the Introduction.

Now the linear decreasing of the SP means that a unstable system should decay exponentially (but for quantum oscillation effects appearing at very large t-values which are neither discussed nor are relevant here) even if it is continuously observed; in other words, *the quantum Zeno effect is not present.* Yet it does not mean that a very frequent "observation" of a decaying system should not affect its "free" lifetime; on the contrary, there could be cases in which the latter should appreciably change when the decaying system is being repeatedly observed. Note that the linear term appearing in the short-time behavior of the survival probability is different from the decay width (compare, for instance, Eq. (10) and (11) in the superrenormalizable case).

As a final comment, we would like to point out that the fact that the SP, in the relativistic case, decreases at very short times linearly (instead of quadratically) with t seems to be be relevant to the discussion of locality and separability in quantum physics [5].

REFERENCES

1. L. Fonda, G.C.Ghirardi, and A. Rimini, *Rep. Prog. Phys.* **41**, 587 (1978).
2. W.H. Itano, D.J. Heinzen, J.J. Bollinger, and D.J. Wineland, *Phys Rev. A* **41**, 2295 (1990).
3. A.D. Panov, *Ann. Phys. (N.Y.)* **249**, 1 (1996).
4. R.F.Alvarez-Estrada and J.L. Sánchez-Gómez, *Rev.Mex.Fis.* **42**, 216 (1996).
5. E. Santos, private communication.

A MINIMAL LOCAL EXTENSION OF THE QUANTUM FORMALISM

Thomas D. Angelidis

Centre for Mathematical Physics
PO Box 3881
London NW8 0PF
United Kingdom

A consistent local theory $Th(\mathcal{G})$ extending the quantum formalism is proposed. As a consequence, a realistic and local ("common cause") interpretation of the Einstein-Podolsky-Rosen-Bohm ideal experiment can now be given.

Key words: locality, extension, quantum formalism, symmetry breaking.

1. INTRODUCTION

Four postulates ($\Pi_1, \Pi_2, \Pi_3, \Pi_4$) are added to the quantum formalism (QF) which may be said to affect only the universe of discourse of QF by extending the range of its variables in a demonstrably consistent way. The proof, given elsewhere [1,2], proceeds by giving a model $\mathcal{G}$ of the extended theory $Th(\mathcal{G})$ in the model-theoretic sense. Other models $\mathcal{G}'$, elementarily equivalent to $\mathcal{G}$, have also been constructed [1,2]. As the adding of the four postulates to QF does not lead to a contradiction within the strengthened formalism (of $Th(\mathcal{G})$), this strengthening of QF seems to be *minimal*.

The suggested impossibility of a consistent local extension of QF is thereby questioned and reconsidered. For, according to what may be called the 'received view,' there is no model provided by quantum theory (as it stands) which permits a local ("common cause") interpretation of the Einstein-Podolsky-Rosen-Bohm (EPRB) *ideal* experiment.

The four postulates added to QF stipulate a *finer* state specification than that given by the quantum state $|\gamma_1, \gamma_2\rangle$, known as the singlet state. This finer state specification describes how the spontaneous annihilation process itself *breaks* the *spherical symmetry* of the singlet state $|\gamma_1, \gamma_2\rangle$ into a

M. Ferrero and A. van der Merwe (eds.),
New Developments on Fundamental Problems in Quantum Physics, 17–29.

pair of back-to-back photons having only *axial symmetry* about the direction of their motion. This symmetry breaking process creates new *local* "elements of reality" not captured by the usual QF.

In the proposed local extension of QF, postulate Π_3 adjoins to QF the variable μ - missing from the specification of the singlet state $|\gamma_1, \gamma_2\rangle$ - whose values characterise these new *local* "elements of reality" created by the spontaneous symmetry breaking process. More precisely, each value of μ specifies the initial ("hidden") direction of the *common* plane of polarization of the two back-to-back photons being born at the instant the singlet state disintegrates.

This minimal extension of QF, in the shape of the theory $Th(\mathcal{G})$, may be a small step towards a realization of Einstein's hope of strengthening quantum theory to obtain a realistic and local theory.

2. THE FORMAL POSTULATES OF LOCALITY

Bell et al. [3,4] enunciated three conditions for locality, which intend to characterise physical locality. Here we shall collect together these conditions, and re-state them in the shape of three formal postulates of locality which any theory T, whose postulates are expressed in terms of some quadruple $\langle \hat{p}_1, \hat{p}_2, \rho, \Lambda \rangle$ of specified functions, must satisfy if T is to qualify as a local theory in the sense of Bell et al.

(L1) Any joint probability function p_{12} must be defined as a specified instance of the syntactical form

$$p_{12}(\alpha, \beta) := \int_{\Lambda} \rho(\lambda) \hat{p}_1(\lambda, \alpha) \hat{p}_2(\lambda, \beta) d\lambda,$$

where any specified function $\hat{p}_1$ must not depend upon the variable β, and where any specified function $\hat{p}_2$ must not depend upon the variable α. [The product form $\hat{p}_{12}(\lambda, \alpha, \beta) = \hat{p}_1(\lambda, \alpha) \hat{p}_2(\lambda, \beta)$ is known as the "*factorizability condition.*"]

(L2) Any specified function ρ must not depend either upon the variable α or upon the variable β.

(L3) Any specified range Λ of the variable λ must not depend either upon the variable α or upon the variable β.

Note well that (L2) does *not* exclude the possibility that the function ρ may be chosen to depend upon some other variable, say μ, provided μ (like λ) is a variable *distinct* from both variables α and β.

3. BELL'S CONJECTURE OF NONLOCALITY

Bell's [3] impossibility proof, which we shall here call Bell's *conjecture* of nonlocality, purports to show that QF cannot be consistently extended by adjoining local "hidden variables" to it.

In words, Bell's conjecture of nonlocality is taken to assert "*the incompatibility of any local hidden variables theory with certain quantum mechanical predictions*" [5].

We shall slightly sharpen Bell's [3] own formulation here. Let T be a theory whose postulates are expressed in terms of the quadruple $\langle \hat{p}_1, \hat{p}_2, \rho, \Lambda \rangle$ where Λ is the range of the variable λ, ρ is a *specified* function defined on Λ, and $\hat{p}_1, \hat{p}_2$ are *specified* functions defined on $\Lambda \times D$, and D is the range of the variables α and β. Let (L1)$\wedge$ (L2)$\wedge$ (L3) be the logical conjunction of the three formal postulates of locality enunciated by Bell et al. (cf. Sec. 2), where the symbol "$\wedge$" stands for the (truth-functional) *conjunction*. Then, Bell's conjecture of nonlocality asserts that:

> There exists NO consistent theory T whose postulates $\langle \hat{p}_1, \hat{p}_2, \rho, \Lambda \rangle$ satisfy (L1)$\wedge$ (L2)$\wedge$(L3) and such that
>
> $$(\forall \alpha, \beta \in D) \left[\frac{1}{2} \cos^2(\alpha - \beta) = \int_\Lambda \rho(\lambda) \hat{p}_1(\lambda, \alpha) \hat{p}_2(\lambda, \beta) d\lambda \right] \qquad (1)$$
>
> holds.

Or, in Bell's [3] own words, the QF probability function p_{12}^{QF}, defined by

$$p_{12}^{\mathrm{QF}} : (\alpha, \beta) \to \frac{1}{2} \cos^2(\alpha - \beta), \quad \forall (\alpha, \beta) \in D^2, \qquad (2)$$

"*cannot be represented, either accurately or arbitrarily closely, in the form (1).*"

However, a consistent theory T has been constructed [1,2] whose postulates $\langle \hat{p}_1, \hat{p}_2, \rho, \Lambda \rangle$ do satisfy (L1)$\wedge$(L2)$\wedge$(L3) and generate a family of functions $\{p_{12}^{\mu} | \mu \in M\}$ which *converges uniformly* to a unique *limit function* identical with the QF function p_{12}^{QF} for $\forall(\alpha, \beta) \in D$, as the *syntactical form* (1) precisely requires. If we understand Bell's own words correctly, the possible existence of a family of functions endowed with this property was denied by him.

Recall that the QF probability function p_{12}^{QF}, defined by (2), describes the quantum-statistical correlations exhibited in the EPRB *ideal* experiment for *all* values assigned to the variables α and β in the range D, that is, $\forall(\alpha, \beta) \in D^2$. Each *value* of the function p_{12}^{QF} is interpreted as the (relative) joint probability for the *coincidence detection (count)* of both photons

γ_1 and γ_2 *after* passing their respective polarisers. Under different value assignments in D, the values (elements of D) $\alpha_1, \alpha_2, \ldots$ and $\beta_1, \beta_2, \ldots$ assigned to the variables α and β are interpreted as the directions of the setting of the polarisers P_1 and P_2 respectively. The photons γ_1 and γ_2, emitted in opposite (back-to-back) directions, are born by the spontaneous annihilation decay of the singlet state $|\gamma_1, \gamma_2\rangle$ prepared by a suitable source.

4. THE POSTULATES OF THE THEORY T_Π

$$(\Pi_1) \qquad \hat{p}_1 : (\lambda, \alpha) \rightarrow \cos^2(\lambda - \alpha) \qquad \forall(\lambda, \alpha) \in \Lambda \times D$$

$$(\Pi_2) \qquad \hat{p}_2 : (\lambda, \beta) \rightarrow \cos^2(\lambda - \beta) \qquad \forall(\lambda, \beta) \in \Lambda \times D$$

$$(\Pi_3) \qquad \rho_p(\lambda, \mu) := \tfrac{1}{2}\left[\delta(\lambda - \mu) + \delta(\lambda - \mu + \tfrac{1}{2}\pi)\right]$$

$$(\Pi_4) \qquad \Lambda := \{\lambda | -\infty < \lambda < +\infty\}$$

- The postulates of the theory T_Π manifestly satisfy (L1)$\wedge$(L2)$\wedge$(L3). Furthermore, the values of $\hat{p}_1$ and $\hat{p}_2$ are bounded by 0 and 1, as probabilities should be. This answers what Feynman [6] has called the "fundamental problem".
- The theory T_Π is *consistent* as $T_\Pi \subseteq Th(\mathcal{G})$ and $Th(\mathcal{G})$, the theory of $\mathcal{G}$, is always a consistent theory (cf. Ref [2] p. 1652).
- From T_Π one deduces a family of functions $\{p_{12}^{\mu} | \mu \in M\}$ which *converges uniformly* to a unique *limit function* identical with the QF function p_{12}^{QF} for $\forall \alpha, \beta \in D$ as the *syntactical form* (1) precisely requires.
- Thus, the consistent theory T_Π *refutes* the conjecture of nonlocality.
- In the presence of Π_3, the *sum* of products of probability amplitudes in QF – corresponding to the two mutually exclusive alternatives involved in the linear superposition (nonfactorisable) state $|\gamma_1, \gamma_2\rangle$ - is being transcribed into the *sum* of products of (relative) probabilities in $Th(\mathcal{G})$ for the alternatives in question. This shows how the formal postulate of locality referred to as *factorizability* is satisfied in $Th(\mathcal{G})$ (cf. Sec. 7).

5. THE PHYSICAL INTERPRETATION OF THE POSTULATES OF T_Π

The first two postulates Π_1 and Π_2 describe the standard probabilistic local interaction between individual photons and polarisers.

The more important postulate Π_3 describes how the spontaneous annihilation process breaks the *spherical symmetry* of the singlet state, defined by

$$|\gamma_1, \gamma_2\rangle = 1/\sqrt{2}\,[|x(\gamma_1)\rangle|x(\gamma_2)\rangle + |y(\gamma_1)\rangle|y(\gamma_2)\rangle],$$

into a pair of back-to-back photons having only *axial symmetry* about the direction of their motion. Π_3 does so by introducing a slightly *finer* description than that given by the singlet state as follows.

To begin with we propose the following realistic interpretation of the singlet state $|\gamma_1, \gamma_2\rangle$ as describing the *exclusive disjunction* (ED):

'both photons are plane-polarized in the x direction OR
both photons are plane-polarized in the y direction'. (ED_1)

Next we shall describe the two characteristic features of our proposed local realistic extension of the singlet state $|\gamma_1, \gamma_2\rangle$ which are incorporated into the definition of ρ_p stipulated by postulate Π_3.

- *The first characteristic feature is this*: There is nothing special about the orthogonal directions x and y in ordinary space (here 'orthogonal' in ordinary space does correspond to 'orthogonal' in the Hilbert space sense). One could equally well choose any other pair of orthogonal directions in ordinary space, say, λ and $\lambda + \frac{1}{2}\pi$ (or, say, λ and $\lambda - \frac{1}{2}\pi$) *all the more so since the singlet state is spherically symmetric.* Thus, the exclusive disjunction (ED_1) now reads:

  'both photons are plane-polarized in the λ direction OR
  both photons are plane-polarized in the $\lambda + \frac{1}{2}\pi$ direction'. (ED_2)

- *The second characteristic feature is this*: *Before* the spontaneous annihilation decay of the spherically symmetric singlet state $|\gamma_1, \gamma_2\rangle$ occurs, *all* choices of directions λ are equally likely. *After* the spontaneous annihilation decay of $|\gamma_1, \gamma_2\rangle$ has taken place, a *specific* direction in ordinary space, given by a value of the variable μ, has been randomly chosen by the spontaneous annihilation process itself.

Postulate Π_3 incorporates these two characteristic features, partly shared by and partly missing from the specification of the singlet state $|\gamma_1, \gamma_2\rangle$, into the definition of ρ_p (the subscript p in ρ_p stands for "photon pair").

Thus, postulate Π_3 stipulates the (relative) probability distribution ρ_p for the spherically symmetric singlet state $|\gamma_1, \gamma_2\rangle$ to spontaneously disintegrate into two back-to-back photons plane-polarized in a *specific* but randomly chosen direction, given by a *value* of μ, out of *all* the equally likely choices of directions given by the range Λ of values of the variable λ.

Furthermore, each value of the variable μ is ***sufficient*** to completely specify the initial direction of the common plane of polarization of the two back-to-back photons emerging at the instant the singlet state disintegrates: For by the definition of ρ_p, the exclusive disjunction (ED_2) becomes

$$\begin{array}{l}\text{`both photons are plane-polarized in the } \mu \text{ direction OR} \\ \text{both photons are plane-polarized in the } \mu - \tfrac{1}{2}\pi \text{ direction,'}\end{array} \qquad (ED_3)$$

where each of the two mutually exclusive alternatives has probability of occurrence equal to $\frac{1}{2}$.

6. THE CONDITIONAL SENTENCE Σ

The realistic and local ("common cause") interpretation of the EPRB *ideal* experiment proposed here is based on the (first-order) conditional sentence Σ of the theory $Th(\mathcal{G})$:

$$(\forall\varepsilon > 0)(\exists\eta > 0)(\forall\mu \in M)(\forall\alpha,\beta \in D)\big[(|\mu - \alpha)| < \eta) \vee (|\mu - \beta| < \eta) \supset \\ |p_{12}^{\mu}(\alpha,\beta) - p_{12}^{\mathrm{QF}}(\alpha,\beta)| < \varepsilon\big], \qquad (\Sigma)$$

and on the structure $\mathcal{G}$ in which the sentence Σ is satisfied so that $\mathcal{G}$ is a *model* of Σ ($\mathcal{G} \models \Sigma$) or, equivalently, Σ is a *theorem* of $Th(\mathcal{G})$ ($Th\ (\mathcal{G}) \vdash \Sigma$) proved in [1,2]. The domain (universe) of $\mathcal{G}$ is the set G.

Some formal remarks concerning Σ (cf. Ref. [2]):

- The sentence Σ expresses the *formal definition* of the convergence of the family of functions $\{p_{12}^{\mu}|\mu \in M\}$ to the function p_{12}^{QF} in the topology of *uniform* convergence.
- This (uniform) topology is assigned to the function set $\{p_{12}^{\mu}|\mu \in M\}$ and *not* to the domain of definition of each member p_{12}^{μ}.
- It is well known that the uniform topology (for a function space) is *independent* of any topology that may be assigned (if one is required) to the domain of any function belonging to the function space which (domain) may be any arbitrary non-empty set.
- Consequently, the proof that p_{12}^{QF} is the uniform limit of $\{p_{12}^{\mu}|\mu \in M\}$ does not assume any topology for the domain of definition of each member p_{12}^{μ}.

Some further formal remarks concerning Σ:

- The formal definition Σ of the 'limiting case' does *not* license the identification of the variable μ either with the variable α or with the variable β, denoted respectively by the equations $\mu = \alpha$ or $\mu = \beta$.
- This can be demonstrated as follows. Since Σ is *true (valid) in* $\mathcal{G}$ ($\mathcal{G} \models \Sigma$) for *all* value assignments s in $\mathcal{G}$, and since there are value assignments s in $\mathcal{G}$ such that both equations $\mu = \alpha$ and $\mu = \beta$ are *false*, it follows that the formal definition of the 'limiting case' does *not* logically imply the identification $\mu = \alpha$ or $\mu = \beta$ in $\mathcal{G}$, *nor* licenses this identification anywhere in our construction.

Question: Could one demonstrate the identification $\mu = \alpha$ or $\mu = \beta$ in $Th(\mathcal{G})$?

In a formal language with equality, denoted by the predicate symbol "=" in boldface (cf. Ref. [2] p. 1640), speaking about identity brings in the Leibniz principle which admits the formulation [7]:

$$\forall xy\, [x = y \Leftrightarrow \forall P(Px \Leftrightarrow Py)]\,. \qquad \text{(LP)}$$

LP says that any two individuals (or attributes or functions) of whatever type are identical if and only if each has all the properties that the other has (identity means agreement in all properties). LP with $\Rightarrow$ states the *interchangeability principle* for identical individuals etc. LP with $\Leftarrow$ states the *indiscernibles principle.*

Thus, in order to establish the identification, say, of μ with α ($\mu = \alpha$) in $Th(\mathcal{G})$ we would be required to demonstrate that $\forall P(P\mu \Leftrightarrow P\alpha)$ for $\forall P \in Th(\mathcal{G})$ which, in the presence of the indiscernibles principle, would establish $\mu = \alpha$ (by *modus ponens*) in $Th(\mathcal{G})$.

Proposition: It is not the case that $\forall P(P\mu \Leftrightarrow P\alpha)$ for $\forall P \in Th(\mathcal{G})$.

Proof: Let M be the set of irrational numbers. Let D be the set of rational numbers ($M \cap D = \emptyset$). Then, it is not the case that the variable μ has all the properties the variable α has, as each variable is assigned values in sets (ranges) characterized by disjoint properties. ■

- The physical interpretation of the variables μ and α (or β) also shows that they have different properties as their assigned values denote objects with different physical properties and space-time locations.
- Even if the variables μ and α (or β) happened to share *some* properties in $Th(\mathcal{G})$, this would not be sufficient to establish their identification in $Th(\mathcal{G})$.

Consequently, the reply to the question posed above is that one cannot establish the identification $\mu = \alpha$ or $\mu = \beta$ in $Th(\mathcal{G})$ (cf. Ref. [2] p. 1648 for a different proof using the downward Löwenheim-Skolem theorem).

7. THE PHYSICAL INTERPRETATION OF THE SENTENCE Σ

We begin by showing how Bell's [8] hypothesis of "*local causality*" (LC), also referred to as *factorizability*, is satisfied in $Th(\mathcal{G})$.

- Each function p_{12}^{μ} belonging to the set $\{p_{12}^{\mu}|\mu \in M\}$ is deduced as a *specified instance* of the syntactical form stipulated by (L1) using the conjunction $\Pi = \Pi_1 \wedge \Pi_2 \wedge \Pi_3 \wedge \Pi_4$ of the postulates of T_Π so that

$$\begin{aligned} p_{12}^{\mu}(\alpha,\beta) &:= \int_{\Lambda} \rho_p(\lambda,\mu)\hat{p}_1(\lambda,\alpha)\hat{p}_2(\lambda,\beta)d\lambda \\ &= \frac{1}{4}\left[1+\cos 2(\mu-\alpha)+\cos 2(\mu-\beta)\right]. \end{aligned} \tag{3}$$

- Each joint (relative) probability function p_{12}^{μ} describes a purely *local* and *probabilistic* interaction between each correlated photon $\gamma_1(\gamma_2)$ and its corresponding polariser $P_1(P_2)$, where each polariser acts independently of the other polariser *precisely* as stipulated by the 'factorizability condition' in (L1). This can be seen from

$$p_{12}^{\mu}(\alpha,\beta) = \frac{1}{2}\left[\hat{p}_1(\mu,\alpha)\hat{p}_2(\mu,\beta) + \hat{p}_1(\mu-\frac{1}{2}\pi,\alpha)\hat{p}_2(\mu-\frac{1}{2}\pi,\beta)\right] \quad \text{(LC)}$$

deduced from (L1) using the postulated distribution ρ_p.
- LC shows that each p_{12}^{μ} can be written as the *sum* of two real-weighted products $\hat{p}_1\hat{p}_2$ of probabilities corresponding to the two mutually exclusive alternatives described by (ED_3).
- Recall that each function $\hat{p}_1(\hat{p}_2)$, given by $\Pi_1(\Pi_2)$, determines the (relative) probability that each photon $\gamma_1(\gamma_2)$ will get through its corresponding polariser $P_1(P_2)$, *given* that photon $\gamma_1(\gamma_2)$ is plane-polarized in the μ direction OR in the $\mu - \frac{1}{2}\pi$ direction.

Thus, in Bell's [8] own words, LC incorporates his "*hypothesis of 'local causality' or 'no action at a distance'*."

Next we explain how the conditional sentence Σ describes the *local* overall response of the apparatus to each and every pair of back-to-back photons emitted by the source. We do this by considering *specified instances* of Σ in $\mathcal{G}$ (a *specified instance* of the first-order sentence Σ in $\mathcal{G}$ is of course a *propositional sentence*; and the operation of deducing specified instances of Σ in $\mathcal{G}$ is called the *instantiation* of Σ in $\mathcal{G}$).

Each *instantiation* of the sentence Σ describes as ONE experiment the *local* overall response of the apparatus to a single pair of back-to-back photons emitted by the source. In order to proceed, write the antecedent of Σ in the equivalent form $S_\alpha \cup S_\beta$ where

$$S_\alpha = \{\mu| - 2\varepsilon + \alpha < \mu < \alpha + 2\varepsilon\}, \quad (4)$$
$$S_\beta = \{\mu| - 2\varepsilon + \beta < \mu < \beta + 2\varepsilon\}, \quad (5)$$

are indexed subsets of M such that $M = \cup_\alpha S_\alpha = \cup_\beta S_\beta$, and

$$M := \{\mu| - \infty < \mu < +\infty\}, \quad (6)$$

is the range of the variable μ (by the rules of substitution the ranges of the variables λ and μ must be the same).

Let s_1 be a value assignment in $\mathcal{G}$ of some instantiation of the sentence Σ, and let μ_1, α_1, β_1 be the values in M and D a ssigned to the variables μ, α, β under s_1.

Whenever the antecedent in Σ is satisfied (true) under s_1, that is

$$\mu_1 \in S_{\alpha 1} \subset M \quad \text{OR} \quad \mu_1 \in S_{\beta 1} \subset M, \quad (7)$$

the consequent in Σ can be deduced from Σ by *modus ponens.* And the so deduced consequent determines the probability of the single pair of back-to-back photons characterized by the (random) value $\mu_1 \in M$ to get through the polarisers and *cause* a coincidence count, this probability being equal to the *value* $\frac{1}{2}\cos^2(\alpha_1 - \beta_1)$ under s_1 of the QF probability function p_{12}^{QF}.

Question: What happens to the single pair of back-to-back photons if its value $\mu_1 \in M$ is such that the antecedent in Σ is *not* satisfied, that is, if

$$\mu_1 \notin S_{\alpha 1} \quad \text{AND} \quad \mu_1 \notin S_{\beta 1}? \quad (8)$$

Reply: If so, then the consequent in Σ *cannot* be deduced from Σ (to presume that such 'deduction' is sound is to commit the common fallacy known in logic as "*denying the antecedent*"). Thus, it cannot be asserted that the single pair of back-to-back photons with $\mu_1 \in M$ causes a coincidence count with probability $\frac{1}{2}\cos^2(\alpha_1 - \beta_1)$.

However, the single pair of back-to-back photons with $\mu_1 \in M$ may fall inside another subset, say, $S_{\alpha 4}$ or $S_{\beta 4}$ of the set M, that is,

$$\mu_1 \in S_{\alpha 4} \subset M \quad \text{OR} \quad \mu_1 \in S_{\beta 4} \subset M, \quad (9)$$

so that it causes a coincidence count with a *different* probability $\frac{1}{2}\cos^2(\alpha_4 - \beta_4)$, determined by the consequent in Σ deduced from Σ (by *modus ponens*) under another value assignment s_4 in $\mathcal{G}$ which agrees with s_1 in $\mathcal{G}$ on the variable μ ($\mu_1 = \mu_4$).

This argument, based upon instantiating Σ in $\mathcal{G}$, also shows how the apparatus responds, as it should, to the whole range M characterising the *ensemble* of pairs of back-to-back photons emitted by the source and *not* only to those pairs of back-to-back photons with values of μ belonging, say, to the "small" subset $S_{\alpha 1}$ or $S_{\beta 1}$. This is what the universal quantifiers $(\forall \mu \in M)$ and $(\forall \alpha, \beta \in D)$ in the prefix of Σ do: They take into account ALL the "small" subsets S_α or S_β of M $(M = \cup_\alpha S_\alpha = \cup_\beta S_\beta)$.

So the universal quantifiers $(\forall \mu \in M)$ and $(\forall \alpha, \beta \in D)$ in the prefix of Σ take into account the *whole* array of such possibilities (by instantiation) so that the detectors register coincidence (and single) counts with the same probabilities as those given by QF for each and every pair of back-to-back photons emitted by the source.

Thus, in the presence of the conditional sentence Σ of $Th(\mathcal{G})$, each value of the QF probability function p_{12}^{QF} (*deducible* in $Th(\mathcal{G})$) can now be interpreted, in purely *local* terms, as the measure of the *chance* of each pair of back-to-back photons, upon being born by the spontaneous annihilation process, to get through the corresponding polarisers and cause a coincidence count in agreement with the EPRB *ideal* experiment.

This realistic and local ("common cause") explanation of the quantum-statistical correlations exhibited in the EPRB *ideal* experiment - in terms of the adjoined *local* variable μ (absent from the specification of the singlet state $|\gamma_1, \gamma_2\rangle$) whose values specify the initial ("hidden") directions of the common planes of polarisation of each and every pair of back-to-back photons born at the instant the singlet states annihilate – is missing from the usual (unextended) QF.

8. THE BELL INEQUALITY

Here we shall briefly outline the main results of the argument which shows that the Bell inequality is *valid* (always satisfied) in the extended theory $Th(\mathcal{G})$. Save for some sketch proofs given below, the whole argument and proofs will be published in a forthcoming paper.

Proposition: $Th(\mathcal{G}) \vdash \tau \ (\Leftrightarrow \mathcal{G} \models \tau)$.

Proof: .... ■

- The Bell sentence (inequality) τ is a *theorem* of $Th(\mathcal{G})$ or, equivalently, $\mathcal{G}$ is a *model of* τ or, equivalently, τ is *valid (true) in* $\mathcal{G}$ for *all* value assignments s in $\mathcal{G}$.
- The (first-order) Bell sentence τ is *deduced* from the conjunction of $\Pi = \Pi_1 \wedge \Pi_2 \wedge \Pi_3 \wedge \Pi_4$ with an arithmetical lemma [4] and reads:

$$(\forall \mu \in M)\big[-2 \leq \cos 2(\mu-\alpha_1)\cos 2(\mu-\beta_1) - \cos 2(\mu-\alpha_1)\cos 2(\mu-\beta_2) + \cos 2(\mu-\alpha_2)\cos 2(\mu-\beta_1) + \cos 2(\mu-\alpha_2)\cos 2(\mu-\beta_2) \leq +2\big], \qquad (\tau)$$

where the symbols $\alpha_1, \alpha_2, \beta_1, \beta_2$ occurring in τ are ***individual constants*** which the structure $\mathcal{G}$ maps to ***designated*** individuals objects (numbers) of the domain G of $\mathcal{G}$ (i.e., $\alpha_1^{\mathcal{G}}$ is a real number, etc.) [9]. This mapping under $\mathcal{G}$ must be distinguished from a value assignment in $\mathcal{G}$ which does not affect the denotation of individual constants [9]. The designated individual objects (numbers) are determined by $\mathcal{G}$ and are here interpreted as the distinct directions of the settings of the polarisers.

- $\neg\tau \notin Th(\mathcal{G})$ since the theory $Th(\mathcal{G})$ is complete and consistent. In other words, the false sentence $\neg\tau$ does not belong to $Th(\mathcal{G})$.

The next proposition shows that in $Th(\mathcal{G})$ all the values of the QF probability function p_{12}^{QF} satisfy the Bell sentence (inequality) τ.

Proposition: $\mathcal{G} \models \Sigma \wedge \tau$.

Proof (sketch): Since a sentence is valid iff it is satisfiable, it suffices to exhibit ONE and the same value assignment s_1 in $\mathcal{G}$ such that the resulting instantiations of Σ and τ are simultaneously satisfied in $\mathcal{G}$, and thus Σ and τ are simultaneously *valid (true) in $\mathcal{G}$* for *all* value assignments s in $\mathcal{G}$. ▪

- The set $\{\Sigma, \tau\}$ is consistent for *all* s in $\mathcal{G}$.
- Whenever the antecedent of Σ is satisfied one deduces (by *modus ponens*) the QF probability function p_{12}^{QF} as the *limit function* of the family $\{p_{12}^{\mu} | \mu \in M\}$.

Question: Is the Bell inequality τ violated in $Th(\mathcal{G})$?

Reply: No. The Bell sentence (inequality) τ is *valid* in $Th(\mathcal{G})$ (always satisfied) for *all* value assignments s in $\mathcal{G}$ since $Th(\mathcal{G}) \vdash \tau$ ($\Leftrightarrow \mathcal{G} \models \tau$).

- Let $\mathcal{G}_m$ be the structure that maps the individual constants $\alpha_1, \alpha_2, \beta_1, \beta_2$ occurring in τ to the designated individual objects (numbers) of the domain of $\mathcal{G}_m$ called [10] the "maximum violation values". Then, in $\mathcal{G}_m$, the false Bell propositional sentence χ ("violated inequality"):

  $$-2 \leq \cos(\alpha_1-\beta_1) - \cos(\alpha_1-\beta_2) + \cos(\alpha_2-\beta_1) + \cos(\alpha_2-\beta_2) \leq +2, \qquad (\chi)$$

  would be deducible from the true Bell sentence (inequality) τ ($\tau \in Th(\mathcal{G}_m)$) *if* there existed a value assignment s in $\mathcal{G}_m$ such that the variable μ is assigned the value $\mu_1 = \alpha_1$ in the first two terms occurring in τ AND μ is assigned the value $\mu_1 = \alpha_2$ in the last two terms occurring in τ with $\alpha_1 \neq \alpha_2$ (due to the distinct settings of each polariser), that is:

$$s: \quad \mu_1 = \alpha_1 \ \text{ AND } \ \mu_1 = \alpha_2 \ \text{ with } \alpha_1 \neq \alpha_2. \tag{10}$$

- Going from right to left in (10), we would then have $\mu_1 \neq \mu_1$ under one and the same s in $\mathcal{G}_m$, which is impossible as no number is different from itself, an obvious contradiction. Thus, no such s in $\mathcal{G}_m$ exists.
- It also follows that $\chi \notin Th(\mathcal{G}_m)$.
- The usual 'substitution' leading to the false Bell sentence χ ("violated inequality"), *distinct* from the true Bell sentence (inequality) $\tau(\tau \in Th(\mathcal{G}_m))$, would amount to the existence of a value assignment s in $\mathcal{G}_m$ under which a number is different from itself.
- We would respectfully submit that such an absurdity has nothing to do with locality or with the quantum formalism.

9. SUMMARY

- The EPR paper left open the question whether or not a finer description exists than that stipulated by any quantum state $|\psi\rangle$, and concluded with the belief that such a theory is possible.
- Postulate Π_3 of the proposed theory $Th(\mathcal{G})$ stipulates a finer state specification than that given by the singlet state $|\gamma_1, \gamma_2\rangle$.
- This *finer* state specification describes how the spontaneous annihilation process *breaks* the *spherical symmetry* of the singlet state $|\gamma_1, \gamma_2\rangle$ into a pair of back-to-back photons having only *axial symmetry* about the direction of their motion.
- Postulate Π_3 adjoins to QF the variable μ whose values characterise the new *local* "elements of reality" created by the symmetry breaking process.
- Each value of the variable μ specifies the *initial* ("hidden") direction of the common plane of polarization of the two back-to-back photons born at the instant the singlet state disintegrates.
- The proposed theory $Th(\mathcal{G})$ provides a realistic and local ("common cause") explanation of the EPRB ideal experiment.
- The proposed local theory $Th(\mathcal{G})$ extends (strengthens) QF by adding to QF four postulates which affect only the universe of discourse of QF by extending it to the universe (domain) of $\mathcal{G}$ to include the range of the variable μ (and of λ) in a consistent way.
- The Bell inequality τ is valid in $Th(\mathcal{G})$.

REFERENCES

1. Th.D. Angelidis, *Proc. Athens Acad.* **66**, 292 (1991).
2. Th.D. Angelidis, *J. Math. Phys.* **34**, 1635 (1993).
3. J.S. Bell, *Physics (N.Y.)* **1**, 195 (1964).
4. J.F. Clauser and M.A. Horne, *Phys. Rev. D* **10**, 526 (1974).
5. A. Shimony, *Found. Phys.* **19**, 1426 (1989).
6. R.P. Feynman, *Intl. J. Theor. Phys.* **21**, 467 (1982).
7. R. Carnap, *Introduction to Symbolic Logic* (Dover, New York, 1958), pp. 88, 111.
8. J.S. Bell, *CERN preprint TH-2926* (1980); published in *J. Phys. (Paris)* **42**, C2, 41 (1981).
9. J.L. Bell and M. Machover, *A Course in Mathematical Logic* (North-Holland, Amsterdam, 1977), pp. 10, 49-50, 162-163.
10. J.F. Clauser and A. Shimony, *Rep. Prog. Phys.* **41**, 1881 (1978).

CHAOS AND DECOHERENCE IN A QUANTUM SYSTEM WITH A REGULAR CLASSICAL COUNTERPART

Y. Ashkenazy,[a] **M. Lewkowicz,**[a,b] **and J. Levitan**[a,b]

[a] *Department of Physics, Bar-Ilan University Ramat-Gan, Israel*
[b] *The Research Institute, The College of Judea and Samaria Ariel, Israel*

We show that chaotic like behavior in a quantum system facilitates decoherence. It appears that the time scale on which decoherence takes place depends on the degree of complexity of the underlying quantum system, i.e., more complex systems decohere relatively faster than less complex ones.

Key words: chaos, decoherence, classical limit.

Bohr's correspondence principle states that quantum mechanical results become 'classical' when Planck's constant becomes very small. Ehrenfest's theorem, that the mean position of a quantum state will follow a classical trajectory, is generally taken as an argumentation for the correspondence principle, at least as far as dynamics is concerned. However, Ehrenfest's theorem is neither sufficient nor necessary to characterize the classical regime [1]. Ehrenfest's theorem breaks down sooner for a chaotic system than for a regular one [1,2]. Zurek and Paz showed [3] that a significantly non linear potential causes a rapid spread of the wave function implying that classical and quantum dynamics begin to differ.

In order to find a quantum mechanical paraphrase to the celebrated definition of classical chaos as being the exponential divergence in time of neighboring points in phase space we recognize the fast spreading of the wave function as an indication for chaotic-like behaviour. This is not a rigorous definition but rests on the fact that the wave function may be thought of as representing an ensemble of points in phase space and that its spread with time is a measure for the divergence of those points.

We have recently utilized a simple one dimensional model of a square barrier embedded in an infinite potential well to demonstrate that tunneling

M. Ferrero and A. van der Merwe (eds.),
New Developments on Fundamental Problems in Quantum Physics, 31–37.

leads to very complex behaviour of the wave function [2]. There are many parallels to the well-known characteristics of classical chaos, e.g., an exponential decreasing correlation function of the peak-to-peak time series [2], a phase space plot of the expectation values $< x > (t)$ and its time derivative $< p > (t)$, revealing a phenomenon similar to period doubling and attractor-like behaviour of a double well chaotic system such as a driven Duffing oscillator [2] and level statistics for levels slightly above the barrier showing signs of Wigner statistics [4].

But most striking is the behaviour of the spatial entropy function $S(t) = -\int |\psi(x,t)|^2 \ln|\psi(x,t)|^2 dx$. This entropy function shows a rapid rise at early times to a non periodic fluctuating function around a smooth almost constant asymptotic value illustrating the early burst of chaotic behaviour.

By displacing the barrier in the double well system to the right or to the left certain positions are passed where the system becomes almost degenerated [5]. It is exactly for those positions that one may find significant tunneling.

Fig. 1 shows $S(t)$ for different positions of the barrier in the well. In Fig. 1a the barrier is placed at the centre of the double well where the highest degree of the 'almost degeneracy' for the levels is found. Changing the site of the barrier and thus going through positions of less and less 'almost degeneracy' a slower and slower approach to equilibrium and a faster and faster approach to recurrence of $S(t)$ is noted (see Fig. 1b-1d and also [5]).

A time dependent adiabatic modulation of the barrier which is tantamount to coupling of the system to the environment enhances the complex, chaotic-like behaviour. A first indication of this is rendered by an increased correspondence to Wigner statistics when an ensemble of slightly different barrier heights is considered.

Considering that chaos facilitates tunneling and vice versa [2,5] and that tunneling is enhanced by a high degree of 'almost degeneracy,' we conclude (see Fig. 1a-1d) that systems with a higher degree of complex, chaotic like, behaviour have a sharper increase of the entropy (higher entropy production) and a faster approach to equilibrium than systems with lower complexity.

One can use the model to illustrate some differences between classical and quantum dynamics. We start with a wave packet initially located on the left hand side of the square barrier placed in the infinite potential well. The parameters are as follows: The barrier height $V = 5$, the half-width of the well $\ell = 55$, the width of the Gaussian wave packet of the form $|\psi|^2 \cong c\exp[-(x - x_0)^2/2\sigma^2], \sigma = 5$, and the barrier width $2a = 2$, and the mass $m = 1/2$. The Gaussian wave packet is constructed from the first 30 energy levels and the standard deviation from an exact (normed) Gaussian is $8.81 \cdot 10^{-5}$. The computation preserves the norm to high accuracy, 0.999997. The problem is solved analytically as a function of time. Note that there is

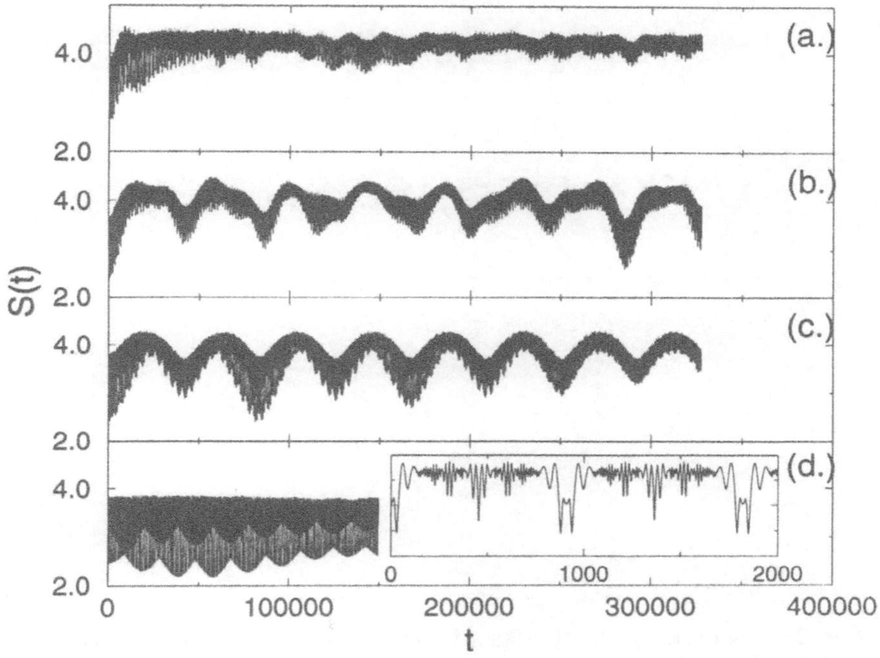

Fig. 1. The entropy function as a function of time for different locations in the well ($X \in [-55, 55]$). (a) the center of the barrier in the middle of the well X=0, (b) the center of the barrier in $X = -10,89$, (c) the center of the barrier in $X = -3,63$, (d) the center of the barrier in $X = -1$. (The inset shows an enlargement of typical periodic oscillations.)

no additional error for large t, since t enters in periodic exponential form.

The motion of this quantum state is compared to that of a classical ensemble whose initial position and momentum distributions are equal to those of the quantum state, the initial phase space distribution being the product of the position and the momentum distributions (this method is introduced and worked out in [1]).

In Figs. 2a-2d we have compared the time evolution of the spatial entropy function $S(t)$ for the quantum case and the classical case for decreasing values of the energy of the incoming particle high relative to the barrier height, $< E >= 5$. In Fig. 3a-3d the same comparison is performed for a low energy of the incoming particle (low relative to the barrier height), $< E >= 0.05$.

For both ratios $< E > /V$ it becomes obvious that decreasing $\hbar$ does not recover the classical behaviour of the entropy by merely letting $\hbar \rightarrow 0$, even though significant tunneling ($< E >\sim V$) leads to some similarity

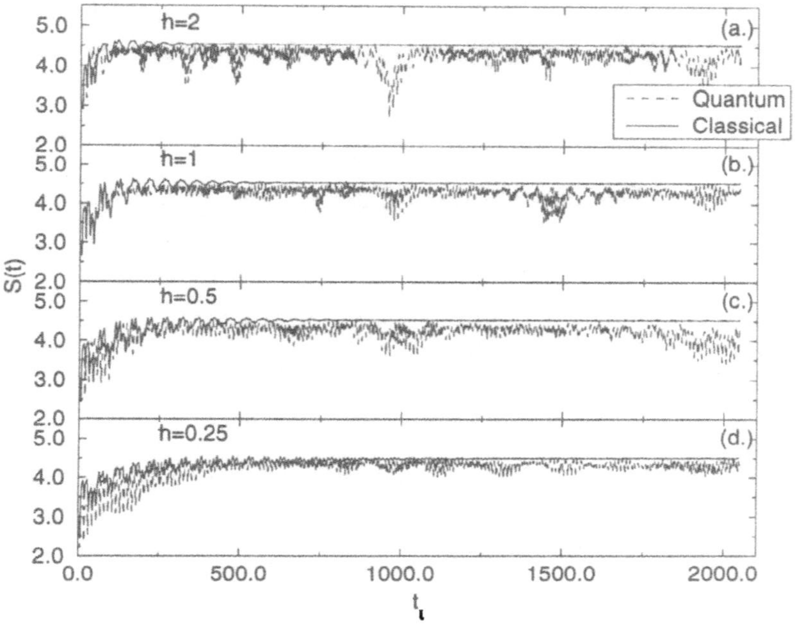

Fig. 2. The entropy function as a function of time for the classical case (solid line) and the quantum case (dotted), for different values of $\hbar$. The barrier is placed in the middle of the potential well and the mean energy of the incoming particle – $< E >$ – is equal to the barrier height $V, V = 5$. (a) $\hbar = 2$, (b) $\hbar = 1$, (c) $\hbar = 0.5$, (d) $\hbar = 0.25$.

between the classical and the quantum case.

Therefore, we stress that:

(1) $\hbar \rightarrow 0$ is a singular transition.
(2) Tunneling and thus chaotic-like behaviour weakens the differences between classical and quantum mechanics.

This second conclusion becomes even more apparent by comparing the classical case with the quantum case in connection with the position of the barrier inside the potential well. In [5] it is shown that placing the barrier at the center of the well results in a strong complex behaviour of the wave-function whereas placing the barrier off centre leads to a less complex, almost regular behaviour. Indeed, by comparing the two positions of the barrier, centered in Figs. 2a-2d and off center in Figs. 4a-4d, one observes that a similarity in the time development between the classical case and the quantum analogue is obtained much faster when the barrier is placed in the center, i.e., when there is a higher degree of 'almost degeneracy' and

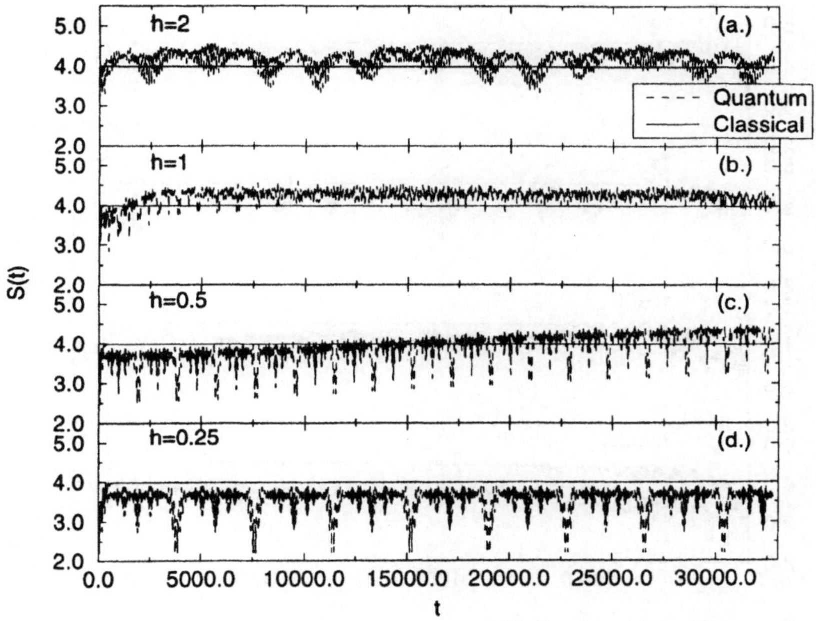

Fig. 3. The entropy function as a function of time for the classical case (solid line) and the quantum case (dotted), for different values of $\hbar$. The barrier is placed in the middle of the potential well and the mean energy of the incoming particle– $< E >= 0.05$– is low relative to the barrier height $V, V = 5$. (a) $\hbar = 2$, (b) $\hbar = 1$, (c) $\hbar = 0.5$, (d) $\hbar = 0.25$.

hence more chaotic like behaviour. We therefore conjecture that the time scale on which decoherence takes place depends on the degree of complexity of the underlying quantum mechanical system, i.e., more complex systems decohere relatively faster than less complex ones.

This is in accordance with the expectation that decoherence effects should suppress the possibility of interference and hence reduce the off diagonal elements of the reduced density matrix. The Schrödinger evolution cannot transform a pure state into a mixture and hence not make the off diagonal elements vanish completely (see also [6,7]). Quantum chaos on the other hand is expected to lead to irreversibility and hence also to manifest itself by implying that the non diagonal elements vanish (or become very small) [6-9], so it is understandable that quantum chaos will facilitate decoherence. This is illustrated in the present model by a fast increasing spatial entropy function (rapid entropy production) and is in accordance with previous results obtained in studies of decohering systems [6-8].

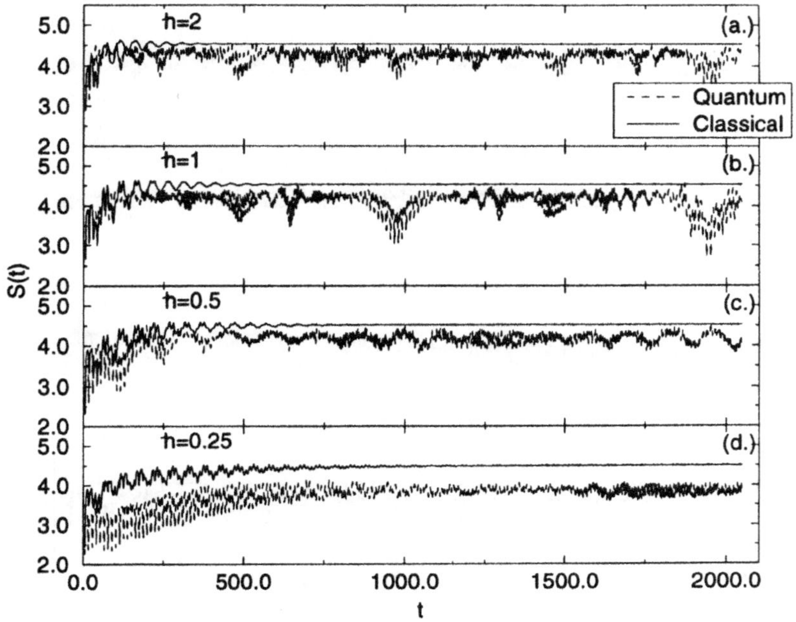

Fig. 4. The entropy function as a function of time for the classical case (solid line) and the quantum case (dotted) for different values of $\hbar$. The barrier is placed off center, $X = -1$, and the mean energy of the incoming particle $- < E > -$ is equal to the barrier height $V, V = 5$. (a) $\hbar = 2$, (b) $\hbar = 1$, (c) $\hbar = 0.5$, (d) $\hbar = 0.25$.

REFERENCES

1. L.E. Ballentine, Yumin Yang and J.P. Zibin, "Inadequacy of Ehrenfest's theorem to characterize the classical regime," *Phys. Rev. A* **50**, 2854-2859 (1994).
2. Y. Ashkenazy, L.P. Horwitz, J. Levitan, M. Lewkowicz, and Y. Rothschild, "Chaoticlike behavior in a quantum system without classical counterpart," *Phys. Rev. Lett.* **75**, 1070-1073 (1995).
3. W.H. Zurek and J.P. Paz, "Decoherence, chaos and the second law," *Phys. Rev. Lett.* **72**, 2508-2511 (1994).
4. R. Berkovits, Y. Ashkenazy, L.P. Horwitz, and J. Levitan, "Chaotic signatures in the spectrum of a quantum double well," *Physica A*, in press.
5. L.P. Horwitz, J. Levitan, and Y. Ashkenazy, "Complexity, tunneling and geometrical symmetry," *Phys. Rev. E*, in press.
6. R. Omnès, *The Interpretation of Quantum Mechanics* (Princeton University Press, Princeton, 1994), pp. 268-273.
7. W.H. Zurek and J.P. Paz, "Quantum chaos: a decoherent definition,"

Physica D **83**, 300-308 (1995).
8. J. Levitan, M. Lewkowicz, and Y. Ashkenazy, "Quantum chaos as a decoherence enhancing agent," *Found Phys.*, in press.
9. L. Bonci, R. Roncaglia, B.J. West, and P. Grigolini, "Quantum irreversibility and chaos," *Phys. Rev. Lett.* **67**, 2593-2596 (1991).

QUANTUM COMPUTATION

A. Barenco, S. F. Huelga,[1] and A. K. Ekert

Clarendon Laboratory, University of Oxford
Oxford OX1 3PU, United Kingdom

We describe how physics of computation determines computational complexity. In particular we show how quantum phenomena lead to qualitatively new modes of computation. An ideal quantum computer would allow to face certain tasks out of reach of classical computers, the factorization problem being the most striking example. Experimentally, however, implementation of quantum computation faces some serious difficulties due to an interaction with the environment which causes decoherence. We mention quantum error-correction as a process which can, to some extent, protect quantum computers from unwelcome effects of dissipation and decoherence. This is a brief, introductory review of quantum computation. It is based on [1].

Key words: quantum computation, error correction.

1. COMPUTATION AND PHYSICS

Computers are physical objects and computations are physical processes. Quantum computers are machines that rely on characteristically quantum phenomena, such as quantum interference and quantum entanglement in order to perform computation.

The classical theory of computation usually does not refer to physics and as the result it is often falsely assumed that its foundations are self-evident and purely abstract. Although in the sixties Landauer [2] pointed out the physical nature of information, it was not until the first works on quantum computation by Deutsch [3] and Feynmann [4] that the fundamental connection between the laws of physics and computation was properly

[1] Permanent Address: Departamento de Fisica, Universidad de Oviedo, Calvo Sotelo s/n, 33007 Oviedo, Spain.

M. Ferrero and A. van der Merwe (eds.),
New Developments on Fundamental Problems in Quantum Physics, 39–54.

emphasised. Recent developments in the theory of quantum computational complexity [5,6,7,8] provide a vivid example of this connection.

Computers solve problems following a precise set of instructions that can be mechanically applied to yield the solution to any given instance of a particular problem. A specification of this set of instruction is called an algorithm. Examples of algorithms are the procedures taught in elementary schools for adding and multiplying whole numbers; when these procedures are mechanically applied, they always yield the correct result for any pair of whole numbers. However, any operation on numbers is performed by physical means and what can be done to a number depends on the physical representation of this number and the underlying physics of computation. For example, when numbers are encoded in quantum states then quantum computers, *i.e.* physical devices whose unitary dynamics can be regarded as the performance of computation, can accept states which represent a coherent superposition of many different numbers (inputs) and evolve them into another superposition of numbers (outputs). In this case computation, *i.e.* a sequence of unitary transformations, affects simultaneously each element of the superposition allowing a massive parallel data processing albeit within one piece of quantum hardware. As the result quantum computers can efficiently solve some problems which are believed to be intractable on any classical computer [3,5,6,7,8].

In this paper we illustrate the relevance of the underlying physics of computation by describing the difference between classical and quantum computation. We have chosen *factorisation* as an example that sets the efficiency of quantum computation ahead of any classical data processing. Indeed the contrast is striking. Shor [8] has recently shown that quantum computers can efficiently factor big integers and compute discrete logarithms. These tasks are believed to be intractable on any classical computer!

Finally let us mention also that factorisation is not of purely academic interest. It is the problem which underpins security of many classical public key cryptosystems, for example, RSA — the most popular public key cryptosystem named after the three inventors, Rivest, Shamir, and Adleman [9] — gets its security from the difficulty of factoring large numbers. Hence for the purpose of cryptoanalysis the experimental realisation of quantum computation is a most interesting issue.

2. COMPLEXITY OF FACTORING

Consider an integer N with L decimal digits. Factoring N means finding its prime factors, *i.e.*, finding whole numbers $\{p\}$ such that any p divides N with the remainder 0.

One way to calculate the prime factors is to try to divide N by

$2, 3, \ldots \sqrt{N}$ and to check the reminder. This method is very time consuming. It requires about $\sqrt{N} \approx 10^{L/2}$ divisions, hence the time it takes to execute this algorithm increases exponentially with L. Even if the computer can perform as much as 10^{10} divisions per second it would take about a second to factor a 20 digit number, about a year to factor a 34 digit number and more than the estimated age of the Universe (10^{17}s) to factor a 60 digit long number!

Although the problem of finding an efficient algorithm for factoring large numbers was worked on by famous mathematicians such as Fermat and Legendre, only recently (since the invention of public key cryptosystems in the seventies) a significant progress has been made in designing good factoring algorithms. The best algorithms such as the Multiple Polynomial Quadratic Sieve [10] and the Number Field Sieve [11]) have an execution time that grows as a subexponential function of L ($\sim \exp[L^{1/3}]$). Although they are much faster than the trial division method still they cannot be regarded as efficient algorithms.

For an algorithm to be efficient (and usable), the time it takes to execute the algorithm must increase no faster than a polynomial function of the size of the input (L in our case). If the best algorithm we know for a particular problem has the execution time (viewed as a function of the size of the input) bounded by a polynomial then we say that the problem belongs to class P. Problems outside class P are known as *hard* problems. Thus we say, for example, that multiplication is in P whereas factoring is not in P and that is why it is a hard problem (for more information about computational complexity see, for example, [12] or [13]).

The snag is that the complexity classes such as P are defined with respect to classical computation. Classical algorithms for factorisation are not known to belong to P, however, there exists an efficient quantum factoring algorithm [8]!

3. QUANTUM REGISTERS

We start our description of quantum computation with the basic unit of information namely a single bit. If a physical object can be put into two different, distinguishable states then this object can represent two different numbers. We call any two–state system a physical bit; when the system is quantum and the two states are two orthogonal *quantum* states, we refer to it as a *quantum bit* or simply a *qubit*. Any two–state quantum system is a potential candidate for a qubit. Both a single classical bit and a qubit can represent at most two different numbers, however, qubits are different because apart from the two orthogonal basis states, which we label as $|0\rangle$ and $|1\rangle$, they can also be put into infinitely many other states of the form

$|\Psi\rangle = c_0|0\rangle + c_1|1\rangle$.

Let us mention in passing that although a qubit can be prepared in an infinite number of different quantum states it cannot be used to transmit more that one bit of information. This is because no detection process can reliably differentiate between nonorthogonal states (c.f. [14, 15, 16, 17]). However, information encoded in nonorthogonal states and in quantum entanglement can be used in systems known as quantum cryptography [18,19,21] or quantum teleportation [22].

Consider now a register composed of m physical qubits. There are 2^m different orthogonal quantum states of this register therefore the register can represent 2^m different numbers, *e.g.* from 0 to $2^m - 1$. The most general (pure) state of this register can be written as

$$|\Psi\rangle = \sum_x c_x|x\rangle, \tag{1}$$

where number x is represented in the register in binary form

$$|x\rangle = |x_{m-1}\rangle \otimes |x_{m-2}\rangle \otimes \ldots |x_1\rangle \otimes |x_0\rangle \tag{2}$$

according to the decomposition

$$x = \sum_{i=0}^{m-1} 2^i x_i, \quad x_i = 0 \text{ or } 1. \tag{3}$$

Note that Eq. 1 describes the state in which several different values of the register are present *simultaneously*; this quantum feature has no classical counterpart. In order to prepare a specific number in the register we have to perform m elementary operations on each qubit to set it into one of the two orthogonal state $|0\rangle$ or $|1\rangle$. However, in quantum computers m elementary unitary transformations performed bit by bit can also prepare the register in a coherent superposition of all 2^m numbers that can be stored in the register. Take the register initially in $|0\rangle \otimes |0\rangle \otimes \ldots |0\rangle$ state and apply the unitary operation

$$A = \frac{1}{\sqrt{2}}\begin{pmatrix} 1 & 1 \\ 1 & -1 \end{pmatrix} \tag{4}$$

to each qubit. The resulting state of the register is an equally weighted superposition of all 2^m numbers,

$$|\Psi\rangle = \overbrace{A|0\rangle \otimes A|0\rangle \otimes \ldots A|0\rangle}^{m \text{ times}} = \frac{1}{2^{m/2}} \sum_{x=0}^{2^m-1} |x\rangle. \tag{5}$$

It is quite remarkable that in quantum registers m elementary operations can generate a state containing all 2^m possible numerical values of the register. In contrast, in classical registers m elementary operations can only prepare one state of the register representing one specific number. It is this ability of creating quantum superpositions which makes the "quantum parallel processing" possible. If after preparing the register in a coherent superposition of several numbers all subsequent computational operations are unitary (*i.e.*, preserve the superpositions of states) then with each computational step the computation is performed simultaneuosly on all the numbers present in the superposition.

This type of computation is particularly useful for problems which, in classical case, involve performing the same computation several times for different input data. Estimating the period of a given function $f(x)$ is a typical example as it requires evaluating function $f(x)$ many times for different x.

4. COMPUTING FUNCTIONS

Let us describe now how quantum computers compute functions. For this we will need two quantum registers of length m and n. Consider a function

$$f : \{0, 1, \ldots 2^m - 1\} \longrightarrow \{0, 1, \ldots 2^n - 1\}, \tag{6}$$

where m and n are natural numbers.

A classical computer computes f by evolving each labeled input, $0, 1, \ldots 2^m - 1$ into a respective labeled output, $f(0), f(1), \ldots f(2^m - 1)$. Quantum computers, due to the unitary (and therefore reversible) nature of their evolution, compute functions in a slightly different way. In order to compute functions which are not one–to–one and to preserve the reversibility of computation, quantum computers have to keep the record of the input. Here is how it is done.

We will use the two quantum registers; the first register to store the input data, the second one for the output data. Each possible input x is represented by $|x\rangle$ — the quantum state of the first register. Analogously, each possible output $y = f(x)$ is represented by $|y\rangle$ — the quantum state of the second register. Vectors $|x\rangle$ belong to the 2^m–dimensional Hilbert space $\mathcal{H}_1$, and vectors $|y\rangle$ belong to the 2^n–dimensional Hilbert space $\mathcal{H}_2$ ($\mathcal{H}_1$ and $\mathcal{H}_2$ are tensorial products of the Hilbert spaces of respectively m and n qubits). States corresponding to different inputs and different outputs are orthogonal, $\langle x|x'\rangle = \delta_{xx'}$, $\langle y|y'\rangle = \delta_{yy'}$. The function evaluation is then determined by the evolution of the two registers,

$$|x\rangle|0\rangle \xrightarrow{U_f} |x\rangle|f(x)\rangle. \tag{7}$$

We can always prepare specific x in the first register and read the value $f(x)$ from the second register. It was shown that as far as the computational complexity is concerned a reversible function evaluation, *i.e.*, the one that keeps track of the input, is as good as a regular, irreversible evaluation [23]. This means that if a given function can be computed in polynomial time it can also be computed in polynomial time using a reversible computation. The computation we are considering here is not only reversible but also quantum and we can do much more than computing values of $f(x)$ one by one. We can prepare a superposition of all input values as a single state and by running the computation U_f *only once*, we can compute *all* of the 2^m values $f(0), \ldots, f(2^m - 1)$,

$$\left(\frac{1}{2^{m/2}} \sum_{x=0}^{2^m-1} |x\rangle\right)|0\rangle \xrightarrow{U_f} \frac{1}{2^{m/2}} \sum_{x=0}^{2^m-1} |x\rangle|f(x)\rangle. \tag{8}$$

It looks too good to be true, so where is the catch? How much information about f does the state

$$|f\rangle = \frac{1}{2^{m/2}}(|0\rangle|f(0)\rangle + |1\rangle|f(1)\rangle + \ldots + |2^m - 1\rangle|f(2^m - 1)\rangle) \tag{9}$$

contain?

Unfortunately no quantum measurement can extract all of the 2^m values $f(0), f(1), \ldots, f(2^m - 1)$ from $|f\rangle$. However, there are measurements that provide us with information about joint properties of all the output values $f(x)$ such as, for example, periodicity. We will see in the following sections, how a periodicity estimation can lead to fast factorisation.

5. FACTORING AND PERIODICITY

The factorisation problem is related to finding periods of certain functions. In particular one can show that finding factors of N is equivalent to finding a period of $f_N(x)$, where

$$f_N(x) = a^x \bmod N \tag{10}$$

and a is *any* randomly chosen number which is coprime with N. The result of this operation is the remainder after the division of a^x by N. The function is periodic and the period r, which depends on a and N, is called the order of a modulo N. For example, the increasing powers of 2 modulo 15 go like $1, 2, 4, 8, 1, 2, 4, 8, 1, \ldots$ and so on — the order of 2 modulo 15 is 4 (for more information see, for example, [24]).

Knowing r, *i.e.*, the order of a modulo N, we can factor N provided r is even and $r \bmod N \neq -1$. When a is chosen randomly the two conditions are satisfied with probability greater than half. To factor N it is enough to calculate the greatest common divisor of $a^{r/2} \pm 1$ and N. Fortunately an easy and very efficient algorithm to compute the greatest common divisor has been known since 300 BC. The algorithm, known as the Euclidean algorithm, is described in Euclid's *Elements*, the oldest Greek treatise in mathematics to reach us in its entirety (try your elementary school textbooks as a reference). The result of taking this greatest common divisor, written as $(a^{r/2} \pm 1, N)$, is a factor of N.

To see how this method works let us consider a very simple example of factoring 15. Firstly we select a, such that $(a, N) = 1$; *i.e.*, a could be any number from the set $\{2, 4, 7, 8, 11, 13, 14\}$. Let us pick up $a = 11$ and let us compute the order of 11 modulo 15. Values of $11^x \bmod 15$ for $x = 1, 2, 3, \ldots$ go as $11, 1, 11, 1, 11, \ldots$ giving $r = 2$. Then we compute $a^{r/2}$ which gives 11 and we find the largest common factor $(11 \pm 1, N)$ *i.e.* $(10, 15)$ and $(12, 15)$ which gives 5 and 3, the two factors of 15. Respective orders modulo 15 of elements $\{2, 4, 7, 8, 11, 13, 14\}$ are $\{4, 2, 4, 4, 2, 4, 2\}$ and in this particular example any choice of a except $a = 14$ leads to the correct result. For $a = 14$ we obtain $r = 2$, $a^{r/2} \equiv -1 \bmod 15$ and the methods fails.

Classically finding r is as time consuming as finding factors of N by the trial divisions, however, if we employ quantum computation r can evaluated very efficiently. Shor [8] describes a quantum algorithm which provides the order r of a randomly chosen a and which runs in polynomial time *i.e.*, requires poly$(\log N)$ steps. Let us now outline the main features of this algorithm.

6. MEASURING PERIODICITY

Suppose you want to find period r of $f_N(x)$ where $x = 0, 1, 2, \ldots M-1$ for some large $M = 2^m$ ($M \approx N^2$ and the values $f(0), f(1), \ldots f(r)$ are all different). Here is an efficient quantum method. First we choose a computational basis (which we label $\{|x\rangle\}$ for the first register and $\{|f_N(x)\rangle\}$ for the second register) and compute function $f_N(x)$ in a quantum way:

$$\frac{1}{\sqrt{M}} \left(\sum_{0}^{M-1} |x\rangle \right) |0\rangle \xrightarrow{U_{f_N}} \frac{1}{\sqrt{M}} \sum_{0}^{M-1} |x\rangle |f_N(x)\rangle. \tag{11}$$

Function $f_N(x) = a^x \bmod N$ can be computed efficiently. Next perform a measurement in the computational basis to determine the bit values in the second register. Suppose the outcome of this measurement is $f_N(l)$ for a least l ($f_N(l) = f_N(jr + l)$ for $j = 0, 1, 2, \ldots$). The post measurement state is

$$|\phi_l\rangle = \frac{1}{\sqrt{A+1}} \sum_{j=0}^{A} |jr+l\rangle |f_N(l)\rangle, \tag{12}$$

where A is the greatest integer less than M/r. Thus in the first register we have a uniform superposition of labeled basis states where the labels have been chosen with period r $(l, l+r, l+2r, \ldots, l+Ar)$. From this state we wish to extract the information about the periodicity r.

The extraction of r will be achieved by applying to the first register the quantum discrete Fourier transform, *i.e.*, the unitary transformation (DFT) which acts on a M dimensional Hilbert space and is defined relative to a chosen basis $|0\rangle, \ldots, |M-1\rangle$ by

$$DFT : |x\rangle \longmapsto \frac{1}{\sqrt{M}} \sum_{y=0}^{M-1} \exp(2\pi i x y/M)\, |y\rangle. \tag{13}$$

The reason for calling this particular unitary transformation the discrete Fourier transform becomes obvious when you notice that in the transformation

$$DFT: \quad |\phi_{in}\rangle = \sum_x c_x |x\rangle \longmapsto |\phi_{out}\rangle = \sum_y c_y |y\rangle \tag{14}$$

the coefficients c_y are the discrete Fourier transforms of c_x's, *i.e.*,

$$c_y = \frac{1}{\sqrt{M}} \sum_x \exp(2\pi i x y/M) c_x. \tag{15}$$

There exists an efficient quantum algorithm for DFT which is a quantum analog of the Fast Fourier Transform algorithm (for details, see [25, 21]).

To see the principle of how this works, we consider first the simplified situation where r divides M exactly. Write $A = M/r - 1$. The final state corresponding to (12) is then

$$|\phi_{in}\rangle = \sqrt{\frac{r}{M}} \sum_{j=0}^{A} |jr+l\rangle = \sum_{x=0}^{M-1} c_x |x\rangle, \tag{16}$$

where $c_x = \sqrt{(r/M)}\delta_{x,jr+l}$ for $j = 0, 1, \ldots A$ is a periodic function of x (see Fig. 1.) Performing DFT on $|\phi_{in}\rangle$ gives

$$|\phi_{out}\rangle = \sum_y c_y |y\rangle, \tag{17}$$

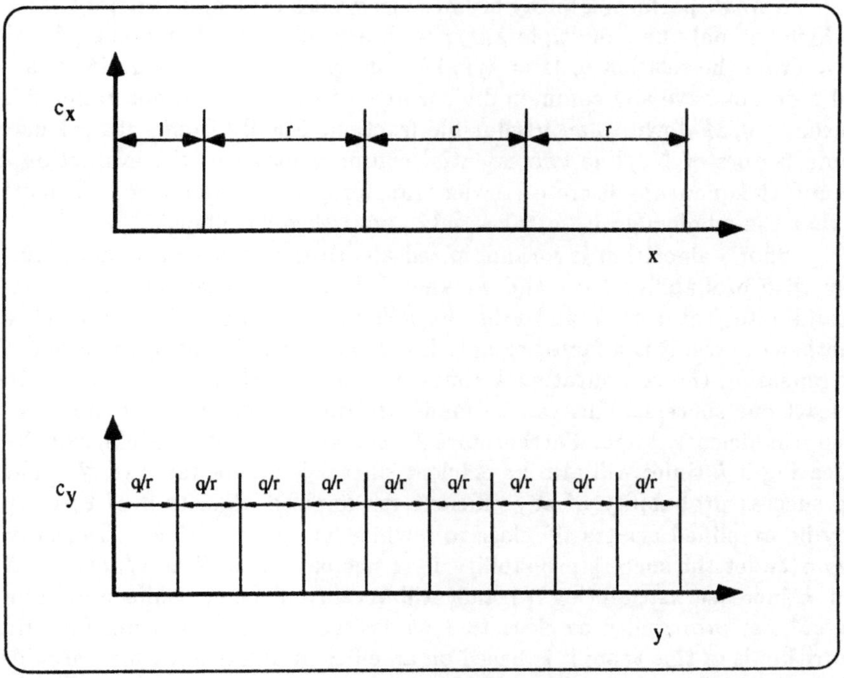

Fig. 1. c_x and the modulus of its Fourier transform c_y.

where the amplitude of c_y is

$$c_y = \frac{\sqrt{r}}{M}\sum_{j=0}^{A}\exp\left(\frac{2\pi i(jr+l)y}{M}\right) = \frac{\sqrt{r}}{M}\left[\sum_{j=0}^{A}\exp\left(2\pi i\frac{jry}{M}\right)\right]\exp\left(2\pi i\frac{ly}{M}\right). \tag{18}$$

The term in the square bracket on the r.h.s. is zero unless y is a multiple of M/r,

$$c_y = \begin{cases} \exp(2\pi i l y/M)/\sqrt{r}, & \text{if } y \text{ is a multiple of } M/r \\ 0, & \text{otherwise,} \end{cases} \tag{19}$$

i.e., the Fourier transform of a state with period r is a state with period M/r:

$$|\phi_{out}\rangle = \frac{1}{\sqrt{r}}\sum_{j=0}^{r-1}\exp(2\pi i l y/M)\,|j\frac{M}{r}\rangle. \tag{20}$$

Note that the Fourier transform "inverts" the periodicity of the input ($r \to M/r$) and has a translation invariance property which washes out the shift l (see Fig.1).

Now we perform a bit by bit measurement on the first register to learn y which can only be a multiple $\lambda M/r$ with $\lambda = 0, \ldots, r-1$ chosen equiprobably. From the relation $y/M = \lambda/r$, knowing y and M and assuming that λ and r do not have any common divisor apart from 1 we can determine r by canceling y/M down to an irreducible fraction. Finally from r we calculate prime factors of N. The two essential computations *i.e.*, the evaluation of f_N and the quantum discrete Fourier transform can be performed efficiently so that the whole algorithm takes only about $(\log N)^3$ steps!

Shor's algorithm is a randomized algorithms which runs successfully only with probability $1 - \epsilon$ and we know when it is successful. It produces a candidate factor of N and must be followed by a trial division to check whether the result is a factor or not. If $\epsilon > 0$ is independent of the input N. By repeating the computation k times, we get probability $1 - \epsilon^k$ of having at least one success. This can be made arbitrarily close to 1 by choosing a fixed k sufficiently large. Furthermore if a single computation is efficient then repeating it k times will also be efficient since k is independent of N . Thus the success probability of any efficient randomized algorithm of this type may be amplified arbitrarily close to 1 while retaining efficiency. Indeed we may even let the success probability $1 - \epsilon$ decrease with N as $1/\text{poly}(\log N)$ and k increase as $\text{poly}(\log N)$ and still retain efficiency while amplifying the success probability as close to 1 as desired. Shor's quantum factoring algorithm is of this type; it is based on an efficient algorithm which provides a factor of the input N with probability which decreases as $1/\text{poly}(\log N)$.

The randomness in the algorithm is due to certain mathematical results concerning the distribution of prime and coprime numbers. For example, for λ being chosen at random it follows from number theory that the probability of λ and r to have no common divisor apart from 1 is greater than $1/\log r$ for largish r (see for example [24] or [27]). Also if we abandon the assumption that r divides M (very unlikely and adopted here only for pedagogical purposes) the Fourier transform of c_x will not produce sharp maxima as in Fig. 1. which may contribute to possible errors while reading y from the register. Subsequent estimation of r is calculated using additional mathematical approximation techniques (continued fraction expansion).

If we try to factor bigger and bigger numbers N it is enough to repeat the computation $\approx$ *poly*$(\log N)$ times to amplify the success probability as close to 1 as we wish. This gives an efficient determination of r and an efficient method of factoring any N!

7. FROM QUANTUM COMPUTATION TO QUANTUM NETWORKS

An open question has been whether it would ever be practical to build physical devices to perform such computations, or whether they would forever remain theoretical curiosities. Like classical computers, quantum computers can be built out of logic gate networks. In the case of a quantum computer, a logic gate can be thought as a unitary operation that acts only on the space of a restricted number of qubits, one for a one–bit gate, two for a two–bit gate, etc. Deutsch [28] described quantum networks composed of elementary logic gates connected together by wires and showed that there exists a universal quantum gate from which any quantum computation can be built. More recently Barenco [29] and independently Sleator and Weinfurter [30] proved that a single two–bit gate suffices to implement the Deutsch gate. Finally it has been shown that almost any non–trivial two–bit gate is universal [31,32].

Quantum logic gates perform elementary unitary operations on qubits. In this section we will illustrate how complex quantum operations, such as the quantum discrete Fourier transform discussed above, can be implemented as a network consisting of only one– and two–bit gates.

Consider the single qubit gate **A** performing the unitary transformation

$$A = \frac{1}{\sqrt{2}} \begin{pmatrix} 1 & 1 \\ 1 & -1 \end{pmatrix}, \tag{21}$$

where the diagram on the right provides a schematic representation of the gate **A** acting on a qubit q. Consider also the two–bit gate $\mathbf{B}(\phi)$ acting on qubits q_1 and q_2 and performing the operation

$$B(\phi) = \begin{pmatrix} 1 & 0 & 0 & 0 \\ 0 & 1 & 0 & 0 \\ 0 & 0 & 1 & 0 \\ 0 & 0 & 0 & e^{i\phi} \end{pmatrix}$$

in the Hilbert space $\mathcal{H} = \mathcal{H}_{q_1} \otimes \mathcal{H}_{q_2}$ of the two qubits with the basis $\{|0\rangle|0\rangle, |0\rangle|1\rangle, |1\rangle|0\rangle, |1\rangle|1\rangle\}$ (the diagram on the right shows the structure of the gate). The gate $\mathbf{B}(\phi)$ performs a conditional phase shift *i.e.* multiplication by phase factor $e^{i\phi}$ but only if the two qubits are both in their $|1\rangle$ states.

The two gates can be used to implement the efficient quantum DFT on a register of any size. For example, consider a four–bit register with qubits $a_0, \ldots a_3$. The network in Fig. 2 follows step by step the classical algorithm of a DFT (see for instance [33]), and perform the operation

$$|a\rangle \longmapsto \frac{1}{\sqrt{2^4}} \sum_{c=0}^{2^4-1} \exp(2\pi i a c/2^4)|b\rangle, \tag{23}$$

where $|b\rangle$ represents the value c read *reversing* the order of the bits, *i.e.*,

$$b = \sum_{i=0}^{3} 2^i c_{3-i}, \quad \text{with } c_k \text{ given by} \quad c = \sum_{k=0}^{3} 2^k c_k. \tag{24}$$

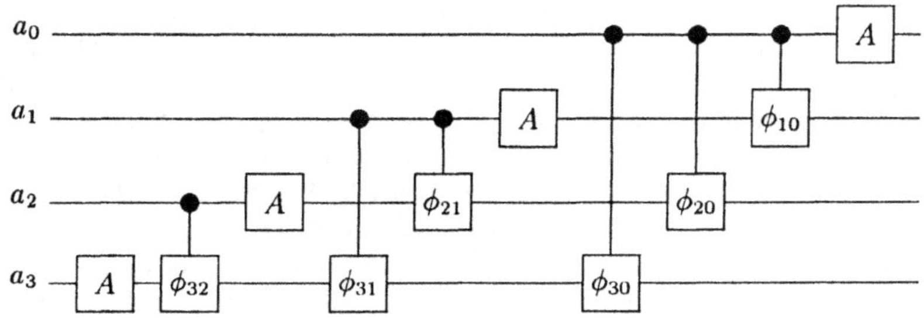

Fig. 2. Network effecting a DFT on a four-bit register, the phases that appear in the operations $\mathbf{B}(\phi_{jk})$ are related to the "distance" of the qubits upon which **B** acts, namely $\phi_{jk} = \pi/2^{j-k}$.

A general case of L qubits requires a trivial extension of the network following the same sequence pattern of gates **A** and **B**.

Each gate operates for a fixed period of time (the clock time of the computer) and the number of gates needed to complete the full quantum DFT grows only as a quadratic function of the size of the register. (The transformation on the L–qubit register requires L operations **A** and $L(L-1)/2$ operations **B**, in total $L(L+1)/2$ elementary operations). Thus the quantum DFT can be performed efficiently. Moreover, it can be even simplified. Note that in the network shown in Fig. 2, the operations $\mathbf{B}(\phi)$ that involve distant qubits a_j and a_k, *i.e.*, qubits for which $|j-k|$ is big (and therefore $\phi = \pi/2^{k-j}$ approaches zero), are close to unity. Therefore when performing the quantum DFT on registers of size L, one can neglect operations **B** on distant qubits (more precisely on qubits a_j and a_k for which $|j-k| > \log_2(L)+2$) and still retrieve the periodicity of coefficients c_x.

The network of gates for the quantum DFT enables the efficient implementation of the second part of Shor's algorithm. The first part requires an efficient quantum evaluation of the function $f_N(x) = a^x \bmod N$. The computation of $f_N(x)$ is "easy," *i.e.*, the number of gates does not grow faster than a polynomial in the size of the input. The respective network is constructed by combining networks which perform addition and multiplication in a reversible and unitary way.

8. PRACTICALITIES

It remains an open question which technology will be employed to build the first quantum computer. In principle, the conditional quantum dynamics which supports quantum logic gates and quantum networks can be implemented in many different ways, ranging from the Ramsey atomic interferometry [34], interacting electrons in quantum dots [35] to ions in ion traps [36] and atoms coupled to high finesse optical cavities [37]. However, when one deals with realistic experimental conditions the operation is no longer error-free, which will prevent to maintain a coherent unitary evolution until the completion of the computation. There are basically two types of errors:

- Errors induced by the imperfection of the operation of the quantum computer (*coherent noise*)

- Errors induced by the coupling to the surronding enviroment, responsible for decay and decoherence processes (*incoherent noise*).

These errors will cause quantum coherence to be lost or partially destroyed, thus reducing the efficiency of the computer.

Simple theoretical models of decoherence [38, 39] show that the probability P of a successful computation in a single run decreases exponentially with the input size L. If $S(L)$ denotes the total number of qubits needed to perform the calculation and $t(L)$ the total time to perform it,

$$P = P_0\, e^{(-S(L)t(L)\tau_{dec})}, \tag{25}$$

τ_{dec} being the typical decoherence time (see [40] for the different orders of magnitude involved in the technogies mentioned at the begining of this section) and P_0 the probability with no errors. Hence, to obtain at least one correct computation one needs to run the computer on average $1/P = 1/(P_0 e^{(-S(L)t(L)\tau_{dec})})$ times and the problem becomes exponentially difficult. In fact, the bound obtained for a coherent computation to be completed within the decoherence time is dramatically reduced when decay processes, such as spontaneous emission, are taken into account (See M. B. Plenio and P. L. Knight in these proceedings). Consequently, to overcome the effect of dissipation/decoherence will require a procedure different from simply increasing the number of runs. What we need is some form of efficient *"quantum error correction"* to stabilise the computation.

The basic idea underlying this strategy is imported from classical error correction codes [41] and it is illustrated in Figure 3. The information that needs to be protected from the influence of noise is encoded in more bits than the required minimum (redundancy). The system is made larger and hence more resistent to perturbations; in this way, even if some random bits experiences random unwanted interactions, the information encoded in the remaining ones is sufficient to reconstruct the initial state through an appropiate decoding procedure.

It is useful to think of the issue of error correction as a problem of communication over a noisy quantum channel where Alice wants to transmit

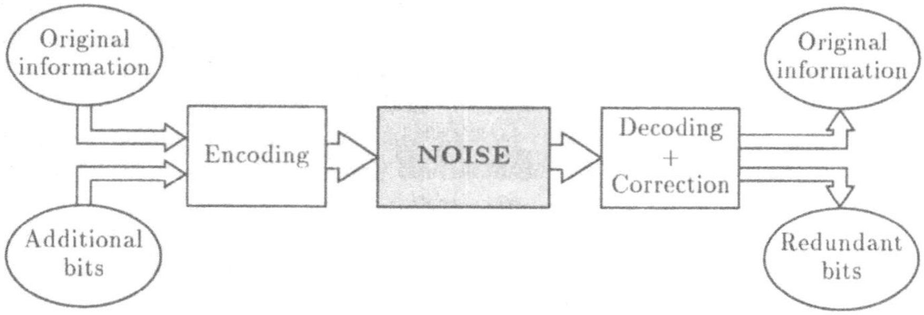

Fig. 3. Basic operation of an error correction code.

Bob a message consisting of l qubits. In order to increase the probability of error-free transmission she encodes the state of l qubits into a set of n qubits. If no more than k errors occur on k (or less) of the n transmitted qubits, Bob can retrieve the initial state via a decoding operation which involves projective measurements onto subspaces that belong to different possible errors. The condition that these subspaces are mutually orthogonal leads to certain conditions on the codewords [48] from which one can derive bounds for n, k and l analogous to the sphere packing bound in classical error correction. Assuming perfect encoding and decoding operations, it has been proven so far that to protect a single qubit ($l = 1$) against one error ($k = 1$), one needs to encode it in five ($n = 5$) qubits [42]. Provided that only dephasing errors occur, then only $n = 3$ are necessary [43].

A problem in the practical implementation of the codes is the occurrence of errors during the error correction which would destroy the error correction procedure completely. However, recent results [44,45] indicate a way to deal with this problem by implementing fault tolerant quantum computation and error correction. As a summary we can say that, although some general results on the construction of particular error-correcting codes have been obtained [46,47,48,49,50], it is still an open and very important question whether quantum error correction can save the exponential power of quantum computation.

9. CONCLUSION

Although the current technologies cannot support even a very simple quantum factorisation we hope that the world-wide experimental efforts will make practical quantum computation possible in a not too distant future. It should be stressed, however, that from the fundamental standpoint it is irrelevant when exactly the first non-trivial quantum computer is built — what matters is that quantum computation tell us about a connection between physics and computation making the two branches of science inseparable.

ACKNOWLEDGEMENTS

This work was partially supported by the Royal Society. A. B. acknowledges the financial support of the Swiss Science Foundation, as well as Worcester College, Oxford. S.F.H acknowledges support from TMR European Community program.

REFERENCES

1. A. Barenco and A. Ekert, *Acta Physica Slovaca* **45**, 205-216 (1995). A. Barenco, *Contemp. Phys.* **37** (5), 375–389 (1996).
2. R. Landauer, *IBM J. Res. Dev.* **5**, 183 (1961).
3. D. Deutsch, *Proc. R. Soc. London A* **400**, 97 (1985).
4. R. P. Feynmann, *Int. J. Theor. Phys.* **21**, 467 (1982).
5. D. Deutsch. and R. Jozsa, *Proc. R. Soc. London A* **439**, 553 (1992).
6. E. Bernstein and U. Vazirani, in *Proceedings, 25th ACM Symposium on the Theory of Computation*, p. 11 (1993).
7. D. S. Simon in *Proceedings of the 35th Annual Symposium on the Foundations of Computer Science*, S. Goldwasser, ed. (IEEE Computer Society Press, Los Alamitos, CA), p. 116 (1994).
8. P. W. Shor in *Proceedings of the 35th Annual Symposium on the Foundations of Computer Science*, S. Goldwasser, ed. (IEEE Computer Society Press, Los Alamitos, CA), p. 124 (1994).
9. R. Rivest, A. Shamir, and L. Adleman, *On Digital Signatures and Public–Key Cryptosystems*, MIT Laboratory for Computer Science, Technical Report, MIT/LCS/TR-212 (January 1979).
10. R. D. Silverman, *Math. Comp.* **48**, 329 (1987).
11. A. K. Lenstra, H. W. Lenstra Jr., M. S. Manasse, and J. M. Pollard, in *Proceedings, 22nd ACM Symposium on the Theory of Computing*, pp. 564-572 (1990).
12. D. Welsh *Codes and Cryptography* (Clarendon Press, Oxford, 1988).
13. C. H. Papadimitriou, *Computational Complexity* (Addison-Wesley, Reading, 1994).
14. A. S. Holevo, *Problemy Peredachi Informatsii* **9**, 3 (1979) (this journal is translated by IEEE under the title *Problems of Information Transfer*).
15. E. B. Davies, *IEEE Trans. Inform. Theory* **IT 24**, 596 (1978).
16. C. A. Fuchs and C. M. Caves, *Phys .Rev. Lett.* **73**, 3047 (1994).
17. R. Jozsa and B. Schumacher, *J. Mod. Optics* **41**, 2343 (1994).
18. S. Wiesner, *Sigact News* **15**(1), 78 (1983).
19. C. H. Bennett, *Phys. Rev. Lett.* **68** 3121 (1982).

20. C. H. Bennett and G. Brassard, in *Proceedings, IEEE international Conference on Computers, Systems, and Signal Processing, Bangalore, India* 175, (IEEE, New York, 1984).
21. A. K. Ekert, *Phys. Rev. Lett.* **71**, 4287 (1993).
22. C. H. Bennett, G. Brassard , C. Crepeau ,R. Jozsa, A. Peres, and W. K. Wootters, *Phys. Rev. Lett.* **70**, 1895 (1993).
23. C. H. Bennett, *SIAM J. Comput.* **18(4)**, 766 (1989).
24. M. R. Schroeder, *Number Theory in Science and Communication* (Springer, New York, 1984).
25. D. Coppersmith, IBM Research Report No. RC19642 (1994).
26. A. K. Ekert and R. Jozsa, *Rev. Mod. Phys.*, to appear (1995).
27. G. H. Hardy and E. M. Wright, *An Introduction to the Theory of Numbers*, 4th edn. (Oxford University Press, Oxford, 1965).
28. D. Deutsch, *Proc. R. Soc. London A* **425**, 73 (1989).
29. A. Barenco, *Proc. R. Soc. London A* **449**, 679 (1995).
30. T. Sleator and H. Weinfurter, *Phys. Rev. Lett.* **74**, 4087 (1995).
31. D. Deutch, A. Barenco, and A. K. Ekert, *Proc. R. Soc. London A*, in press (1995).
32. S. Lloyd, Los Alamos National Laboratory preprint (1994).
33. D. E. Knuth, *The Art of Computer Programming, Volume 2: Seminumerical Algorithms* (Addison-Wesley, Reading, 1981).
34. M. Brune, P. Nussenzveig, F. Schmidt-Kaler, F. Bernardot, A. Maali, J. M. Raimond, and S. Haroche, *Phys. Rev. Lett.* **72**, 3339 (1994).
35. A. Barenco, D. Deutsch, R. Jozsa, and A. Ekert, *Phys. Rev. Lett.* **74**, 4083 (1995).
36. J. I. Cirac and P. Zoller, *Phys. Rev. Lett.* **74**, 4091 (1995).
37. Q.A. Turchette, C. J. Hood, C. J. Lange, H. Mabuchi, and H. J. Kimble, *Phys. Rev. Lett.* **75**, 4710 (1995).
38. W. Unruh, *Phys. Rev. A* **51**, 992 (1994).
39. G. M. Palma, K. A. Suominen, and A. Ekert, *Proc. R. Soc. London A* **452**, 567 (1996).
40. D. DiVincenzo, *Phys. Rev. A* **50**, 1015 (1995).
41. F. J. MacWilliams and N. J. A. Sloane, *The Theory of Error-Correcting Codes* (North-Holland, Amsterdam, 1977).
42. R. Laflamme, C. Miguel, J. P. Paz, and W. H. Zurek, *Phys. Rev. Lett.* **77**, 198 (1996).
43. P. W. Shor, *Phys. Rev. A* **52**, R2493 (1995).
44. P. W. Shor, quant-ph/9605011.
45. D. P. DiVicenzo and P. W. Shor, *Phys. Rev. Lett.* **77**, 3260 (1996).
46. A. R. Calderbank and P. W. Shor, *Phys. Rev. A* **54**, 1098 (1996).
47. A. M. Steane, *Proc. R. Soc. A* **452**, 2551 (1996).
48. A. Ekert and C. Macchiavello, *Phys. Rev. Lett.* **77**, 2585 (1996).
49. E. Knill and R. Laflamme, quant-ph/9604034.
50. A. R. Calderbank, E. M. Rains, P. W. Shor, and N. J. A. Sloane, quant-ph/9605005.

SUPERDIFFUSIVE BEHAVIOR IN WEAKLY CHAOTIC SYSTEMS: A NEW AVENUE TO QUANTUM MACROSCOPIC MANIFESTATIONS

Luca Bonci

Dipartimento di Fisica dell'Università di Pisa
Piazza Torricelli 2, 56127 Pisa, Italy
bonci@ipifidpt.difi.unipi.it

We study the anomalous diffusion resulting from the standard map. The quantum mechanical breakdown of this property occurs in a time depending logarithmically on the inverse of the Planck constant and implies a coherence among so small regions of phase-space as to be robust against environment fluctuations.

Key words: correspondence principle, quantum chaos, anomalous diffusion.

The Wigner formulation of quantum mechanics [1] is a suitable framework to the study of the quantum-classical correspondence. In this framework a quantum state is represented by a distribution in the usual phase-space and the time evolution is described by a differential operator which can be divided in a classical term, L_{Cl}, coinciding with the classical Liouville operator, plus a term, $L_Q = \sum_{r=3}^{\infty}(\frac{\partial^r}{\partial q^r}V(q,t))(\frac{\hbar}{2i})^{r-1}\frac{1}{r!}\frac{\partial^r}{\partial p^r}$, describing the quantum corrections to the classical motion [2]:

$$\frac{\partial}{\partial t}\rho_W(q,p;t) = [L_{Cl} + L_Q]\rho_W(q,p;t) \,. \tag{1}$$

The quantum distribution ρ_W must satisfy the Heisenberg uncertainty relations and thus a quantum state cannot be represented by a delta-like distribution. This means that a proper quantum-classical comparison can be cast in a statistical sense, comparing the time evolution of ρ_W with the classical evolution of a corresponding statistical distribution, namely,

$$\frac{\partial}{\partial t}\rho_L(q,p;t) = L_{Cl}\rho_L(q,p;t) \,. \tag{2}$$

M. Ferrero and A. van der Merwe (eds.),
New Developments on Fundamental Problems in Quantum Physics, 55–58.

If we choose a suitable initial condition, i.e., a quantum state with a plausible classical counterpart, we can argue that the validity of the classical picture is broken when $\rho_W(q,p;t)$ significantly departs from $\rho_L(q,p;t)$. In the case of ordinary classical motion the time required for this breakdown is

$$t_B \sim \frac{1}{\hbar^\alpha} , \tag{3}$$

where $\hbar$ is a dimensional constant, corresponding to the ratio between the real Planck constant and the characteristic action of the system, and α is of the order of unity. This means, for a macroscopic body, a time so astronomically large as to result in a completely classical evolution.

If the classical motion is chaotic a different situation occurs. The Liouville distribution undergoes a fast process of fragmentation and quickly develops fine scale structures in phase-space. This process has the effect of enhancing the role of the operator L_Q whose strength depends on the distribution gradients, leading to a quantum-classical breakdown in a time depending logarithmically on $\hbar$ [3]

$$t_\chi \sim \frac{1}{\lambda} \ln \frac{1}{\hbar}, \tag{4}$$

where λ is the largest Lyapunov exponent of the chaotic classical system.

There is a strong difference between the two predictions; in fact t_χ increases much more slowly than t_B as the system becomes macroscopic, making plausible an experimental observation of quantum effects at macroscopic level.

Despite the wide agreement on the existence of this phenomenon, the logarithmic breakdown had not been observed till now. On the contrary, in all the cases studied so far the classical and quantum expectation values have been shown to diverge in a time depending on $\hbar$ as a power law, like in the non chaotic case.

To illustrate this behavior let us take in consideration the so called Kicked Rotator (KR), a one-dimensional model system whose classical discrete dynamics is described by the well known standard map [4]:

$$\begin{aligned} p_{t+1} &= p_t + K \sin \theta_t \\ \theta_{t+1} &= \theta_t + p_{t+1} \ , \quad \bmod(2\pi) \ . \end{aligned} \tag{5}$$

The classical evolution of this system can be chaotic, depending on the value of the nonlinearity parameter K; in that case the KR shows a diffusive behavior of the mean energy $\langle p_t^2 \rangle$ which increases linearly. In the quantum evolution this behavior is eventually destroyed by the so called dynamical localization of quantum chaos [5]. Nevertheless, to observe this quantum-classical discrepancy we have to wait a time $t_L \sim 1/\hbar^\mu$ with μ of the order of the unity.

This lack of breakdown at the time t_χ can be explained by looking at the evolved ρ_W. At the logarithmic time the quantum term L_Q becomes effective, preventing the development of the finer and finer structures characterizing the chaotic classical evolution: ρ_W is no longer identical to ρ_L, but they differ on a microscopic scale, which cannot be revealed by looking at the mean value of a smooth observable like the energy.

This seems to be the general behavior which makes the time t_χ unobservable at the macroscopic level. To reveal this time we should find a macroscopic observable depending on the microscopic structure of the phase-space distribution. We reached this result by studying the same KR model, but in a different region of parameters. There exist, in fact, particular values of K which produce a situation of weak chaos: embedded in the chaotic sea there exist fractal regions called accelerator islands. Due to this small regions the system can exhibit a superdiffusive behavior, i.e., a faster than linear energy increase [6] which is strongly related to the self-similar structure of the islands [7]. This is just a macroscopic behavior which depends on the fine phase-space structure which we expect to be destroyed by the quantum corrections stemming from L_Q. We studied the system and found that the classical behavior is lost in the quantum case and that a standard (linear) diffusion is recovered in a time which depends on $\hbar$ like t_χ [8].

The quantum breakdown of anomalous diffusion (QBAD) of the KR seems to be a good candidate for the observation of quantum macroscopic manifestations, even because there exist experimental realizations of this model [9], but the previous analysis is not exhaustive because we did not take into account the effects of the environment.

The unavoidable interaction with the environment, destroying the quantum coherence, prevents the appearance of quantum effects [10,11], and this destruction is more effective as we reach the macroscopic level. A general expression for the decoherence time can, in fact, be written as

$$t_{dec} = \frac{\hbar^2}{2M\sigma(\Delta Q)^2}\,, \tag{6}$$

where σ measures the temperature of the environment and ΔQ the characteristic size of the system. From this point of view QBAD has been shown to be more insensitive to decoherence than other quantum effects like, for example, dynamical localization of time t_L [12]. In fact QBAD rests on the impossibility of the quantum evolution to reproduce the fractal classical dynamics, but this concerns coherence on so small regions in phase-space as to make plausible to find a decoherence time larger than t_χ.

We can thus conclude that QBAD could represent a new avenue to quantum macroscopic manifestations for two main reasons:

- its characteristic time t_χ depends logarithmically on the Planck constant;

- this breakdown is robust against quantum environment decoherence.

REFERENCES

1. M. Hillery, R.F. O'Connell, M.O. Scully, and E.P. Wigner, *Phys. Rep.* **106**, 121 (1984).
2. For the sake of simplicity we restrict ourself to the case of a particle moving in a one-dimensional potential $V(q,t)$.
3. G.P. Berman and G.M. Zaslavsky, *Physica A* **91**, 450 (1978); M.V. Berry, *Ann. Phys.* **131**, 163 (1981); G.M. Zaslavsky, *Phys. Rep.* **80**, 157 (1981);W.H. Zurek and J.P. Paz, *Phys. Rev. Lett.* **72**, 2508 (1994).
4. M. Tabor, *Chaos and Integrability in Nonlinear Dynamics* (Wiley, New York, 1989).
5. G. Casati, B.V. Chirikov, F.M. Izrailev, and J. Ford in *Stochastic Behavior in Classical and Hamiltonian Systems* (Lectures Notes in Physics, Vol. 3), G.Casati and J. Ford, eds. (Springer, Berlin, 1979).
6. R. Ishizaki, H. Hata. T. Horita, and H. Mori, *Prog. Theor. Phys.* **84**, 179 (1990).
7. G. Zumofen and J. Klafter, *Europhys. Lett.* **25**, 565 (1994); Gy. Trefán, E. Floriani, B.J. West, and P. Grigolini, *Phys. Rev. E* **50**, 2564 (1994).
8. R. Roncaglia, L. Bonci, B.J. West, and P. Grigolini, *Phys. Rev. E* **51**, 5524 (1995).
9. J.C. Robinson, C. Bharucha, F.L. Moore, R.Jahnke, G.A. Georgakis, Q. Niu, M.G. Raizen, B.Sundaram, *Phys. Rev. Lett.* **74**, 3963 (1995).
10. W.H. Zurek, *Phys. Rev. D* **24**, 1516 (1981); E. Joos and H.D. Zeh, *Z. Phys. B* **59**, 223 (1985); A.O. Caldeira and A.J. Leggett, *Phys. Rev. A* **31**, 1059 (1985); W.G. Unruh and W.H. Zurek, *Phys. Rev. D* **40**, 1071 (1989).
11. P. Grigolini, *Quantum Irreversibility and Measurement* (World Scientific, Singapore, 1993).
12. L. Bonci, P. Grigolini, A. Laux, and R. Roncaglia, *Phys. Rev. A.* **54**, 112 (1996).

A PROOF WITH 18 VECTORS OF THE BELL-KOCHEN-SPECKER THEOREM

Adán Cabello

Departamento de Física Teórica
Universidad Complutense
28040 Madrid, Spain

A state-independent, additive proof of the Bell-Kochen-Specker theorem with 18 four-dimensional vectors is presented. These vectors can be physically interpreted as a set of 18 propositions concerning a system of two spin-1/2 particles. Geometrically they can be interpreted as some relevant directions of a tesseract (a four-dimensional hypercube).

Key words: Bell-Kochen-Specker theorem, Kochen-Specker theorem, non-contextual hidden-variables.

1. THE PROOF

Recently [1], we have proposed what to our knowledge is the most economic state-independent additive proof of the Bell-Kochen-Specker (BKS) theorem [2-4]. In this poster I review this proof, dealing in more detail with its physical and geometrical interpretations.

Given an individual physical system, let $v(\mathbf{u})$ denote the values ($1 =$ true, $0 =$ false) in said system of the proposition P_u (mathematically represented by the projector $|\mathbf{u}\rangle\langle\mathbf{u}|$) in a non-contextual hidden-variables theory.

The premises in the BKS theorem can be formulated as follows:

(a) *In an individual system, each proposition P_u has a unique value, 0 or 1, which is independent of which other observables are being considered jointly (non-contextuality).*

(b) *For each set of one-dimensional commuting projectors (representing a set of mutually compatible propositions) whose sum is the unit matrix*

M. Ferrero and A. van der Merwe (eds.),
New Developments on Fundamental Problems in Quantum Physics, 59–62.

in the n-dimensional Hilbert space of the states of the system, the value to one and only one of the projectors is 1, and the values to the other $n-1$ *are 0.*

Following these rules, if the value of the projector over a given vector is 1, the values of the projectors over all orthogonal vectors must be zero. Let us consider the values to the projectors over the following 9 sets of orthogonal four-dimensional vectors:

$$v(0,0,0,1)+v(0,0,1,0)+v(1,1,0,0)+v(1,-1,0,0)=1, \quad (1)$$

$$v(\mathit{0,0,0,1})+v(0,1,0,0)+v(1,0,1,0)+v(1,0,-1,0)=1, \quad (2)$$

$$v(1,-1,1,-1)+v(1,-1,-1,1)+v(\mathit{1,1,0,0,})+v(0,0,1,1)=1, \quad (3)$$

$$v(\mathit{1,-1,1,-1})+v(1,1,1,1)+v(\mathit{1,0,-1,0})+v(0,1,0,-1)=1, \quad (4)$$

$$v(\mathit{0,0,1,0})+v(\mathit{0,1,0,0})+v(1,0,0,1)+v(1,0,0,-1)=1, \quad (5)$$

$$v(\mathit{1,-1,-1,1})+v(\mathit{1,1,1,1})+v(\mathit{1,0,0,-1})+v(0,1,-1,0)=1, \quad (6)$$

$$v(1,1,-1,1)+v(1,1,1-1)+v(\mathit{1,-1,0,0})+v(\mathit{0,0,1,1})=1, \quad (7)$$

$$v(\mathit{1,1,-1,1})+v(-1,1,1,1)+v(\mathit{1,0,1,0})+v(\mathit{0,1,0,-1})=1, \quad (8)$$

$$v(\mathit{1,1,1,-1})+v(\mathit{-1,1,1,1})+v(\mathit{1,0,0,1})+v(\mathit{0,1-1,0})=1. \quad (9)$$

Theorem: There is no set of values satisfying (a) and (b) for the set of 18 propositions appearing in (1)-(9).

Proof: The sum of the right-hand sides of (1)-(9) is *odd* (it is 9), whereas the sum of the left-hand sides is necessarily *even*, because each value is 0 or 1 and appears twice in (1)-(9) (vectors in italics have appeared in a previous equation).

2. PHYSICAL MEANING

The 18 vectors in (1)-(9) can be interpreted as propositions of a system of two spin-1/2 particles. There are two kinds of vectors and therefore two kinds of propositions:

(a) The 12 vectors in (1)-(4) are *factorizable* (i.e., of the form $(a,b)^{(1)}\otimes(c,d)^{(2)}$), and the corresponding projectors are products of local observables. For instance, in the usual Pauli representation, the (non-normalized) vector

$(1,-1,0,0)$ represents the state $|\sigma_z = +1\rangle^{(1)} \otimes |\sigma_x = -1\rangle^{(2)}$, and its corresponding projector represents the proposition: "the observable $\hat{\sigma}_z^{(1)}$ has a well-defined (hidden) value $+1$ *and* the observable a well-defined (hidden) value -1," with $v(1,-1,0,0) = 1$ if it is "true," or $v(1,-1,0,0) = 0$ if it is "false."

(b) The remaining six vectors in (5)-(9) are *entangled*, and the corresponding propositions cannot be factorized in terms of local observables. Each one can be expressed in terms of a pair of the observables $\hat{\sigma}_z^{(1)} \otimes \hat{\sigma}_z^{(2)}, \hat{\sigma}_z^{(1)} \otimes \hat{\sigma}_x^{(2)}, \hat{\sigma}_x^{(1)} \otimes \hat{\sigma}_z^{(2)}$ and $\hat{\sigma}_x^{(1)} \otimes \hat{\sigma}_x^{(2)}$. For instance, $(1,-1,1,1)$ is an eigenvector of $\hat{\sigma}_z^{(1)} \otimes \hat{\sigma}_x^{(2)}$ and $\hat{\sigma}_x^{(1)} \otimes \hat{\sigma}_z^{(2)}$ with eigenvalues -1 and $+1$, respectively, and therefore can be associated with the proposition: "the observables $\hat{\sigma}_z^{(1)} \otimes \hat{\sigma}_x^{(2)}$ *and* $\hat{\sigma}_x^{(1)} \otimes \hat{\sigma}_z^{(2)}$ have well-defined (hidden) values -1 *and* $+1$, respectively". Each of the Eqs. (5)-(8) involves a pair of these entangled vectors, whereas (9) involves four.

3. GEOMETRICAL INTERPRETATION

The set of 18 vectors used in the proof is a subset of one of 24 vectors previously considered by Peres [5]. Peres' set of 24 vectors can be geometrically interpreted [1] as the directions obtained joining the center of a tesseract (a four-dimensional hypercube [6]) with some "relevant" points of the tesseract:

(a) The 8 centers of hyperfaces of the tesseract. These give rise to *4 directions* that can be associated with the vectors: (1, 0, 0, 0), (0, 1, 0, 0), (0, 0, 1, 0), (0, 0, 0, 1).

(b) The 24 centers of the faces. These give rise to *12 directions* represented by the vectors: (1, 1, 0, 0), (1, −1, 0, 0), (0, 0, 1, 1), (0, 0, 1, −1), (1, 0, 1, 0), (1, 0, −1, 0), (0, 1, 0, 1), (0, 1, 0, −1), (1, 0, 0, 1), (1, 0, 0, −1), (0, 1, 1, 0), (0, 1, −1, 0).

(c) The 16 vertices of the tesseract. These give rise to *8 directions* represented by the vectors: $(1,1,1,1), (1,-1,1,-1), (1,1,-1,-1), (1,-1,-1,1), (1,1,1,-1), (1,1,-1,1), (1,-1,1,1), (-1,1,1,1)$.

The 18 vectors appearing in our proof can be obtained by removing one direction (a), three (b)'s, and two (c)'s in a particular way; see Fig. 1.

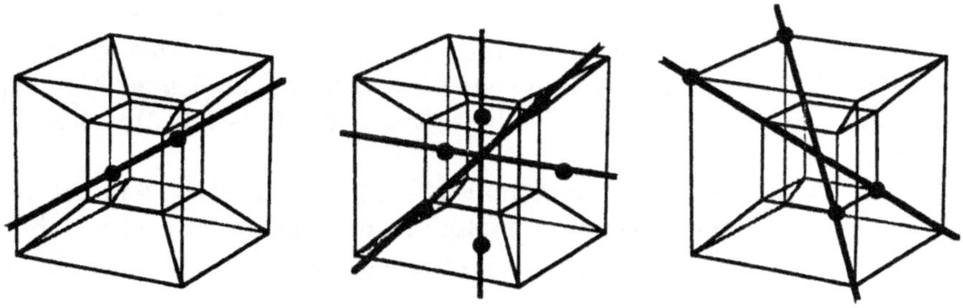

Fig. 1. "Relevant" directions that must be removed from the set of 24 of Peres to obtain the set of 18. Left: one direction (a), center: three directions (b), and right: two directions (c). There are 16 different ways of removing 6 directions from Peres' set of 24 with this symmetry.

ACKNOWLEDGMENTS

I like to thank the organizers for inviting me to the symposium and to Susana G. Fdez. Huelga for letting me use her computer to write this poster.

REFERENCES

1. A. Cabello, J. M. Estebaranz, and G. García Alcaine, "Bell-Kochen-Specker theorem: A proof with 18 vectors," *Phys. Lett. A* **212**, 183-187 (1996).
2. E. P. Specker, "Die Logik nicht gleichzeitig entscheidbarer Aussagen," *Dialectica* **14**, 239-246 (1960).
3. J. S. Bell, "On the problem of hidden variables in quantum mechanics," *Rev. Mod. Phys.* **38**, 447-452 (1966).
4. S. Kochen and E. P. Specker, "The problem of hidden variables in quantum mechanics," *J. Math. Mech.* **17**, 59-87 (1967).
5. A. Peres, "Two simple proofs of the Kochen-Specker theorem," *J. Phys. A* **24** , L175-L178 (1991).
6. H. S. M. Coxeter, *Regular polytopes* (Methuen, London, 1948).

TEMPORAL BELL INEQUALITIES AND THE UNCERTAINTY PRINCIPLE

T. Calarco

Dipartimento di Fisica, Università di Ferrara
Via Paradiso 12, Ferrara, Italy 44100

Various schemes have been proposed to test temporal Bell inequalities either by using macroscopic degrees of freedom as the magnetic flux of a rf-SQUID or microscopic degrees as the occupation number of a two level system. We analyze such proposals and the quantum predictions for the violations of temporal Bell inequalities taking into account the consequences of the Heisenberg principle. These inequalities involve correlation probabilities between subsequent measurements of a dichotomic observable. A criterion for the distinguishability of two states, required to define such correlations, is introduced and linked to the effect of the measurement process and the consequent modification of the state.

Key words: Bell's inequalities, quantum measurement theory, macroscopic realism, SQuIDs.

1. INTRODUCTION

Joint probabilities for measurements of a dichotomic observable O can be defined for both a classical or a quantum system: for example, let

$$P^{ab}_{+-} \equiv \text{Prob}\,\{O_a+; O_b-\} \tag{1}$$

be the probability of obtaining respectively the outcome + and − in two measurements labelled a and b. Classically such probabilities are additive: thus they satisfy inequalities such as

$$P^{bc}_{+-} = P^{abc}_{++-} + P^{abc}_{-+-} \leq P^{ab}_{++} + P^{ac}_{--}. \tag{2}$$

Quantum mechanically they are not additive and in some cases lead to violation of the inequalities fulfilled by classical physics. Detecting such violations in quantum systems can disprove some classical assumptions about

M. Ferrero and A. van der Merwe (eds.),
New Developments on Fundamental Problems in Quantum Physics, 63–69.

Table 1. Spatial and temporal Bell inequalities.

TYPE OF BI	**spatial**	**temporal**	
RELATED RELATED	EPR [1]	Schrödinger's cat [2]	
CLASSICAL ASSUMPTION TESTED	*realism locality*	*realism noninvasive measurability* at the *macroscopic* level	*realism* at the *mesoscropic/* microscopic level
PHYSICAL SYSTEM	Pair of 100% correlated particles	SQuID (Super-conducting Quantum Interference Device	atom oscillating between two levels
MEASURED OBSERVABLE	particle spins along *directions* $\boldsymbol{a}, \boldsymbol{b}, \boldsymbol{c}$	trapped magnetic flux at *times* t_a, t_b, t_c	occupation of energy levels at *times* t_a, t_b, t_c
PROPOSED EXPERIMENT	Aspect et al. [7]	Tesche [8]	?
DETECTION OF VIOLATIONS	YES [7]	Cosmelli et al. [9]?	?

their behavior. Different experiments were recently proposed and realised aimed at testing in this way some paradoxical aspects of the quantum measurement theory (Table 1). The crucial point here is that in the temporal version, unlike the spatial one, the same system is monitored twice. According to quantum mechanics, in this case the first measurement affects the subsequent dynamical evolution of the system and in particular the second measurement outcome. On the other hand, before starting an experimental effort, it is needed to know if quantum mechanics actually predicts *detectable* violations of temporal Bell inequalities (hereafter TBI). Therefore the measurement effect must be taken into account.

2. THE MEASUREMENT EFFECT

The effect of a measurement of an observable ϕ at time t_a with outcome Φ and instrumental uncertainty $\Delta\Phi$ can be represented by a filtering weight [10]:

$$\psi_{\Phi}(\phi, t_a^+) = \exp\left\{-\frac{(\phi-\Phi)^2}{2\Delta\Phi^2}\right\}\psi(\phi, t_a^-). \tag{3}$$

Such a Gaussian weight, smoother than the usual perfect (double-step) filter [11], has however the same effect, the collapse of the wavefunction around the measurement outcome. The squared modulus of Eq. (3) gives a probability distribution for the outcomes

$$P^a(\Phi) = \frac{\|\psi_\Phi(t_a^+)\|^2}{\int \|\psi_\Phi(t_a^+)\|^2 d\Phi} = \frac{1}{\sqrt{\pi}\Delta\Phi\|\psi_\Phi(t_a^+)\|^2}. \tag{4}$$

whose dispersion is named effective uncertainty:

$$\Delta\Phi^a_{\text{eff}} = \left[2\int(\Phi-\tilde{\Phi})^2 P^a(\Phi)d\Phi\right]^{\frac{1}{2}}, \tag{5}$$

where $\tilde{\Phi}$ is the most probable outcome.

This formalism allows to compute in terms of the eigenfunctions of the system all the relevant quantities, in particular the probabilities entering Eq. (2). Each term there corresponds to a specific couple of measurements, performed at particular times with particular signs of the outcomes Φ. Such measurements are represented by means of Eq. (3). Evaluation of Eq. (4) on the proper wavefunctions gives directly the desired correlation probabilities, with the associated effective uncertainties.

3. TEMPORAL BELL INEQUALITIES IN SQUIDS

An rf-SQuID is a superconducting ring interrupted by a weak link (a thin insulating layer). It has a dynamics schematized through the motion of the magnetic flux coordinate ϕ in the effective double-well potential

$$V(\phi) = -\frac{1}{2}\mu\phi^2 + \frac{1}{4}\lambda\phi^4, \tag{6}$$

with μ, λ positive constants linked to the parameters (capacitance, inductance and external magnetic flux) of the superconducting junction. Under certain experimental conditions, the two wells of the potential correspond to states of flux $|\phi_+\rangle$ and $|\phi_-\rangle$ of opposite signs – *i.e.*, to a clockwise and a counterclockwise current of equal intensity in the ring. The main advantage of such a device is that ϕ is a macroscopic degree of freedom which behaves quantum mechanically. Thus one can test quantum mechanics against a realistic theory in the macroscopic domain [4].

The method sketched in Sec. 2 allows to check the fulfillment of Eq. (2) in the presence of the measurements. Violations to the inequality are actually found in a certain range of the measurement times ($t_{ab} \equiv t_b - t_a$ and $t_{bc} \equiv t_c - t_b$), call it the *violation regime.*

However, for detecting these violations in a meaningful experiment, it is required that the two states $|\phi_+\rangle$ and $|\phi_-\rangle$ could be considered as macroscopically distinct: this is expressed as

$$\Delta\Phi_{\text{eff}} > \phi_+ - \phi_-, \tag{7}$$

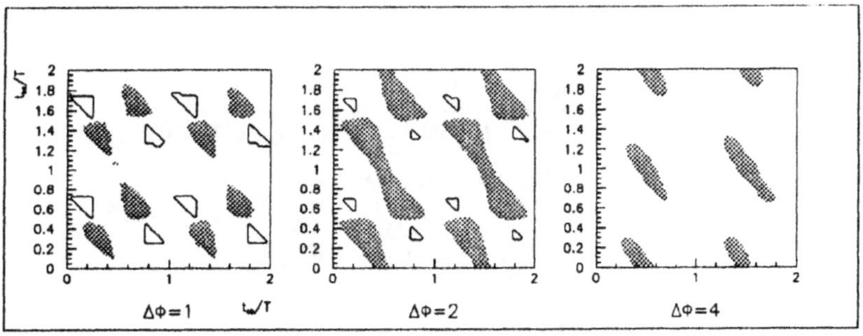

Fig. 1. Violation regime (shaded areas) and distinguishability regime (small triangular zones) for the temporal Bell inequality (2) T is the tunnelling time.

which in turn is fulfilled for some values of (t_{ab}, t_{bc}), call it the *distinguishability regime*. Fig. 1 shows the two regimes in the $t_{ab} - t_{bc}$ plane, for various values of the instrumental uncertainty $\Delta\Phi$ [12]. No intersection exists: violations of TBI in an rf-SQuID seem not to be compatible with distinguishability of the macroscopic states.

4. TEMPORAL BELL INEQUALITIES WITH CORRELATION FUNCTIONS

The distinguishability criterion stated in Eq. (7) can be made more general and its relationship with the detectability of violations to TBI more transparent by means of another formalism proposed by Paz *et al.* [5]. Accordingly, the time-dependent correlation function for a dichotomic observable with eigenvalues ± 1 is written

$$K(t_1, t_2) \stackrel{\text{def}}{\equiv} \int \mathcal{D}[Q(t)]P[Q(t)]Q(t_1)Q(t_2), \tag{8}$$

where $Q(t) = \langle\psi(t)|\hat{Q}|\psi(t)\rangle$ and $\hat{Q} = \hat{P}_+ - \hat{P}_-$ in terms of the projectors $\hat{P}_\pm$ on the two eigenspaces. In a realistic theory, TBI are derived such as [6]

$$\Delta K \equiv -K(2t) - 2K(t) \leq 1. \tag{9}$$

If at time t_1 a measurement is performed, $Q(t)$ changes as

$$Q_M(t > t_1) = \sum_{s', s''=+,-} s's'' \left|\langle s''|U(t-t_1)P_{s'}|\psi(t_1)\rangle\right|^2 . \tag{10}$$

When taking into account only the outcomes of measurements in which the state is left unchanged within some confidence level ε, the propagator becomes [13]

$$K_\varepsilon(t_1,t_2) = \int \mathcal{D}[Q_M(t)]P_\varepsilon[Q_M(t)]Q_M(t_1)Q_M(t_2), \tag{11}$$

where $P_\varepsilon[Q_M(t)]$ imposes the restriction

$$\langle\psi_\pm(t_1)|\psi_\pm(t_1)\rangle \geq \varepsilon.$$

The maximal temporal Bell inequality violation ΔK_{max}, if any, depends upon the distinguishability of the states ε.

The dichotomic observable can in general be derived from another one, possibly continuous, which is actually measured. For example, in the previous Section $\hat{Q} = \hat{\phi}/|\hat{\phi}|$: in this case $\hat{P}_\pm = \Theta(\pm\hat{\phi})$. On the other hand, experiments can be performed by measuring correlations of an observable intrinsically dichotomic, as the state of a two-level system: then $\hat{P}_\pm = |\pm\rangle\langle\pm|$, as for a Rydberg atom interacting with a superconducting cavity [6], or an atom undergoing optically driven Rabi oscillations [5].

4.1. TBI in Two-Level Systems

It is convenient to choose systems in which the state oscillates between the two levels $|+\rangle$ and $|-\rangle$:

$$|+\rangle \longrightarrow \cos[\Omega(t_1-t_2)]|+\rangle + \sin[\Omega(t_1-t_2)]|-\rangle, \tag{12}$$

where $\Omega = \Omega_{JC}\sqrt{n+1}$ for a Rydberg atom in a resonant cavity [6]. (Ω_{JC} being the Jaynes-Cummings angular frequency and n the number of photons in the cavity); $\Omega = \Omega_R$ (the Rabi frequency) for an atom driven by an external force [5]. In both these cases Eq. (11) can be evaluated explicitly (see the continuous curve in Fig. 2):

$$K_\varepsilon(t_1,t_2) = \frac{1}{\pi}\left[2\sqrt{\varepsilon(1-\varepsilon)} + \arccos(2\varepsilon-1)\right]K(t_2-t_1). \tag{13}$$

4.2. SQuID as a Two-Level System

Unlike the discrete case, the eigenspaces of the projectors $P_\pm \equiv \Theta(\pm\hat{\phi})$ are not one-dimensional: an average over the possible initial states is required. The most simple parametrization, in terms of the first two energy eigenstates, is $\psi^\theta(\phi,t_0) = \cos\theta\,\psi_0(\phi) + \sin\theta\,\psi_1(\phi)$. The "cat state"

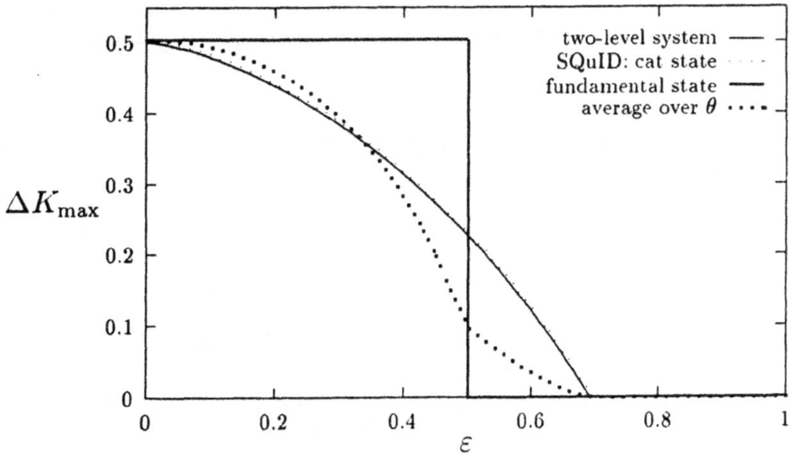

Fig. 2. Dependence of the maximal TBI violation $\Delta K_{\max}$ on the distinguishability level ε for a SQuID in various initial states, compared with that for a two-level system.

($\theta = \pi/4$) behaves as the discrete two-level system. On the other hand the oscillations in $Q(t)$, needed to observe violations to TBI, are absent in a stationary situation such as a single energy eigenstate, corresponding to θ integer multiple of $\pi/2$: in this case no meaningful violation is present. The averaged behavior (Fig. 2) is at the borderline between observing and not observing the violations, depending on the confidence level required for attributing a definite flux sign. In other words, only a probabilistic test of the violations to TBI is possible.

5. CONCLUSIONS

Temporal Bell inequalities (TBI) involve correlation probabilities between consecutive measurements on the same system. In the spatial case two distant subsystems are observed only once. According to quantum mechanics, the first measurement affects the evolution of the state and the subsequent outcomes. The entity of violations to TBI will depend upon the measurement strength. Strong measurements destroy the coherence of the state, from which violations could arise; weak measurements have not enough resolution power. For appropriate measurement strengths, quantum mechanics including the measurement effect predicts violations to TBI within a certain range of correlation times, the violation regime. Violation regime and distiguishability regime are in competition in all the investigated systems, rf-SQuIDs and two-level-like systems. Only a probabilistic violation can be estabilished in an actual experiment.

REFERENCES

1. A. Einstein, B. Podolsky, N. Rosen, *Phys. Rev.* **47**, 777 (1935).
2. E. Schrödinger, *Naturwiss.* **23**, 807 (1935); **23**, 823 (1935); **23**, 844 (1935).
3. J. S. Bell, *Physics* **1**, 195 (1964).
4. A. J. Leggett and A. Garg, *Phys. Rev. Lett.* **54**, 857 (1985).
5. J. P. Paz and G. Mahler, *Phys. Rev. Lett.* **71**, 3235 (1993).
6. S. F. Huelga, T. W. Marshall, and E. Santos, *Phys. Rev.* **A52**, R2497 (1995).
7. A. Aspect *et al.*, *Phys. Rev. Lett.* **47**, 460 (1981); **49**, 91, 180 (1982).
8. C. D. Tesche, *Phys. Rev. Lett.* **64**, 2358 (1990).
9. C. Cosmelli, G. Diambrini-Palazzi, G. Di Cosimo, A. Di Domenico, M. G. Castellano, R. Leoni, P. Carelli, proposal presented to CNR, Rome, 18/2/1993.
10. M. B. Mensky, R. Onofrio, and C. Presilla, *Phys. Rev. Lett.* **70**, 2825 (1993).
11. J. V. von Neumann, *Mathematical Foundations of Quantum Mechanics* (Princeton University Press, Princeton, 1955).
12. R. Onofrio and T. Calarco, *Phys. Lett.* **A208**, 40 (1995).
13. T. Calarco and R. Onofrio, *Appl. Phys. B* **64**, 137 (1997).

PARAMETRIC DOWN-CONVERSION EXPERIMENTS IN THE WIGNER REPRESENTATION

Alberto Casado

Departamento de Física Aplicada
Escuela Superior de Ingenieros
Avda. Reina Mercedes s/n
41012 *Sevilla, Spain*

Parametric down-conversion experiments are studied within the framework of the Wigner representation. After the description of signal and idler beams and the theory of detection in the Wigner formalism, we calculate the auto- and crosscorrelations which are necessary in order to compute the detection probabilities in second- and fourth-order interference experiments.

Key words: parametric down-conversion, Wigner function, vacuum field.

1. THE DESCRIPTION OF A SIGNAL AND IDLER BEAMS IN THE WIGNER REPRESENTATION

We shall consider electromagnetic fields corresponding to narrow light beams and we shall ignore polarization. We assume that the light beam contains frequencies within a range between ω_{min} and ω_{max} and wave vectors whose transverse components are limited by a small upper value, that is,

$$\omega_{\mathrm{min}} < \omega_{\mathbf{k}} < \omega_{\mathrm{max}}, \quad |\mathbf{k}^{\mathrm{tr}}| \ll \frac{\omega_{\mathrm{min}}}{c}. \tag{1}$$

Hence the "relevant" component of the electric vector in the Wigner formalism is

$$E^{(+)}(\mathbf{r},t) = i\sum_{[\mathbf{k}]} \left(\frac{\hbar\omega_{\mathbf{k}}}{L^3}\right)^{\frac{1}{2}} \alpha_{\mathbf{k}}(t)\mathrm{e}^{i\mathbf{k}\cdot\mathbf{r}}. \tag{2}$$

In order to obtain the evolution equations of amplitudes α, we use the fact

M. Ferrero and A. van der Merwe (eds.),
New Developments on Fundamental Problems in Quantum Physics, 71–76.

that the interaction Hamiltonian is quadratic:

$$\hat{H} = \sum_{\mathbf{k}} \hbar\omega_{\mathbf{k}} \left(\hat{a}^{\dagger}_{\mathbf{k}}\hat{a}_{\mathbf{k}} + \frac{1}{2} \right) + \left(i\hbar g' V \sum_{\mathbf{k},\mathbf{k}'} f(\mathbf{k},\mathbf{k}') e^{-i\omega_0 t} \hat{a}^{\dagger}_{\mathbf{k}} \hat{a}^{\dagger}_{\mathbf{k}'} + \text{h.c.} \right). \quad (3)$$

Due to this fact, the evolution equation for $\alpha_{\mathbf{k}}(t)$ is the same as the corresponding for the destruction operator $\hat{a}_{\mathbf{k}}(t)$:

$$\dot{\alpha}_{\mathbf{k}} = -i\omega_{\mathbf{k}}\alpha_{\mathbf{k}} + g'V \sum_{\mathbf{k}'} f(\mathbf{k},\mathbf{k}') e^{-i\omega_0 t} \alpha^{*}_{\mathbf{k}'}. \quad (4)$$

After integrating to second order in the coupling constant g' and by substituting in (2), we find:

$$E_s^{(+)}(\mathbf{r},t) = E_{0s}^{(+)}(\mathbf{r},t) + gVGE_{0i}^{(-)}(\mathbf{r},t) + g^2|V|^2 J E_{0s}^{(+)}(\mathbf{r},t), \quad (5)$$

$$E_i^{(+)}(\mathbf{r},t) = E_{0i}^{(+)}(\mathbf{r},t) + gVGE_{0s}^{(-)}(\mathbf{r},t) + g^2|V|^2 J E_{0i}^{(+)}(\mathbf{r},t), \quad (6)$$

$E_s^{(+)}$ and $E_I^{(+)}$ being the signal and idler beams with average frequencies ω_s, ω_i, and wave vectors $\mathbf{k}_s$, $\mathbf{k}_i$ respectively, fulfilling the matching conditions

$$\omega_s + \omega_i = \omega_0, \quad \mathbf{k}_s + \mathbf{k}_i = \mathbf{k}_0. \quad (7)$$

The quantity $g \equiv g'\Delta t$ is the effective coupling constant, Δt being the time taken for the radiation to cross the crystal. G and J are linear operators related with the interaction into the crystal between the laser and the zeropoint field. For instance, the outgoing signal field can be expressed as a sum which consists in three parts (5):

1. A zeropoint radiation which passes through the crystal without any change. This is the zeroth order contribution.
2. The first order term is a radiation produced by the nonlinear interaction between the laser beam and the zeropoint radiation incoming to the crystal in the direction of the idler beam.
3. The second order term modifies to order g^2 the radiation of the incoming zeropoint in the direction of the signal beam.

2. QUANTUM THEORY OF DETECTION IN THE WIGNER REPRESENTATION

We now formulate the quantum theory of detection in the Wigner representation. The essential point is to express the normal ordering in terms of symmetrical ordering, due to the following property. For any polynomial $P(\alpha, \alpha^*)$

$$\langle P(\alpha,\alpha^*)\rangle \equiv \int P(\alpha,\alpha^*)W(\alpha,\alpha^*)d^{2M}\alpha = \mathrm{Tr}\{\rho S[P(\hat{a},\hat{a}^\dagger)]\}. \tag{8}$$

We now show the results obtained for the single and joint probabilities expressions in the Wigner representation:

2.1. Single Probabilities

$$P_a(\mathbf{r}_a,t) \propto \langle I(\mathbf{r}_a,t) - I_0(\mathbf{r}_a)\rangle, \tag{9}$$

"*The single detection probability is proportional to the mean value of the difference between the actual intensity and the intensity of the vacuum field at the detector.*"

2.2. Joint Probabilities

In the case of PDC experiments we have:

$$P_{ab}(\mathbf{r}_a,t;\mathbf{r}_b,t') \propto \langle\{I(\mathbf{r}_a,t) - I_0(\mathbf{r}_a)\}\{I(\mathbf{r}_b,t') - I_0(\mathbf{r}_b)\}\rangle. \tag{10}$$

"*The joint detection probability in two separate detectors is proportional to the mean value of the product of two such terms.*"

3. SIGNAL AND IDLER FIELD AUTO AND CROSSCORRELATIONS

It is convenient to work with slowly varying amplitudes:

$$F^{(+)}(\mathbf{r},t) = e^{i\omega_a t}E^{(+)}(\mathbf{r},t) = i\sum_{[\mathbf{k}]}\left(\frac{\hbar\omega_{\mathbf{k}}}{L^3}\right)^{\frac{1}{2}} \alpha_{\mathbf{k}}(0)e^{i\mathbf{k}\cdot\mathbf{r}}e^{i(\omega_a-\omega_{\mathbf{k}})t}. \tag{11}$$

It is easy to obtain the amplitude $F^{(+)}(\mathbf{r}_B,t)$ in terms of the amplitude $F^{(+)}(\mathbf{r}_A,t)$ at another point of the light beam. By propagating the field in the usual way we have

$$F^{(+)}(\mathbf{r}_B,t) = F^{(+)}(\mathbf{r}_A,t-\frac{r_{AB}}{c})e^{i\omega_a\frac{r_{AB}}{c}}, \tag{12}$$

where $\mathbf{r}_{AB} = \mathbf{r}_B - \mathbf{r}_A$ and $r_{AB} = |\mathbf{r}_{AB}|$.

3.1. Autocorrelations

For the field at a given point $\mathbf{r}$ and two different times t and t' we have

$$\begin{aligned}
&\langle F_s^{(+)}(\mathbf{r},t)F_s^{(-)}(\mathbf{r},t')\rangle - \langle F_{0s}^{(+)}(\mathbf{r},t)F_{0s}^{(-)}(\mathbf{r},t')\rangle \\
&= g^2|V|^2\{\langle GF_{0i}^{(-)}(\mathbf{r},t)G^*F_{0i}^{(+)}(\mathbf{r},t')\rangle \\
&+\langle F_{0s}^{(+)}(\mathbf{r},t)J^*F_{0s}^{(-)}(\mathbf{r},t')\rangle + \langle F_{0s}^{(-)}(\mathbf{r},t')
\end{aligned} \tag{13}$$

There are contributions of all terms of the expansion (5) of the field. Now, we are going to relate these contributions by using the preservation of commutation rules. If we consider the signal beam emerging from the crystal at different times t and t', then

$$\left[\hat{F}_s^{(+)}(\mathbf{r},t), \hat{F}_s^{(-)}(\mathbf{r},t')\right] = \left[\hat{F}_{0s}^{(+)}(\mathbf{r},t), \hat{F}_{0s}^{(-)}(\mathbf{r},t')\right], \tag{14}$$

where $\hat{F}_s^{(+)}$ is the field *operator* in the Heisenberg picture. Taking the expectation value in the vacuum for the latter expression, and working in the Wigner representation, we obtain

$$\begin{aligned}
&\langle F_{0s}^{(+)}(\mathbf{r},t)J^*F_{0s}^{(-)}(\mathbf{r},t')\rangle + \langle F_{0s}^{(-)}(\mathbf{r},t')JF_{0s}^{(+)}(\mathbf{r},t)\rangle \\
&= \langle GF_{0i}^{(-)}(\mathbf{r},t)G^*F_{0i}^{(+)}(\mathbf{r},t')\rangle.
\end{aligned} \tag{15}$$

Therefore

$$\begin{aligned}
&\langle F_s^{(+)}(\mathbf{r},t)F_s^{(-)}(\mathbf{r},t')\rangle - \langle F_{0s}^{(+)}(\mathbf{r},t)F_{0s}^{(-)}(\mathbf{r},t')\rangle \\
&= 2g^2|V|^2\langle GF_{0i}^{(-)}(\mathbf{r},t)G^*F_{0i}^{(+)}(\mathbf{r},t')\rangle \equiv g^2|V|^2\mu_s(t'-t),
\end{aligned} \tag{16}$$

being $\mu_s(t-t')$ a correlation function which goes to zero when $t'-t$ is greater than the correlation time of the signal, τ_s. Similarly,

$$\langle F_i^{(+)}(\mathbf{r},t)F_i^{(-)}(\mathbf{r},t')\rangle - \langle F_{0i}^{(+)}(\mathbf{r},t)F_{0i}^{(-)}(\mathbf{r},t')\rangle = g^2|V|^2\mu_i(t'-t). \tag{17}$$

The remaining autocorrelations are zero:

$$\langle F_s^{(+)}(\mathbf{r},t)F_s^{(+)}(\mathbf{r},t')\rangle = \langle F_i^{(+)}(\mathbf{r},t)F_i^{(+)}(\mathbf{r},t')\rangle = 0. \tag{18}$$

3.2. Crosscorrelations

Taking the signal and idler fields at two points $(\mathbf{r}, t)$ and $(\mathbf{r}', t')$ respectively, we have:

$$\langle F_s^{(+)}(\mathbf{r},t)F_i^{(+)}(\mathbf{r}',t')\rangle = \langle\{(1+g^2|V|^2J)F_{0s}^{(+)}(\mathbf{r},t) + gVGF_{0i}^{(-)}(\mathbf{r},t)\}$$
$$\times\{(1+g^2|V|^2J)F_{0i}^{(+)}(\mathbf{r}',t') + gVGF_{0s}^{(-)}(\mathbf{r}',t')\}\rangle$$
$$= gV[\langle F_{0s}^{(+)}(\mathbf{r},t)GF_{0s}^{(-)}(\mathbf{r}',t')\rangle + \langle F_{0i}^{(+)}(\mathbf{r}',t')GF_{0i}^{(-)}(\mathbf{r},t)\rangle]. \quad (19)$$

Now we are going to apply again the conservation of commutation rules, but by considering the signal and idler field operators at different times. By taking into account the same considerations as above, we have

$$\left[\hat{F}_s^{(+)}(\mathbf{r},t), \hat{F}_i^{(+)}(\mathbf{r}',t')\right] = \left[\hat{F}_{0s}^{(+)}(\mathbf{r},t), \hat{F}_{0i}^{(+)}(\mathbf{r}',t')\right] = 0 \;\rightarrow$$

$$\langle F_{0s}^{(+)}(\mathbf{r},t)GF_{0s}^{(-)}(\mathbf{r}',t')\rangle = \langle F_{0i}^{(+)}(\mathbf{r}',t')GF_{0i}^{(-)}(\mathbf{r},t)\rangle \equiv \frac{1}{2}\nu(t'-t). \quad (20)$$

The second conmutator is zero because it contains only destruction operators. The function $\nu(\tau)$ defines a coherence time τ_{si} between signal and idler. From (20) and (21), it follows that

$$\langle F_s^{(+)}(\mathbf{r},t)F_i^{(+)}(\mathbf{r}',t')\rangle = gV\nu(t'-t). \quad (21)$$

By a similar argument, it can be proved that

$$\langle F_s^{(+)}(\mathbf{r},t)F_i^{(-)}(\mathbf{r}',t')\rangle = \langle F_s^{(-)}(\mathbf{r},t)F_i^{(+)}(\mathbf{r}',t')\rangle = 0. \quad (22)$$

4. CONCLUSIONS

We have developed the Wigner representation formalism in a fashion suitable for the study of experiments involving PDC photon pairs [1], [2]. The results agree with those obtained using the more common Hilbert-space formulation, both being just two equivalent forms of quantum optics.

The Wigner representation is specially suited for PDC because the Wigner function is Gaussian and positive definite in this case. The Gaussian character simplifies notably the calculation of auto- and crosscorrelations for the various fields involved. On the other hand positivity makes possible a picture in terms of pure waves during the propagation, which may aid the intuition in the study of these experiments.

REFERENCES

1. Alberto Casado, Trevor W. Marshall, and Emilio Santos, *J. Opt. Soc. Am. B.*, in press.
2. Alberto Casado, Agustín Fernández, Trevor Marshall, Ramón Risco-Delgado, and Emilio Santos, "Fourth-order interference in the Wigner representation for parametric down-conversion experiments," preprint, University of Sevilla.

NEGATIVE ENTROPY IN QUANTUM INFORMATION THEORY

Nicolas J. Cerf and Chris Adami

Kellogg Radiation Laboratory, 106-38
California Institute of Technology
Pasadena, California 91125

We present a quantum information theory that allows for the consistent description of quantum entanglement. It parallels classical (Shannon) information theory but is based entirely on density matrices, rather than probability distributions, for the description of quantum ensembles. We find that, unlike in Shannon theory, conditional entropies can be negative when considering quantum entangled systems such as an Einstein-Podolsky-Rosen pair, which leads to a violation of well-known bounds of classical information theory. Negative quantum entropy can be traced back to "conditional" density matrices which admit eigenvalues larger than unity. A straightforward definition of mutual quantum entropy, or "mutual entanglement," can also be constructed using a "mutual" density matrix. Such a unified information-theoretic description of classical correlation and quantum entanglement clarifies the link between them: the latter can be viewed as "super-correlation" which can induce classical correlation when considering a ternary or larger system.

Key words: quantum information theory, quantum entropy, quantum entanglement.

1. INTRODUCTION

Quantum information theory [1] is a new field with potential implications for the conceptual foundations of quantum mechanics. It appears to be the basis for a proper understanding of the emerging fields of quantum computation [2], quantum communication [3], and quantum cryptography [4]. Although some fundamental results have been obtained recently such as the quantum noiseless coding theorem [5] or the rules governing the extraction of classical

M. Ferrero and A. van der Merwe (eds.),
New Developments on Fundamental Problems in Quantum Physics, 77–84.

information from quantum entropy, it is still puzzling in many respects. Quantum information processing basically deals with quantum bits (qubits) [5] rather than bits, the former obeying quantum laws quite different from the classical physics of bits that we are used to. Most importantly, qubits can exist in quantum *superpositions*, a notion completely foreign to classical mechanics, or even classical thinking. To accommodate the relative phases in quantum superpositions, quantum information theory must be based on mathematical constructions which reflect these: the density matrices. The central object of information theory, the entropy, has been introduced in quantum mechanics by von Neumann [6]

$$S(\rho) = -\mathrm{Tr}\,\rho\log\rho\,. \tag{1}$$

Its relationship to the Boltzmann-Gibbs-Shannon entropy

$$H(\boldsymbol{p}) = -\sum_i p_i \log p_i \tag{2}$$

is obvious when considering the von Neumann entropy of a mixture of orthogonal states, in which case the density matrix ρ in (1) contains classical probabilities p_i on its diagonal, and $S(\rho) = H(\boldsymbol{p})$. In general, however, quantum mechanical density matrices have off-diagonal terms, which reflect the relative quantum phases in superpositions.

In classical (Shannon) information theory [7] the concept of conditional probabilities has given rise to the definition of conditional and mutual entropies. These can be used to elegantly describe the trade-off between entropy and information in measurement, as well as the characteristics of a transmission channel. For example, for two systems A and B, the measurement of A by B is expressed by the equation for the entropies

$$H(A) = H(A|B) + H(A{:}B)\,. \tag{3}$$

Here, $H(A|B)$ is the entropy of A after having measured those pieces that become correlated in B, while $H(A{:}B)$ is the information gained about A via the measurement of B. Mathematically, $H(A|B)$ is a *conditional* entropy, and is defined using the conditional probability $p_{i|j}$ and the joint probability p_{ij} describing random variables from ensembles A and B:

$$H(A|B) = -\sum_{ij} p_{ij} \log p_{i|j}\,. \tag{4}$$

The mutual entropy or information $H(A{:}B)$, on the other hand, is defined via the mutual probability $p_{i:j} = p_i\, p_j / p_{ij}$ as

$$H(A{:}B) = -\sum_{ij} p_{ij} \log p_{i:j}\,. \tag{5}$$

Simple relations such as $p_{ij} = p_{i|j}\, p_j$ imply equations such as $H(A|B) = H(AB) - H(B)$ and all the other usual relations of classical information theory [e.g., Eq. (3)]. Curiously, a quantum information theory paralleling these constructions has never been attempted. Rather, a "hybrid" theory was used in which quantum probabilities are inserted in the classical formulae given above, thereby losing the quantum phase crucial to density matrices (see, e.g., [8]). Below in Sec. 2 we show that a consistent quantum information theory can be developed that parallels the construction outlined above, while based entirely on matrices [9].

2. QUANTUM INFORMATION THEORY

Let us consider the information-theoretic description of a composite quantum system AB. A straightforward quantum generalization of Eq. (4) suggests the definition

$$S(A|B) = -\mathrm{Tr}_{AB}[\rho_{AB} \log \rho_{A|B}] \tag{6}$$

for the quantum conditional entropy. In order for such an expression to hold, we define the concept of a "conditional" density matrix,

$$\rho_{A|B} = \left[\rho_{AB}^{1/n}(\mathbf{1}_A \otimes \rho_B)^{-1/n}\right]^n \qquad n \to \infty\,, \tag{7}$$

the analog of the conditional probability $p_{i|j}$. Here, $\mathbf{1}_A$ is the unit matrix in the Hilbert space for A, $\otimes$ stands for the tensor product in the joint Hilbert space, and $\rho_B = \mathrm{Tr}_A[\rho_{AB}]$ denotes a "marginal" density matrix, analogous to the marginal probability $p_j = \sum_i p_{ij}$. The peculiar form involving the infinite limit in Eq. (7) is necessary because joint and marginal density matrices do not commute in general. However, the definition implies that the standard relation

$$S(A|B) = S(AB) - S(B) \tag{8}$$

holds for the quantum entropies and that $S(A|B)$ is invariant under any unitary transformation of the product form $U_A \otimes U_B$. More precisely, it is easy to see that $\rho_{A|B}$ is a positive Hermitian operator (in the joint Hilbert space) whose spectrum is invariant under $U_A \otimes U_B$. Despite the apparent similarity between the quantum definition for $S(A|B)$ and the standard classical one for $H(A|B)$, dealing with matrices rather than scalars opens up a quantum realm for information theory exceeding the classical one. The crucial point is that, while $p_{i|j}$ is a probability distribution in i (in particular $0 \le p_{i|j} \le 1$), its quantum analog $\rho_{A|B}$ is *not* a density operator: it can have eigenvalues *larger* than one, and, consequently, the associated conditional entropy $S(A|B)$ can be *negative.* Only such a matrix-based quantum formalism consistently accounts for the well-known non-monotonicity of quantum entropies (see, e.g., [10]). This means that it is acceptable, in quantum information theory, to have $S(AB) < S(B)$, i.e., the entropy of the entire system AB can be smaller than the entropy of one of its subparts B, a situation which is of course forbidden in classical information theory. This happens for example in the case of quantum *entanglement* between A and B, and will be illustrated below for an EPR pair. Note that, as a consequence of the concavity of $S(A|B)$, a property related to strong subadditivity (see, e.g., [10]) any separable state (i.e., a mixture of product states) is associated with non-negative $S(A|B)$. (The converse is not true.) Therefore, the non-negativity of conditional entropies can be viewed as a necessary condition for separability, and we have shown that this condition can be related to entropic Bell inequalities [11].

Similarly, the quantum analog of the mutual entropy can be constructed, defining a "mutual" density matrix

$$\rho_{A:B} = \left[(\rho_A \otimes \rho_B)^{1/n} \rho_{AB}^{-1/n}\right]^n \qquad n \to \infty\,, \tag{9}$$

the analog of the mutual probability $p_{i:j}$. As previously, this definition implies the standard relation

$$S(A{:}B) = S(A) + S(B) - S(AB) \tag{10}$$

between the quantum entropies. This definition extends the classical notion of mutual or *correlation* entropy $H(A{:}B)$ to the quantum notion of mutual *entanglement* $S(A{:}B)$ and applies to pure as well as mixed states; $S(A{:}B)$ is a general measure of correlations *and* "super-correlations" in information theory. In fact, all the above quantum definitions reduce to the classical ones for a diagonal ρ_{AB}, which suggests that Eqs. (7) and (9) are very reasonable assumptions. It is possible that other definitions of $\rho_{A|B}$ and $\rho_{A:B}$ could be proposed, but we believe this choice is simplest. This formalism suggests that all the relations between classical entropies (e.g., the chain rules for entropies and mutual entropies) also have a quantum analog, and we make use of it in [11,12].

The relations between entropies are conveniently summarized by a Venn-like entropy diagram, as shown in Fig. 1a. The important difference between classical and quantum entropy diagrams is that the basic inequalities relating the entropies are "weaker" in the quantum case, allowing for negative conditional entropies and "excessive" mutual entropies [9]. For example, the upper bound for the mutual entropy (which is directly related to the channel capacity) is $H(A{:}B) \leq \min[H(A), H(B)]$ in classical information theory, while it can reach twice the classical upper bound

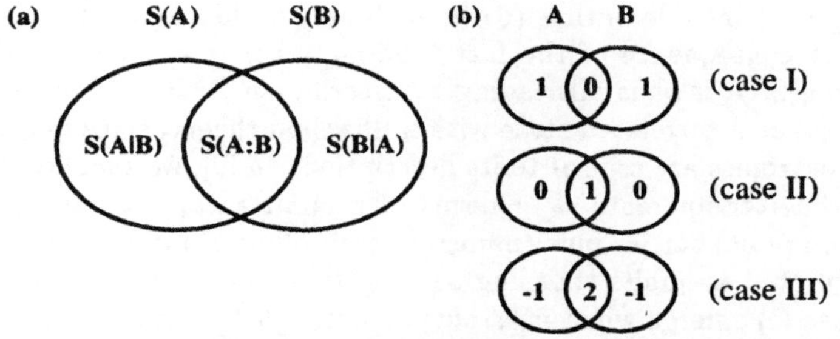

Fig. 1. (a) General entropy diagram for a quantum composite system AB. (b) Entropy diagrams for three cases of a system of 2 qubits: (I) independent, (II) classically correlated, (III) quantum entangled.

$S(A{:}B) \leq 2\min[S(A), S(B)]$ in quantum information theory as a consequence of the Araki-Lieb inequality (see, e.g., [10]). In Fig. 1b, we show the entropy diagram corresponding to three limiting cases of a composite system of two dichotomic variables (e.g., 2 qubits): independent variables (case I), classically correlated variables (case II), and quantum entangled variables (case III). In all three cases, each subsystem taken separately is in a mixed state of entropy $S(A) = S(B) = 1$ bit. Cases I and II correspond to classical situations (which can of course be described in our formalism with density matrices as well), while case III is a purely quantum situation which violates the bounds of classical information theory [9]. It corresponds to an EPR pair, characterized by the pure state $|\psi_{AB}\rangle = 2^{-1/2}(|01\rangle - |10\rangle)$, and, accordingly, it is associated with a vanishing combined entropy $S(AB) = 0$. Using $\rho_{AB} = |\psi_{AB}\rangle\langle\psi_{AB}|$, we see that subpart A (or B) has the marginal density matrix $\rho_A = \mathrm{Tr}_B[\rho_{AB}] = \frac{1}{2}(|0\rangle\langle 0| + |1\rangle\langle 1|)$, and is therefore in a mixed state of positive entropy. This purely quantum situation corresponds to the unusual entropy diagram (−1,2,−1) shown in Fig. 1b. That the EPR situation cannot be described classically is immediately apparent when considering the conditional density matrix[1]: indeed the latter can be written as

$$\rho_{A|B} = \rho_{AB}(\mathbf{1}_A \otimes \rho_B)^{-1} = \begin{pmatrix} 0 & 0 & 0 & 0 \\ 0 & 1 & -1 & 0 \\ 0 & -1 & 1 & 0 \\ 0 & 0 & 0 & 0 \end{pmatrix}. \qquad (11)$$

[1] Note that for Bell states, joint and marginal density matrices commute, simplifying definitions (7) and (9).

Plugging (11) into definition (6) immediately yields $S(A|B) = -1$. This is a direct consequence of the fact that $\rho_{A|B}$ has one "unclassical" (> 1) eigenvalue, 2. It is thus misleading to describe an EPR pair (or any of the Bell states) as a correlated state within Shannon theory, since negative conditional entropies are crucial to its description. In [9], we suggest that EPR pairs are better understood in terms of a qubit-antiqubit pair, where the qubit (antiqubit) carries plus (minus) one bit of information, and antiqubits are interpreted as qubits traveling *backwards* in time. Still, classical *correlations* (case II) emerge when *observing* an entangled EPR pair. Indeed, after measuring A, the outcome of the measurement of B is known with 100% certainty. The key to this discrepancy lies in the information-theoretic description of the measurement process [12] and will be briefly addressed in the next section.

3. CORRELATION VERSUS ENTANGLEMENT

The concept of negative conditional entropy turns out to be very useful to describe n-body composite quantum systems, and it sheds new light on the creation of classical correlations from quantum entanglement. Consider for example a 3-body system ABC in a GHZ state (or an "EPR-triplet"), $|\psi_{ABC}\rangle = 2^{-1/2}(|000\rangle + |111\rangle)$. As it is a pure (entangled) state, the combined entropy is $S(ABC) = 0$. The corresponding ternary entropy diagram of ABC is shown in Fig. 2a. Note that the *ternary* mutual entropy $S(A{:}B{:}C) = S(A)+S(B)+S(C)-S(AB)-S(AC)-S(BC)+S(ABC)$ vanishes (see the center of the diagram); this is generic to any fully entangled three-body system. When tracing over the degree of freedom associated with C, say, the resulting marginal density matrix for subsystem AB is $\rho_{AB} = \mathrm{Tr}_C[\rho_{ABC}] = \frac{1}{2}(|00\rangle\langle 00| + |11\rangle\langle 11|)$, corresponding to a classically correlated system (case II). As the density matrix *fully* characterizes a quantum system, subsystem AB (unconditional on C, i.e., ignoring the existence of C) is in this case physically *indistinguishable* from a statistical ensemble prepared with an equal number of $|00\rangle$ and $|11\rangle$ states. Thus, A and B are correlated in the sense of Shannon theory if C is ignored. The "tracing over" operation depicted in Fig. 2b illustrates this creation of classical correlation from quantum entanglement. This feature is central to the description of the measurement process that we propose in [12], where A and B represent two parts of the measurement device, while C is the measured quantum system. The subsystem AB unconditional on C has a positive entropy $S(AB) = 1$ bit, and is indistinguishable from a classical correlated mixture (this corresponds to the generation of random numbers). On the other hand, the entropy of C conditional on AB, $S(C|AB)$, is negative and equal to -1 bit, thereby counterbalancing $S(AB)$ to yield a vanishing combined entropy

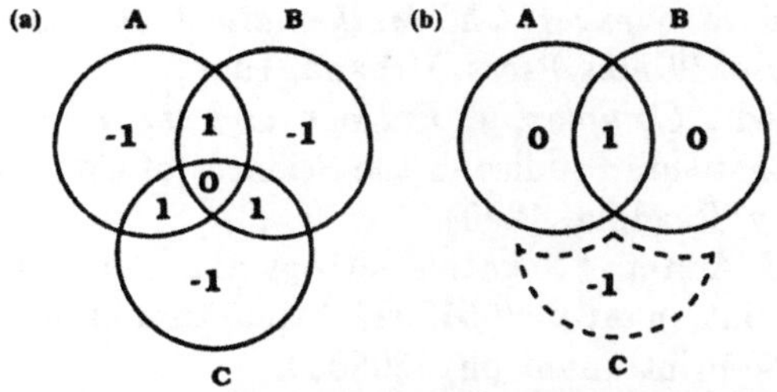

Fig. 2. (a) Ternary entropy diagram for an "EPR-triplet". (b) Entropy diagram for subsystem AB unconditional on C.

$$S(ABC) = S(AB) + S(C|AB) = 0\ , \tag{12}$$

as expected in view of the quantum entanglement between AB and C. We suggest in [12] that this information-theoretic interpretation of entanglement paves the way to a natural, unitary, and causal model of the measurement process, devoid of any assumption of a wave-function collapse, while implying all the well-known results of conventional probabilistic quantum mechanics. The same framework can also be used to interpret the observation of classical correlation between the measurement devices that occurs in the measurement of an EPR pair, and sheds new light on quantum paradoxes [13].

We would like to thank H. Bethe and A. Peres for very useful discussions. This work was supported in part by the National Science Foundation Grant Nos. PHY91-15574 and PHY94-12818.

REFERENCES

1. C. H. Bennett, *Physics Today* **48**(10), 24 (1995).
2. C. H. Bennett and D. P. DiVincenzo, *Nature* **377**, 389 (1995); D. P. DiVincenzo, *Science* **270**, 255 (1995).
3. C. H. Bennett and S. J. Wiesner, *Phys. Rev. Lett.* **69**, 2881 (1992); C. H. Bennett *et al.*, *Phys. Rev. Lett.* **70**, 1895 (1993).
4. A. Ekert, *Nature* **358**, 14 (1992); C. H. Bennett, G. Brassard, and N. D. Mermin, *Phys. Rev. Lett.* **68**, 557 (1992).
5. B. Schumacher, *Phys. Rev. A* **51**, 2738 (1995); R. Jozsa and B. Schumacher, *J. Mod. Opt.* **41**, 2343 (1994).
6. J. von Neumann, *Mathematische Grundlagen der Quantenmechanik* (Springer, Berlin, 1932).

7. C. Shannon and W. Weaver, *The Mathematical Theory of Communication* (University of Illinois Press, Urbana, 1949).
8. W. H. Zurek, ed., *Complexity, Entropy and the Physics of Information*, Santa Fe Institute Studies in the Sciences of Complexity Vol. VIII (Addison-Wesley, Reading, 1990).
9. N. J. Cerf and C. Adami, "Negative entropy and information in quantum mechanics," e-print quant-ph/9512022; "Quantum information theory of entanglement," e-print quant-ph/9605039.
10. A. Wehrl, *Rev. Mod. Phys.* **50**, 221 (1978).
11. N. J. Cerf and C. Adami, "Entropic Bell inequalities," e-print quant-ph/9608047.
12. N. J. Cerf and C. Adami, "Quantum mechanics of measurement," e-print quant-ph/9605002.
13. C. Adami and N. J. Cerf, Caltech preprint KRL–MAP–204 (1996).

STOCHASTIC APPROACH TO THE TUNNEL EFFECT

A. M. Cetto and L. de la Peña

Instituto de Física, UNAM
Apartado postal 20-364, 01000 México, D.F.

There is a growing interest in the investigation about the time required by particles to tunnel accross potential barriers, due both to important technological advances on the subject and to concerns of a conceptual nature. The fact that the various proposed versions of tunneling times are mutually inconsistent and give widely different outcomes, reveals that there is still much to be learned from such an apparently simple problem.

In this paper we approach the problem by using the phenomenological theory of stochastic quantum mechanics to obtain semi-quantitative information about the tunneling times. The results are in good agreement with recent numerical calculations performed within the framework of Nelson's stochastic theory. In particular, the most important effect of an opaque barrier on the tunneling particles is shown to occur before they actually enter the barrier. Further, a very simple statistical approach is shown to give valuable information on the average time spent by all particles inside the barrier.

The stochasticity of the particle dynamics, combined with the obviously wavelike features exhibited in the tunnel effect, call for an analysis within the framework of stochastic electrodynamics. As a first entry to this program, we briefly discuss the possible role of the zeropoint field in determining the behavior of tunneling particles, with the purpose of ultimately obtaining a satisfactory understanding of this intriguing quantum phenomenon.

Key words: potential barrier, tunneling time, stochastic quantum mechanics, zeropoint field, de Broglie wave, stochastic electrodynamics.

1. INTRODUCTION

In this day and age, quantum mechanics still puzzles us with some of its

M. Ferrero and A. van der Merwe (eds.),
New Developments on Fundamental Problems in Quantum Physics, 85–101.

oldest mysteries. The many successful theoretical developments and experimental confirmations, and innumerable analyses and debates on the interpretation of its results, have not yet resulted in a convincing physical picture for the theory—neither in classical terms nor in any other. Some of the most basic quantum phenomena remain to be explained, specifically those where the wave nature of particles seems to play a decisive role. One good example is given of course by the wide variety of interference experiments that have been performed since 1974 [1] in which beams of particles are separated and recombined producing an intensity pattern that can only be explained in terms of waves. Then there is the whole series of Bell-related experiments that do not seem to find a satisfactory explanation purely in terms of particles [2]. An analogous situation is met in relation with problems of scattering of beams of particles by potential barriers and wells. It is amazing to see how deeply the wave–language is involved in the description of these phenomena, without, however, there being a universally accepted image of the physical nature of the wave component in quantum systems.

A well-known instance that lends itself to a discussion of these issues, without being obscured by mathematical complexity, is the old problem of the tunnel effect. In this case, of course, there is the additional complication arising from the classical preclusion of particles moving in a region where the potential energy is greater than the total energy. So the particle+ wave–vocabulary has to be complemented with the magic word 'tunneling' for a complete 'description' of the phenomenon. These words, however, do not help to conjure away the big mystery of what is really going on inside the barrier during tunneling.

In recent years there has been a remarkable revival of the investigations on the tunnel effect, motivated to a great extent by the development of high-speed devices based on tunneling processes in semiconductors (see, e.g., [3]) which open up the possibility to obtain reliable experimental data. There are nowadays many applications of the tunnel effect in such distinct fields as electronics, nuclear physics, cosmology or nanoscopy; however, an analysis of the growing literature reveals that we still do not have a clearcut answer to such basic questions as how long does it take a particle to tunnel.

There are a number of recent reviews where the various proposals for defining the tunneling times are analysed in detail; see, e.g., [4], [5] and [6]. It is clear from these reviews that the different versions of tunneling times proposed up to now not only continue to be controversial; some of them even produce results that are difficult to accept, or at least to understand. As follows from a comparison of the results, one can group these times into two broad categories, according to their behavior for very thick barriers. One category comprises the (asymptotic or extrapolated) phase-shift times [4–7] the dwell time [5], [8], [9] and some of those grouped under the common name of Larmor times [4–6], [10], [11]; all these times become independent of the barrier width a as $a \to \infty$. A tunneling time δt that is independent of a for opaque barriers is equivalent to arbitrarily high effective tunneling velocities $\bar{v}_{\rm eff} \equiv a/\delta t$, which gives rise to the so-called Hartman-Fletcher effect [5], [7], [12]. Other times, such as the Büttiker-Landauer time [4], [6], [9], one of the Larmor times [4], [9], [13], and the mean passage time recently introduced in [14], become proportional to a for very wide or opaque barriers; we group all these under the second category. Further, there is the Sokolovski-Baskin time which has the additional complication of being a complex quantity.

Our motivation for addressing the tunneling problem is a theoretical concern: we take the diversity of tunneling times as a symptom of the lack of understanding of what goes on during tunneling and of what determines the dynamics of the particles in the barrier region. A more thorough study of this apparently simple problem, from as many angles as possible, should hopefully help us get a better understanding of the underlying physics. As a contribution to this needed clarification, we here take a slightly different view on the problem from a statistical angle, and present a more qualitative approach but appealing to some elements provided by stochastic quantum mechanics (SQM) and stochastic electrodynamics (SED).

In the first part of the paper we follow a line of thought recently introduced by Imafuku et al. [14], and apply a simple phenomenological treatment using the tools of SQM in order to evaluate the times spent by tunneling particles within and in the close neighborhood of the potential barrier. In fact, the most interesting physics seems to occur in front of the barrier, before any crossing happens and where the particles apparently spend (on the average) a most important part of the 'tunneling' time. This point, to which little attention had been paid in the literature, turns out to be of great significance. It had already been recognized for some time [4] that the interference of the incoming part and the reflected part of the wave function implies that free-particle motion cannot be assumed all the way up to the barrier. However, only recently specific attention was given to this phenomenon in [14], where the authors study the motions of *individual* electrons using a stochastic model of the system, and thus are able to discriminate between reflected and transmitted particles and to perform averages over the respective subensembles. The interesting results thus obtained are somewhat obscured by the numerical treatment of the problem, which makes it difficult to establish a connection with other known results. Here we suggest a couple of definitions that lead to closed, algebraic expressions which seem to be in close correspondence with those used in [14].

On the other hand, a simple statistical approach to the problem allows to define a quantity representing the average time spent by *all* particles within a barrier of any thickness. The quantity thus defined, called here *sojourn time*, turns out to be in disagreement with all the other times mentioned above; this is however expected, since one is dealing with a different concept, as becomes clear from the discussion. What is surprising is that such elementary definition—in fact, the simplest and most natural definition possible for the average time spent by particles inside the barrier—has not been used to date despite the fact that it leads to intuitively clear and sensible results; this is perhaps due to the prevailing views, that tend to inhibit thinking of particles moving within a classically forbidden region.

Finally, in an attempt to advance in the clarification of the meaning of the quantum wave, we discuss some qualitative aspects of the tunneling problem from the perspective of SED, a theory according to which the zeropoint field is the ultimate source for the conspicuous wave element that accompanies quantum particles.

The calculations that follow are restricted to the one-dimensional stationary problem, as is customary in the literature. Also, since we want to focus on appropriate definitions rather than precise specific calculations, and it is convenient to have explicit results, we consider only square barriers, as is also frequent in the related literature. The definitions are generalizable to more complex situations, which need not be discussed here in detail.

2. STOCHASTIC MODEL FOR THE TUNNELING TIMES

As is well known, quantum mechanics affords a good description of the statistical properties of physical systems but it does not provide a description of the actual individual events of experiments; so, in particular, it is unable to describe the dynamics of particles in terms of individual trajectories. There is, however, a family of phenomenological theories grouped under the heading of stochastic quantum mechanics, that is based on the idea that particles are constantly subjected to a random force which makes them perform a stochastic motion. Such theories afford a natural way to describe an individual sample path, taken as representative of the ensemble, and indeed they have been used for this purpose in a variety of circumstances (see, e.g., [15]). It is therefore interesting to investigate what SQM can tell us about the individual dynamics of tunneling particles.

Apparently the first work in this direction is the one by Imafuku et al. [14], who for their purpose use Nelson's stochastic model of quantum mechanics [16] and perform a statistical study of the trajectories followed by electrons that approach a square tunneling barrier. In the framework of Nelson's theory, the motion of a quantum particle is subjected to a (frictionless) Markov process described by a stochastic Langevin-type equation. The authors use a numerical simulation based on this stochastic equation, and apply it to study the motion of the particles represented by a wave packet incident on the barrier. One first significant outcome is that they can analyse separately the individual sample paths for the subensembles of

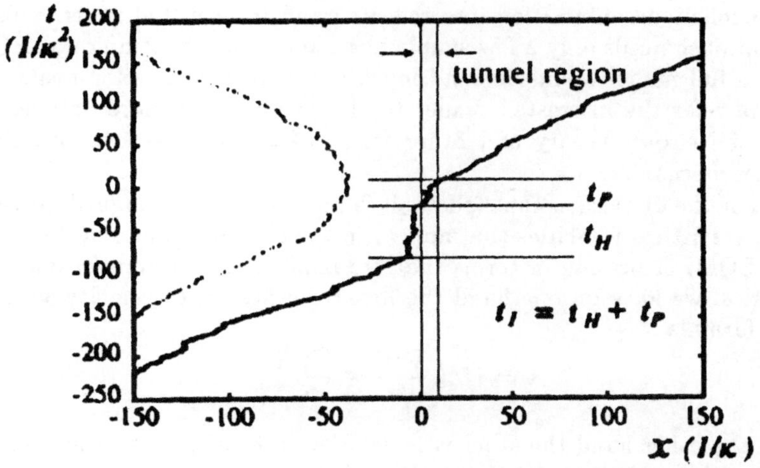

Fig. 1. Two typical sample paths: a transmission path and a reflection one. The potential width is $a = 10/\kappa$ and the potential height is $V = 2E$. Taken from [14], Fig. 4.

reflected and transmitted particles.

A study of the subensemble of transmitted electrons shows, somewhat surprisingly, that these particles wander a long time (they *hesitate*, in the language of Ref. [14]) before the barrier. This wandering (hesitation) time increases with qa, where $q^2\hbar^2 = 2m(V_0 - E)$, and for large values of q it becomes dominant over the passage time spent by the particles inside the barrier. In Fig. 1 we reproduce two typical sample paths obtained in Ref. [14] for different barrier widths, to illustrate the kind of behavor referred to above for a reflected and a transmitted particle.

A practical disadvantage of such a numerical study is of course that the number of sample trajectories pertaining to the subensemble of transmitted particles becomes drastically reduced for high or wide barriers, which produce the most interesting dynamical effects. Moreover, the results of numerical simulations sometimes give a poor idea of the functional depedence of the dynamics on the various parameters involved in the process. It is therefore convenient to develop an algebraic method that allows us to study the effects of the barrier, particularly of an opaque barrier, on the transmitted particles.

We therefore propose to construct expressions for tunneling times similar to those defined in [14], based also on SQM but avoiding a detailed, individual study of the random trajectories. We formulate a heuristic approach to estimate the passage and similar times, without however requiring an explicit decomposition into reflected and transmitted subensembles. This approach needs only a few simple ideas and concepts common to SQM, though a full model would demand much more involved developments. We therefore refer the interested reader to [16] and [17] for more detailed accounts of Nelson's theory and other forms of the stochastic approach to quantum mechanics.

For the discussion that follows it is important to distinguish between the two statistical velocities that appear in the configuration–space description of SQM. Expressed in terms that are familiar in Schrödinger quantum mechanics, we have on one hand the flux (or systematic) velocity $\mathbf{v}$, given by the formula

$$\mathbf{v} = iD\left(\frac{\nabla\psi^*}{\psi^*} - \frac{\nabla\psi}{\psi}\right), \tag{1}$$

and on the other hand the stochastic velocity $\mathbf{u}$, generated by the diffusion that goes along with any inhomogeneous distribution of particles, and given by

$$\mathbf{u} = D\frac{\nabla\rho}{\rho}, \tag{2}$$

where $\rho = \psi^*\psi$ and D is the diffusion constant, which takes the value $D = \hbar/2m$ for quantum stochastic processes (a detailed discussion of this formula can be found in [17], chapter 2). Note that the two velocities are associated with a given quantum state. The flux velocity appears naturally in the continuity equation with the probability current given by $\mathbf{j} = \rho\mathbf{v}$:

$$\frac{\partial\rho}{\partial t} + \nabla\cdot\rho\mathbf{v} = 0, \tag{3}$$

and the stochastic velocity can be introduced in a straightforward way by recasting this equation into a Fokker-Planck equation that accounts for diffusions taking place in the system; by adding to both sides a diffusion term, one gets

$$\frac{\partial\rho}{\partial t} + \nabla\cdot\rho\mathbf{v} + D\nabla^2\rho = D\nabla^2\rho, \tag{4}$$

and the extra term on the left hand side can be rewritten with the help of (2) as

$$D\nabla^2\rho = D\nabla\cdot\nabla\rho = D\nabla\cdot\left(\rho\frac{\nabla\rho}{\rho}\right) = \nabla\cdot\rho\mathbf{u}. \tag{5}$$

With $\mathbf{b}_+ = \mathbf{v} + \mathbf{u}$ representing the forward velocity, Eq. (4) transforms into

$$\frac{\partial\rho}{\partial t} + \nabla\cdot\rho\mathbf{b}_+ = D\nabla^2\rho. \tag{6}$$

This is the *forward* Fokker-Planck equation, a fundamental equation within the framework of the stochastic theory; physical arguments for it are given in detail in the cited literature.

With an equivalent analysis it is possible to obtain a *backward* Fokker-Planck equation,

$$-\frac{\partial \rho}{\partial t} - \nabla \cdot \rho \mathbf{b}_- = D\nabla^2 \rho, \quad \mathbf{b}_- = \mathbf{v} - \mathbf{u}, \tag{7}$$

with $\mathbf{b}_-$ the backward velocity. This equation can be understood by taking into account that going backwards in time reverses the sign of the systematic velocity, but leaves the stochastic velocity given by (2) unchanged.

Now assume a one-dimensional potential barrier that extends from $x = 0$ to $x = a$, and consider the particles that stream toward it from the left and arrive very close to the border. As usual, we call I, II and III, the regions located to the left ($x < 0$), inside ($0 < x < a$) and to the right of the barrier ($x > 0$), respectively. We shall study the stationary situation only, and assume for simplicity that the incident beam is monoenergetic. If one pays attention to the region just in front of the barrier at $x = 0^-$, one will see essentially those particles that wander in front before entering—indeed, as follows from the results of Imafuku et al. [14], the particles that are reflected do not reach in general the zone of very small values of x, tending to reverse their direction of motion much before they can reach it. Now, these particles are wandering near $x = 0$ because they are strongly diffussed towards the left of the repulsive barrier. This suggests using as a measure of the *wandering* (hesitation) time the expression

$$\delta t^{\mathrm{w}} = \int \frac{dx}{v-u} = \int \frac{\rho dx}{j - \rho u} = \int \frac{\rho dx}{j - D\rho'}. \tag{8}$$

Here the region of integration must be restricted to those values of $x < 0$ that are in the vicinity of the barrier and for which $v - u > 0$. The use of the velocity b_- in this definition gives a high probability that we are dealing mainly with the (eventually) transmitted particles. A similar reasoning can be used to estimate the *passage* time, spent by the transmitted particles inside the barrier:

$$\delta t^{\mathrm{p}} = \int_{II} \frac{dx}{v-u} = \int_{II} \frac{\rho dx}{j - D\rho'}. \tag{9}$$

Let us apply the above definitions to a square barrier of height $V_0 > 0$; the incoming particles stream from the left with well-defined momentum $p_0 = mv_0 = \hbar k$ for $x < 0$, and $j = j_1(x) = j_2(x) = j_3(x)$. Writing the stationary solution of the Schrödinger equation in the form $\psi_i(x)$, $i = 1, 2, 3$, one has for sub-barrier energies $0 < E < V_0$,

$$\psi_1 = e^{ikx} + B_1 e^{-ikx}, \tag{10}$$

$$\psi_2 = A_2 e^{-qx} + B_2 e^{qx}, \tag{11}$$

$$\psi_3 = A_3 e^{ik(x-a)}, \tag{12}$$

where

$$k^2 = \frac{2mE}{\hbar^2}, \quad \kappa^2 = \frac{2mV_0}{\hbar^2}, \quad q^2 = \frac{2m(V_0 - E)}{\hbar^2} = \kappa^2 - k^2. \tag{13}$$

The coefficients are given by

$$B_1 = -i\frac{\kappa^2}{2kq} A_3 \sinh qa, \tag{14}$$

$$A_2 = \frac{q - ik}{2q} e^{qa} A_3, \tag{15}$$

$$B_2 = \frac{q + ik}{2q} e^{-qa} A_3, \tag{16}$$

$$A_3 = \frac{2kq}{2kq \cosh qa + i(q^2 - k^2) \sinh qa} \tag{17}$$

and for the transmission coefficient and the current density one obtains, respectively,

$$T = |A_3|^2 = \frac{4k^2q^2}{4k^2q^2 \cosh^2 qa + (k^2 - q^2)^2 \sinh^2 qa}$$

$$= \frac{1 - \tanh^2 qa}{1 + \frac{(k^2-q^2)^2}{4k^2q^2} \tanh^2 qa}, \tag{18}$$

$$j = j_3 = \frac{\hbar k}{m} |A_3|^2 = \frac{\hbar k}{m} T = j_0 T, \tag{19}$$

where $j_0 = \hbar k/m$ is the incident flux.

We begin by applying the above formulas to region II, to study the passage time, which from (9) and (19) is given by

$$\delta t^{\mathrm{P}} = \frac{1}{D} \int_{II} \frac{\rho dx}{2kT - \rho'}. \tag{20}$$

From now on we will write the constant probability current j in the form $j = 2DkT$, with the diffusion coefficient $D = \hbar/2m$, as noted before. From equations (11) and (13)-(19) inside the barrier

$$\rho_2 = T\frac{\kappa^2}{2q^2} \left[\cosh 2q(a - x) - \frac{k^2 - q^2}{\kappa^2}\right]. \tag{21}$$

It is easily seen that everywhere in region II the integrand in (20) is positive. One thus gets

$$\delta t^{\mathrm{P}} = \frac{1}{2Dq} \int_0^a \frac{A + \cosh 2q(a-x)}{B + \sinh 2q(a-x)} dx, \tag{22}$$

$$A = \frac{q^2 - k^2}{\kappa^2}, \quad B = \frac{2kq}{\kappa^2}. \tag{23}$$

The integration gives

$$\delta t^{\mathrm{P}} = \frac{1}{4Dq^2} \left[\ln\left(1 + \frac{1}{B}\sinh 2qa\right) + \frac{A}{\sqrt{1+B^2}} \ln \frac{1 - C_+ + (1 + C_+)e^{-2qa}}{1 - C_- + (1 + C_-)e^{-2qa}} \right], \tag{24}$$

where

$$C_\pm = -\frac{1 \pm \sqrt{1+B^2}}{B}. \tag{25}$$

The limiting values of this passage time are as follows:

$$\delta t^{\mathrm{P}} = \frac{a}{2Dk} = \frac{a}{v_0}, \quad qa \to 0; \tag{26}$$

$$\delta t^{\mathrm{P}} = \frac{a}{2Dq} = \frac{a}{v_0}\frac{k}{q}, \quad qa \to \infty. \tag{27}$$

According to (26), very thin or transparent barriers produce almost no delay, as should be expected. Eq. (27), on the other hand, coincides with the Larmor time $\tau_{z\mathrm{T}}^{\mathrm{La}}$ for opaque barriers (see, e.g., [5], Eq. (19)), but differs of course from the times of the first category, which become independent of a as $qa \to \infty$. Also, although its definition is different from that introduced in [14] (where it is defined as the time interval in which $x(t)$ is in region II), it is interesting to note that (27) describes correctly the results of the numerical evaluation reported in Fig. 5 of Ref. [14] within the specified range of validity. Specifically, for a fixed width a, the passage time varies as the inverse square root of the potential height and thus becomes arbitrarily small for very high barriers (for those very few particles that succeed in tunneling).

As was said before, the most interesting physics seems to occur in region I, where the interference phenomenon leads to a complex behavior of the electrons. This is most easily seen by writing the amplitude of the transmitted wave in the form $A_3 = \sqrt{T} e^{i\varphi_3}$, with the phase given by

$$\varphi_3 = \arctan\left(\frac{k^2 - q^2}{2kq} \tanh qa\right), \tag{28}$$

and multiplying the full wave function (10)-(12) by the phase factor $e^{-i\varphi_3}$; then only the *incoming* wave contains φ_3, which thus represents the phase relative to both the reflected wave (except for an i-factor characteristic of a reflexion) and the transmitted wave (except for an additional phase shift ka corresponding to the separation between regions I and III). Since this phase φ_3 carries nonlocal information, we see that for the stationary states under discussion the behavior of particles in region I depends explicitly on their behavior at any other place. This kind of nonlocality, characteristic of the stationary quantum states, is widely recognized though hardly understood within the quantum mechanical context; we shall come back to this point in the final part of the paper. Note, incidentally, that all the remaining nontrivial phases appearing in the coefficients (14)-(17) are independent of the barrier width.

Owing to the complex dynamics in region I, the wandering time δt^{w} given by (8) can become much larger than δt^{p}. Unfortunately its general expression is quite complicated; we get, with $\varphi_1 = \varphi_3 - \pi/2$,

$$\delta t^{\mathrm{w}} = \frac{1}{2Dk}\int_{-L}^{0} \frac{1 + R + 2\sqrt{R}\cos(2kx - \varphi_1)}{T + 2\sqrt{R}\sin(2kx - \varphi_1)}dx. \tag{29}$$

The selection of the point $x = -L$ to the left of the barrier at which one starts to compute this time is conventional, although it should be taken so as to ensure that Eq. (29) is meaningful, which implies, in particular, that the denominator in the integrand should not be allowed to vanish. Performing the integration one gets

$$\delta t^{\mathrm{w}} = \frac{1}{4Dk^2}\left[\ln\frac{T - 2\sqrt{R}\sin\varphi_1}{T - 2\sqrt{R}\sin(2kL + \varphi_1)} + (1 + R)\mathcal{I}\right] \tag{30}$$

where for $T^2 - 4R = S^2 > 0$,

$$\mathcal{I} = \frac{2}{S}\left[\arctan\frac{2\sqrt{R} - T\tan(\varphi_1/2)}{S} - \arctan\frac{2\sqrt{R} - T\tan(kL + \varphi_1/2)}{S}\right] \tag{31}$$

and for $4R - T^2 = Q^2 > 0$,

$$\mathcal{I} = \frac{1}{Q}\ln\left[\frac{-T\tan(\varphi_1/2) + 2\sqrt{R} - Q}{-T\tan(\varphi_1/2) + 2\sqrt{R} + Q}\right.$$
$$\left.\times\frac{-T\tan(kL + \varphi_1/2) + 2\sqrt{R} + Q}{-T\tan(kL + \varphi_1/2) + 2\sqrt{R} - Q}\right]. \tag{32}$$

For nearly transparent barriers, $qa \to 0$, one can take $T \simeq 1, R \simeq 0$, and (31) is reduced in this instance to $\mathcal{I} = \frac{2}{T}kL$. Eq. (30) gives then

$$\delta t^{\mathrm{w}} \approx \frac{L}{2Dk}, \quad qa \to 0, \tag{33}$$

which means simply that the transparent barrier is approached by the particle at its normal speed $v_0 = 2Dk = \hbar k/m$.

The opposite case of opaque barriers is more complicated, since Eq. (30) holds only in a restricted range of values for L, those for which the quantity $T - 2\sqrt{R}\sin(2kL + \varphi_1)$ is greater than zero. Since $\sin\varphi_1 = -\sqrt{T}\cosh qa$ and $\cos\varphi_1 = [(k^2 - q^2)/2kq]\sqrt{T}\sinh qa$, the angle φ_1 is negative and lies either in the third or fourth quadrant. For opaque barriers such that $qa \gg 1$ and $T = 1 - R \ll 1$, one gets from (30) and (32)

$$\delta t^{\mathrm{w}} = \frac{1}{4Dk^2}\left[\ln\frac{T - 2\sin\varphi_1}{T - 2\sin(2kL + \varphi_1)} + \right.$$

$$\left. \ln\frac{\tan(\varphi_1/2) - T/4}{\tan(kL + \varphi_1/2) - T/4} \cdot \frac{1 - (T/4)\tan(kL + \varphi_1/2)}{1 - (T/4)\tan(\varphi_1/2)}\right] \tag{34}$$

with $\sin\varphi_1 \simeq -2kq/\kappa^2$ and $\cos\varphi_1 \simeq (k^2 - q^2)/\kappa^2$. Let us define the maximum possible value of L as that for which

$$T - 2\sqrt{R}\sin(2kL + \varphi_1) = \epsilon, \quad 0 < \epsilon \ll 1. \tag{35}$$

Then $\tan(kL + \varphi_1/2) - T/4 \simeq -\epsilon/4$, $1 - (T/4)\tan(kL + \varphi_1/2) \simeq 1$, $L \simeq -\varphi_1/2k \simeq q/\kappa^2$, and (34) simplifies into

$$\delta t^{\mathrm{w}} \simeq \frac{1}{4Dk^2}\left(\ln\sin\varphi_1\tan\frac{\varphi_1}{2} - \ln\frac{\varepsilon^2}{8}\right) = \frac{1}{4Dk^2}\left(\ln\frac{q^2}{\kappa^2} - \ln\frac{\varepsilon^2}{16}\right). \tag{36}$$

So for wide barriers the wandering time becomes independent of the width, and if the barrier is also very high ($q^2/\kappa^2 \simeq 1$), it approaches the (maximum) asymptotic value

$$\delta t^{\mathrm{w}}_{\mathrm{asymp}} = \frac{1}{4Dk^2}\ln\frac{16}{\epsilon^2}, \tag{37}$$

which can become very large since $\epsilon \ll 1$. Because the parameter ϵ has been arbitrarily defined, this result cannot be compared easily with the numerical results of [14]. However, taken as a function of $V_0/E = \kappa^2/k^2$, δt^{w} has a behavior that is qualitatively very similar to that of the hesitation time reported in Fig. 6 of [14] for wide barriers. Further, the semiquantitative relation $L \simeq q/\kappa^2$ tells us that the wandering takes place in the close neighborhood of the barrier, within a distance $L \simeq q^{-1}$ in the asymptotic limit $q^2/\kappa^2 \simeq 1$. These results confirm the observation that most of the retarding effect produced by an opaque barrier on the tunneling particles is produced to the left, before the particles actually enter.

3. SOJOURN TIME

According to the statistical view of quantum mechanics (and also to more orthodox views), the definition of the flux velocity, given conventionally by the expression $v(x,t) = j(x,t)/\rho(x,t)$ with $j(x,t)$ the local probability current, can be legitimately applied everywhere. The time spent on the average by particles moving at a given instant through an element of length dx centered at x can therefore be estimated as $dx/v(x)$. This allows us to calculate the *sojourn* time, that is, the time spent on the average by the particles moving through the barrier, as the sum of the mean times required to move step by step inside the barrier, i.e., as

$$\delta t^{\text{soj}} = \int_B \frac{dx}{v(x,t)} = \int_B \frac{\rho(x,t)}{j(x,t)} dx, \tag{38}$$

where B is the domain of the barrier. In the stationary case $j(x)$ is the same constant everywhere, and the sojourn time is given by

$$\delta t^{\text{soj}} = \frac{1}{j} \int_B \rho(x) dx. \tag{39}$$

For the square potential barrier one gets, using (21), (39) and (19),

$$\begin{aligned} \delta t^{\text{soj}} &= \frac{1}{j} \int_0^a \rho_2(x) dx = \frac{1}{j_0} \frac{\kappa^2}{4q^3} \left[\sinh 2qa + \frac{q^2 - k^2}{\kappa^2} 2qa \right] \\ &= \frac{ma}{\hbar k} \left[1 + \frac{1}{2} \left(1 + \frac{k^2}{q^2} \right) \left(\frac{\sinh 2qa}{2qa} - 1 \right) \right]. \end{aligned} \tag{40}$$

Note that the factor T appears both in ρ_2 and in j, and therefore cancels out. This feature distinguishes δt^{soj} from other times proposed to measure the average time spent inside the barrier. For instance, the dwell time is normally defined as in (39), but with the incident flux j_0 replacing the actually circulating current density j [4], [6]; thus,

$$\delta t^{\text{dw}} = \frac{1}{j_0} \int_0^a \rho_2(x) dx = T \delta t^{\text{soj}}. \tag{41}$$

The properties of δt^{soj} seem to be well suited to measure the average time spent by the particles within the barrier, even though they give no detailed information on the real time spent by a given particle that effectively crosses it. In particular, for very thin or transparent barriers, such that $qa \to 0, T \simeq 1$,

$$\delta t^{\text{soj}} \simeq \delta t^{\text{dw}} \simeq \frac{a}{j_0} = \frac{ma}{\hbar k}, \tag{42}$$

which means that a thin barrier does not alter the average velocity $\hbar k/m$ with which the particles move outside the potential region. Further, note that in this limit δt^{soj} coincides with the passage time given by (26); this is consistent with the fact that for $T \simeq 1$ there is practically no contribution from reflected particles to the sojourn time.

For $qa \to \infty$ one gets instead

$$T = \frac{16k^2q^2}{\kappa^4}e^{-2qa} \tag{43}$$

and

$$\delta t^{\text{soj}} = \frac{m}{\hbar k} \cdot \frac{\kappa^2 e^{2qa}}{8q^3}, \tag{44}$$

whereas the dwell time becomes independent of a,

$$\delta t^{\text{dw}} = \frac{2mk}{\hbar \kappa^2 q}. \tag{45}$$

According to (44) the particles spend a very long time inside an opaque barrier before they eventually exit. The sojourn time can acquire such high values because for $T \ll 1$ there is still a nonneglibible density of particles within the barrier—or at least in some parts of it, according to (21)—, but the net *flow* of particles is very low. In fact, the effective velocity of these particles is vanishingly small, $v_{\text{eff}} = a/\delta t^{\text{soj}} \sim qae^{-2qa}$, in contrast to the effective velocity associated with the dwell time and other times belonging to the first category, which grows indefinitely as $qa \to \infty$.

On comparing the times given by the last two equations it is interesting to note their different dependency on the particle energy, represented by the wave number k. According to (44), very slow particles tend to spend a larger time within the barrier, which is intuitively acceptable. In contrast, the vanishing of δt^{dw} for small values of k is not so understandeable, if, as discussed in [6], the dwell time as defined in (41) represents the time the incident flux has to be turned on, to provide the accumulated particle storage in the barrier.

In trying to get some understanding of the origin of the wide differences between the various times just discussed, it might be convenient to recall that they have been calculated for a time–independent solution of the tunneling problem, characterized by a stationary flow of particles. Within the usual framework of quantum mechanics it is impossible to separate the particles that are transmitted, from those that remain within the barrier and from those that are backscattered.

A few time–dependent studies have been performed, however, that can give some intuition about the behavior of the particles inside the potential barrier. In particular, an illustrative result is obtained in Ref. [18]

by studying the transient dynamics of a plane wave confined initially in the half-space $x < 0$ and allowed to propagate towards a potential barrier for $t > 0$ (after the sudden opening of a shutter). The potential considered is a double barrier resonant structure, which gives the possibility to analyse (numerically, at least) the buildup process that leads to the stationary wave solution both in the internal region (resonant states) and outside the barrier. The flow of particles that enter is mainly used up in the construction of the density $\rho_2(x,t)$ until it reaches its stationary value $\rho_2(x)$ and only an insignificant part cross the barrier to exit from the right. It is most interesting to note that even for small times—of the order of the lifetime of the first resonance—the shape of $\rho_2(x,t)$ already resembles the shape of the final $\rho_2(x)$.These results support the introduction of the sojourn time as the total time spent on the average by all particles inside the barrier; the sojourn time can indeed become very large for wide or opaque barriers.

Quantum mechanics, however, does not give us any hint about the specific mechanism responsible for the transmission of particles from left to right, and we are thus left with an ambiguous picture. For consider the stationary situation: do the newly incoming particles cross the entire barrier from left to right, or else do they just enter the barrier from the left, thereby creating a 'pressure' that results in particles from the interior moving towards the right? This latter alternative, if it is correct, would mean that the tunneling time can be much smaller than the sojourn time, and smaller even than the passage time, being limited only by the finite velocity of transmission of disturbances through the barrier. Otherwise, if every individual particle must cross the entire barrier, it is the velocity of the transmitted particles that determines the tunneling time. These various alternatives correspond perhaps, at least in part, to the different tunneling times proposed and reviewed in the literature.

In this context it is pertinent to recall a similar ambiguity, relating to electric conduction in a small wire without impurities [19]. Very fine, recent experiments with quantum point contacts seem to confirm that electric current in such tiny metals is a result of ballistic transmission of electrons, which is most conveniently described by invoking the optical analog in terms of transmission of light [20]. One would expect, similarly, that very fine experiments with tunneling barriers can provide information on the specific transport mechanism involved in electron tunneling.

4. A HEURISTIC VIEW FROM STOCHASTIC ELECTRODYNAMICS

Within the framework of SED, the random zeropoint electromagnetic radiation field of energy $\frac{1}{2}\hbar\omega$ per normal mode is assumed to constitute an essential ingredient in determining the behavior of matter at the quantum level. Now, the background field that is in permanent interaction with the radiating particle is, in its turn, strongly modified by it through radiation reaction, to the extent that the stability of quantum systems should be understood as an outcome of this tight complicity between particles and field. This is the point proposed in Ref. [17].

In more recent work (see, e.g., [17] and [21]) it has been shown that the radiating electron acquires an effective structure that gives rise in its turn to a fine vibration with a very high frequency, $\omega_c = mc^2/\hbar$—considered in this theory as the nonrelativistic version of the zitterbewegung. The vibrating electron becomes thus attached to the corresponding field modes, which form a standing wave in the particle frame of reference, and an amplitude modulated wave in the laboratory frame. When the electron moves with a flow velocity v, the wavelength of this modulation is $\lambda_B = \hbar/mv$, i.e., de Broglie's wavelength, and its group velocity coincides with the particle velocity. A heuristic analysis has been shown in fact to lead to the (Schrödinger) wave equation for the modulation amplitude [17], [21]. Unfortunately, the specific mechanism of wave–particle interaction is so far not studied, so as to enable a detailed account of the particle behavior; we hope to report some advances on this point in a future paper [22].

This theory offers thus in principle a physical basis for the undulatory behavior of quantum particles, and should be helpful in explaining specific wave phenomena, as those mentioned in the introduction. In particular, it is clear that the spatial extension of the modulated field wave—even if it is associated to an essentially point particle—is responsible for apparently nonlocal effects on particles. Such effects are visible when the zeropoint field has adjusted to the experimental setup and the modulated field wave acquires a stable pattern. It is under these conditions, for instance, that the behavior of particles to the left of the tunneling barrier depends explicitly on their behavior at any other place, as discussed in Sec. 2. Also, the fact that the background field can be drastically affected by the reflected particles in the close neighborhood of an opaque potential barrier, should be sufficient to explain the large wandering or hesitation times obtained in this case, according to Eq. (37). For a phenomenon of this sort to be effective, a sufficient time must elapse from the beginning of the experiment to allow for a stationary situation to establish, in which matter and background field have adjusted themselves to the entire setup. It is important to note that at the initial instant the boundary conditions affect the particles only, but they in their turn influence the field, and eventually the modified field signals the existence of those boundary conditions. A similar phenomenon is expected to occur in quantum mechanical problems characterized by boundary conditions on the electromagnetic field itself: after a short transient time, necessary for the establishment of matter-field equilibrium, the particles signal the existence of those boundary conditions. Apparently, quantum mechanics only describes this situation in which the so-called quantum regime has been reached; the transient phenomena that take place in the very brief initial time (smaller than 10^{-21}s for electrons) are outside the realm of the quantum description.

REFERENCES

1. H. Rauch, W. Treimer and U. Bonse, *Phys. Lett. A* **47**, 369 (1974).
2. See, e.g., J.S. Bell, *Speakable and Unspeakable in Quantum Mechanics* (Cambridge University Press, Cambridge, 1987). F. Selleri, *Quantum Paradoxes and Physical Reality* (Kluwer Academic, Dordrecht, 1990). T.W. Marshall and E. Santos, *Phys. Rev. A* **39**, 6271 (1989).
3. F. Capasso, K. Mohammed and A.Y. Cho, *IEEE J. Quantum Electron.* **QE-22**, 1853 (1986).
4. E.H. Hauge and J.A. Støvneng, *Rev. Mod. Phys.* **61**, 917 (1989).
5. V.S. Olkhovsky and E. Recami, *Phys. Rep.* **214**, 339 (1992).
6. R. Landauer and Th. Martin, *Rev. Mod. Phys.* **66**, 217 (1994).
7. Th. E. Hartman, *J. Applied Phys.* **33**, 3427 (1962).
8. F.T. Smith, *Phys. Rev.* **118**, 349 (1960).
9. M. Büttiker, *Phys. Rev. B* **27**, 6178 (1983).
10. A.I. Baz', *Yad. Fiz.* **4**, 252 (1966).
11. V.F. Rybachenko, *Yad. Fiz.* **5**, 895 (1967).
12. J.R. Fletcher, *J. Phys. C* **18**, L55 (1985).
13. M. Büttiker and R. Landauer, *J. Phys. C* **21**, 6207 (1988).
14. K. Imafuku, I. Ohba and Y. Yamanaka, *Phys. Lett. A* **204**, 329 (1995).
15. M McClendon and H. Rabitz, *Phys. Rev. A* **37**, 3479 (1988).
16. E. Nelson, *Phys. Rev.* **150**, 1079 (1966), *Quantum Fluctuations* (Princeton University Press, Princeton, New Jersey, 1985), and references therein. L. de la Peña, *J. Math. Phys.* **10**, 1620 (1969). L. de la Peña and A.M. Cetto, *Found. Phys.* **12**, 1017 (1982), reprinted in: *Quantum, Space and Time: The Quest Continues*, A.O. Barut, A. van der Merwe, and J.-P. Vigier, eds. (Cambridge University Press, Cambridge, 1984). F. Guerra, *Phys. Rep.* **77**, 263 (1981).
17. L. de la Peña and A.M. Cetto, *The Quantum Dice. An introduction to stochastic electrodynamics* (Kluwer Academic, Dordrecht, 1966), and references therein.
18. G. García-Calderón and A. Rubio, "Transient effects and delay time in the dynamics of resonant tunneling," preprint SC96-10, IFUNAM, 1996.
19. R. Landauer, *IBM J.Res.Dev.* **32**, 306 (1988); A.D. Stone and A. Szafer, *IBM J.Res.Dev.* **32**, 384 (1988).

20. See, e.g., H. van Houten and C. Beenakker, *Phys. Today*, July 1996, p. 22, and references therein.
21. A.M. Cetto and L. de la Peña, in *Fundamental Problems in Quantum Physics*, M. Ferrero and A. van der Merwe, eds. (Kluwer Academic, Dordrecht, 1995), p. 47; in *Chaos—The Interplay Between Stochastic and Deterministic Behaviour*, P. Garbaczewski, M. Wolf and A. Weron, eds. (Springer, Berlin, 1996), p. 51.
22. L. de la Peña and A.M. Cetto, "Dynamical effect of de Broglie waves on quantum particles," under preparation.

CLASSICAL PHYSICAL ENTITIES WITH A QUANTUM DESCRIPTION

Bob Coecke, Bart D'Hooghe, and Frank Valckenborgh

TENA, Free University of Brussels
Pleinlaan 2, B-1050 Brussels, Belgium

For general quantum-like systems we introduce classical representations. We show why these representations do not conflict with the Jauch-Piron no-go theorem for quantum mechanics. Moreover, we are able to impose the Hilbert space structure for these representations.

Key words: hidden measurements, propositional structure.

1. INTRODUCTION

In [1], Aerts introduced the 'hidden measurement approach' to quantum mechanics: *with every quantum measurement corresponds a collection of classical measurements (called hidden measurements), and there exists a lack of knowledge concerning which measurement is actually performed.* For a general background on hidden measurements we refer to [1–4] and [7]. In this paper, we investigate how it is possible to avoid the Jauch-Piron no-go theorem. We also illustrate how the propositional structure of quantum mechanics can be imposed in a natural way on the hidden measurement representation. Thus, the classical representations that we consider respect the symmetries of the given entity.

2. ASSUMPTIONS OF THE APPROACH

In this section we consider a situation with a lack of knowledge concerning the interaction of the entity under study with its measurement context; in this case the state of the entity is not sufficient anymore to determine the measurement outcome. With the following reasonable assumptions, we are able to construct a framework to study these situations:

M. Ferrero and A. van der Merwe (eds.),
New Developments on Fundamental Problems in Quantum Physics, 103–107.

Assumption 1. *The entity is in a pure state, which is the complete representation of all its properties.*

Assumption 2. *There exists[1] a set of possible descriptions (='relevant' parameters) for the measurement context during the measurement. We call this set of relevant parameters the 'states of the measurement context' (denoted as Λ).*

Assumption 3. *The result of a measurement is completely determined by the initial conditions: the state of the entity and the state of the measurement context.*

Assumption 4. *There exists a statistical description for the relative frequency of occurrence of the states of the measurement context during the measurement.*

For a fixed state of the measurement context $\lambda \in \Lambda$, the measurement process is strictly classical (because of the deterministic dependence), so for every such strictly classical *hidden measurement* there exists a strictly classical observable:

$$\varphi_\lambda : \Sigma \to O, \tag{1}$$

where Σ is the set of pure states of the physical entity and O is the set of possible outcomes of the measurement. Thus, we have the following set of strictly classical observables that correspond with the different possible states of the measurement context:

$$\Phi_\Lambda = \{\varphi_\lambda | \lambda \in \Lambda\} \tag{2}$$

Since there exists a relative frequency of occurrence for the states of the measurement context, there exists a probability measure:

$$\mu_\Lambda : \mathcal{B}_\Lambda \to [0,1], \tag{3}$$

where $\mathcal{B}_\Lambda$ is a σ-field of subsets of Λ. We are able to compute probabilities defined on subsets of the set of outcomes for every given initial state p and every measurement e. Thus, we have represented the 'unknown but relevant information' for the measurement process in a couple consisting of: a set of strictly classical observables Φ_Λ and a probability measure μ_Λ defined on these observables. For more details on this definition of a hidden measurement representation, we refer to [7]. A theorem on the existence of such a hidden measurement representation for finite dimensional quantum mechanics was contained in [1]. A generalization of this theorem to more general finite dimensional entities can be found in [2]. The general proof for the existence of a hidden measurement representation for infinite dimensional entities can be found in [6]. In [7] it is proved that there always exists a hidden measurement representation with $[0,1]$ as its set of states of the measurement context and in [8] we identified all possible hidden measurement representations.

3. HOW THE NO-GO THEOREM IS AVOIDED

We will refer to the Jauch-Piron improvement (see [10]) of the von Neumann theorem (see [12]). Following [9] and [11], the structure on the collection of measurements $\mathcal{E}$ is represented by a propositional lattice $\mathcal{L}$ (every possible outcome of a measurement $e \in \mathcal{E}$ defines a proposition). Two propositions a and b in $\mathcal{L}$ are compatible (denoted $a \leftrightarrow b$) if the sublattice generated by $\{a, a', b, b'\}$ is distributive (a' and b' are the orthocomplements of a and b). A state (pure or mixed) $p : \mathcal{L} \to [0,1]$ is defined by

$$\begin{aligned} p(\emptyset) = 0,\, p(I) = 1, \\ a \leftrightarrow b \Rightarrow p(a) + p(b) = p(a \wedge b) + p(a \vee b), \\ p(a) = p(b) = 1 \Rightarrow p(a \wedge b) = 1. \end{aligned} \tag{4}$$

A theory is said to admit non-contextual hidden variables if we can add extra variables (not dependent on a) Λ_p to every state p such that it can be written as the sum of dispersion free states (denoted as (p, λ), where $\lambda \in \Lambda_p$ is an additional parameter[2]), i.e.:

$$\forall p \in \Sigma, \exists \Lambda_p, \forall a \in \mathcal{L} : p(a) = \sum_{\lambda \in \Lambda_p} \mu_p(\lambda)(p, \lambda)(a) \tag{5}$$

In the hidden-measurement approach we have the following (we relate with every proposition a a measurement e_a, which itself corresponds with a collection of states of the measurement context Λ_a):

$$\forall a \in \Sigma, \exists \Lambda_a, \forall p \in \mathcal{L} : p(a) = \sum_{\lambda \in \Lambda_a} \mu_a(\lambda)(p, \lambda)(a). \tag{6}$$

For the specific case of quantum mechanics it is possible to find hidden measurement representations such that Λ and μ do not depend on the propositions (see [1] and [5]). Thus, at first sight it seems that we have a contradiction since there exist hidden measurement representations for quantum mechanics that do satisfy (5), which conflicts with the no-go theorems. But, as we will show in the next section, Eq. (4) is not fulfilled for the dispersion-free components.

4. IMPOSING THE PROPOSITIONAL STRUCTURE

In this section we show how the propositional structure of quantum mechanics can be implemented in a straightforward way within this framework using the unitary structure of Hilbert space. We will show that it suffices to have a hidden measurement representation for only one of the measurements to obtain a representation for all measurements. Let Σ be the rays of a n-dimensional Hilbert space. If $\mathcal{E}$ consists of all measurements with n outcomes, we can represent such a measurement by a self-adjoint operator which is decomposable in n orthogonal projectors on the eigenvectors $p_{e,1}, \ldots, p_{e,n}$ and n corresponding eigenvalues $o_{e,1}, \ldots, o_{e,n}$. Consider a given measurement e_0 (with $p_{0,1}, \ldots, p_{0,n}$ as eigenvectors and $o_{0,1}, \ldots, o_{0,n}$ as respective eigenvalues) for which we have a hidden measurement representation, i.e., there exist:

$$\begin{aligned} &\Phi_{\Lambda,0} = \{\varphi_{0,\lambda} : \Sigma \rightarrow \{p_{0,1}, \ldots, p_{0,n}\} | \lambda \in \Lambda\}, \\ &\mu_{\Lambda,0} : \mathcal{B}_\Lambda \rightarrow [0,1] \end{aligned} \tag{7}$$

that characterize this hidden-measurement representation. Then we can define a representation for every $e \in \mathcal{E}$ in the following way:

$$\begin{aligned} &\Phi_{\Lambda,e} = \{\varphi_{e,\lambda} : \Sigma \rightarrow \{p_{e,1}, \ldots, p_{e,n}\} : p \mapsto U_e \circ \varphi_{0,\lambda} \circ U_e^{-1}(p) | \lambda \in \Lambda\}, \\ &\mu_{\Lambda,e} = \mu_{\Lambda,0}, \end{aligned} \tag{8}$$

where U_e is the unitary transformation defined by $\forall i : p_{e,i} = U_e(p_{0,i})$. In this way, the hidden-measurement representation clearly 'respects' the structure that characterizes this quantum entity. Now we are ready to prove what we have claimed at the end of the previous section. Let us consider the lattice of propositions of a two-dimensional Hilbert space. Let e_p be the test of the proposition *the state is p* and $e_{p'}$ the test of the proposition *the state is p'*. Let U be a unitary transformation that exchanges p and p'. If we have a hidden measurement representation for e_p, according to the above procedure, we also have one for $e_{p'}$ through U. Let $q = 1/\sqrt{2}(p + p')$. For the initial condition (q, λ) we have:

$$\begin{aligned} (q,\lambda)(p') &= (1/\sqrt{2}(U(p') + U(p)), \lambda)(U(p)) = (1/\sqrt{2}(p' + p), \lambda)(p) \\ &= (q,\lambda)(p) \end{aligned}$$

and thus $(q,\lambda)(p) + (q,\lambda)(p')$ is either 0 or 2. This conflicts with $(q,\lambda)(p \vee p') = 1$ and $(q,\lambda)(p \wedge p') = 0$. Since this situation can easily be embedded in higher dimensional Hilbert spaces, the same is valid for every dimension. This result is related to what has been called 'contextuality', but it is not entirely the same. A definition for context dependence along the lines of the hidden-measurement approach can be found in [7].

5. ACKNOWLEDGEMENTS

This work is supported by the IUAP-III n°9. Bart D'Hooghe and Frank Valckenborgh are Research Assistants of the Fund for Scientific Research.

6. REFERENCES

1. D. Aerts, *J. Math. Phys.* **27**, 202 (1986).
2. D. Aerts, *Found. Phys.* **24**, 1227 (1994).
3. D. Aerts and S. Aerts, "The hidden-measurement approach to quantum mechanics," these proceedings.
4. D. Aerts, S. Aerts, B. Coecke, B. D.Hooghe, T.Durt, and F. Valckenborgh, "A model with varying fluctuations in the measurement context," these proceedings.
5. B. Coecke, *Found. Phys.* **25** 203 (1995).
6. B. Coecke, *Found. Phys. Lett.* **8**, 437 (1995).
7. B. Coecke, "Classical representations for quantum-like systems through an axiomatics for context dependence," *Helv. Phys. Acta* **69** (1996), to appear.
8. B. Coecke, "A classification of classical representations for quantum-like systems," *Helv. Phys. Acta* **69** (1996), to appear.
9. J.M. Jauch, *Foundations of Quantum Mechanics* (Addison-Wesley, London, 1968).
10. J.M. Jauch and C. Piron, *Helv. Phys. Acta* **36**, 827 (1963).
11. C. Piron, *Foundations of Quantum Physics* (Benjamin, Reading, 1976).
12. J. von Neumann, *The Mathematical Foundations of Quantum Mechanics* (Princeton University Press, Princeton, 1955).

7. NOTES

1. We remark that "existence is not equivalent with "knowledge."
2. This slightly different way of formulating dispersion free states by means of additional parameters to every state is equivalent with the Jauch-Piron way. It only makes the bridge to the hidden-measurement approach easier.

DYNAMICAL EFFECTS GENERATED BY THE CURRENT FLUCTUATIONS IN A LONG SOLENOID

K. Dechoum and H.M. França

Instituto de Física, Universidade de São Paulo
C.P. 66318, 05315-970 São Paulo, SP, Brazil

We discuss the interaction between a microscopic electric dipole oscillator and a long solenoid which are separated by a small distance. The solenoid belongs to an RLC circuit and the zeropoint and thermal current fluctuations within the solenoid coils are taken into account. We show that there is an enhancement in the spontaneous emission of the dipole oxcillator and an anisotropy which is related to the dipole orientation. We suggest experiments which can exhibit these effects related to the electromagnetic interactions between electrons, atoms or molecules and the solenoid.

Key words: spontaneous emission, zeropoint energy, stochastic electrodynamics.

1. THE PHYSICAL SYSTEM

We have considered a very simple device that can show new dynamical effects generated by the existence of electromagnetic fluctuations (thermal and zeropoint) [1-3]. This system is composed by a long solenoid which is part of a simple RLC circuit. The solenoid (inductance L) extends from $-l/2$ to $l/2$ along the z axis and is composed by a large number N of circular coils with radius a [4]. At a distance y from the solenoid axis there is an electric dipole oscillator (frequency ω_0) which is oriented in the x direction (see Fig. 1). Therefore, if the dipole oscillator is close enough to the solenoid, it is expected a rich interaction between the electric dipole and the circuit (macroscopic oscillator). Notice that the zeropoint energy of the microscopic oscillator, $\hbar\omega_0/2$, is very often bigger than the zeropoint energy of the circuit, namely $\hbar\Omega/2$, since $\Omega^2 = 1/LC$ and C is the capacitance.

The non-relativistic equation of motion for the oscillating particle (charge e and mass m) is [5]

M. Ferrero and A. van der Merwe (eds.),
New Developments on Fundamental Problems in Quantum Physics, 109–114.

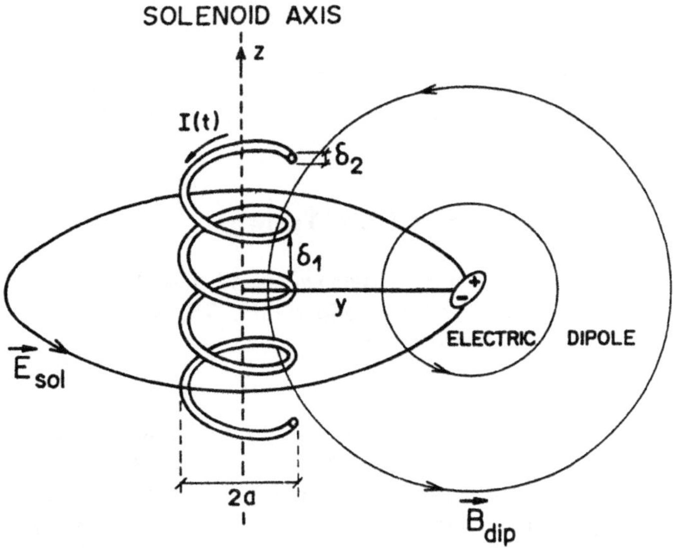

Fig. 1. Schematic picture of the electric dipole at a distance y from the solenoid axis. The relevant fields generated by the solenoid ($\boldsymbol{E}_{sol}$) and the oscillating dipole ($\boldsymbol{B}_{dip}$) are indicated. The solenoids used by Möllenstedt and Bayh [4] are such that $1.5 \times 10^{-4} cm \leq \delta_1 \leq 4 \times 10^{-4} cm$, $\delta_2 \simeq 3 \times 10^{-4} cm$ and $a \simeq 7 \times 10^{-4} cm$. The number of coils is $N \simeq 10^3$ and the solenoids length l varies from $0.5 cm$ to $0.7 cm$.

$$m\left(\ddot{x} + \gamma \dot{x} + \omega_0^2\, x\right) = e\left[E_{VF}(t) - \frac{2\pi N a^2}{c^2 l y}\dot{I}(t)\right] , \tag{1}$$

where $\gamma = 2e^2\omega_0^2/3mc^3$ is the free space damping constant. The damping force is generated by the radiation reaction process. The electric field associated with the vacuum fluctuations is denoted by $E_{VF}(t)$ in (1). Moreover, the second term on the right hand side of (1) is the x component of the electric field generated by the fluctuating current $I(t)$ existing within the solenoid coils ($c\boldsymbol{E}_{sol} = -\partial \boldsymbol{A}_{sol}/\partial t$).

The RLC circuit is supposed to be disconnected from any power source (like batteries) and, therefore, the fluctuating current $I(t)$ obeys the equation [2,5]

$$L\dot{I} + RI + \frac{1}{C}\int I\, dt = \mathcal{E}_N(t) + \frac{2\pi N a^2}{c^2 l y} e\ddot{x}(t) , \tag{2}$$

where $\mathcal{E}_N$ is the (spontaneous) Nyquist e.m.f. acting on the circuit. The second term in (2) is the e.m.f. generated by the fluctuating magnetic field ($\boldsymbol{B}_{dip}$) caused by the oscillatory motion of the electric dipole [see Fig. 1]. This magnetic field generates a time varying flux in each coil of the solenoid.

The coupled equations of motion (1) and (2) are linear and can be solved exactly by the Fourier transform method as was shown recently by Blanco, Dechoum, França, and Santos [5]. These authors have used the approach based on classical stochastic electrodynamics (SED) which is very convenient for the study of the electromagnetic fluctuation phenomena [6-8].

2. INPUT

The necessary input to obtain physical quantities from (1) and (2) are the spectral distributions $S_N(\omega)$, associated with the Nyquist noise, and $S_{VF}(\omega)$, associated with the vacuum electromagnetic fluctuations. Their explicit expressions are [2,3]

$$S_N(\omega) = \frac{R\hbar\omega}{\pi}\left[\frac{1}{2} + \frac{1}{\exp(\frac{\hbar\omega}{kT}) - 1}\right], \tag{3}$$

and

$$S_{VF}(\omega) = \frac{2\hbar\omega^3}{3\pi c^3}\left[\frac{1}{2} + \frac{1}{\exp(\frac{\hbar\omega}{kT}) - 1}\right], \tag{4}$$

where ω is the frequency associated to the Fourier decomposition of the spectrum and T is the temperature of the system.

3. NEW PREDICTIONS

(a) Oscillator. From the exact solutions of the equations (1) and (2), and also using (3) and (4), it is possible to show that, in equilibrium, the coordinate x of the microscopic oscillator is such that its mean square value will be [5],

$$\langle x^2\rangle = 2\frac{e^2}{m^2}\int_0^\infty d\omega \frac{S_{VF}(\omega)\Gamma(\omega)}{(\omega^2-\omega_0^2)^2 + \tau^2\omega^6\Gamma^2(\omega)} \simeq \frac{\hbar}{m\omega_0}\left[\frac{1}{2} + \frac{1}{\exp(\frac{\hbar\omega_0}{kT}) - 1}\right], \tag{5}$$

where $\tau = 2e^2/3mc^3$ and the function $\Gamma(\omega)$ is related to the parameters that characterizes the RLC circuit. This function is defined in Eq. (7) below.

The final result for $\langle x^2\rangle$ is equal to the free space value. This means that, at zero temperature, the energy of the mechanical oscillator will be given by $\hbar\omega_0/2$. This is a remarkable result because the electric dipole is immersed in the noise generated by the circuit, which does not appear in the final expression (5). The reason for this, explained in great detail by Blanco et al. [5], is that the dissipation by the oscillator (which is strongly altered) compensates the effect of the modification in the fluctuating fields by the presence of the solenoid. Therefore, the result of the integral (5) does not depend on the function $\Gamma(\omega)$. However, the oscillator exhibits an enhancement in spontaneous emission after excitation in such a way that, if displaced from the equilibrium position ($\langle x\rangle = 0$), the average coordinate of the charged particle will be given by [5]

$$\langle x\rangle = const \times \sin(\omega_0 t)\exp\left[-\gamma\Gamma(\omega_0)\frac{t}{2}\right] , \tag{6}$$

where

$$\Gamma(\omega_0) \equiv 1 + \frac{3}{2}\frac{R}{c|Z(\omega_0)|^2}\left(\frac{2\pi N a^2}{\ell y}\right)^2 . \tag{7}$$

In (7) the function $Z(\omega) = R - i(\omega L - 1/\omega C)$ is the impedance of RLC circuit. It should be remarked that $\Gamma(\omega_0)$ can be much greater than 1, namely, $\Gamma \simeq 1 + 10a^2/y^2$ if $\omega_0^2 = 1/LC$ and $R \simeq 4 \times 10^{-10}$sec/cm and $L \simeq 5 \times 10^{-20}$sec^2/cm, for instance. These values for R and L correspond to the long solenoid used by Möllenstedt and Bayh [4] for the observation of the Aharonov-Bohm effect.

(b) *RLC* circuit From Eq. (2) one can see that there are also two sources of noise acting on the circuit as random e.m.f.. Therefore, from the exact solution of the coupled equations (1) and (2), and also from (3) and (4), it is possible to show that the complete expression for the spectral distribution is [5]

$$S(\omega) \simeq \frac{\hbar\omega}{2\pi}\coth\left(\frac{\hbar\omega}{2kT}\right)\left\{R + L\frac{3\pi}{4}(\tau\omega_0)\frac{\omega_0 c a^2}{l y^2}\delta(\omega - \omega_0)\right\} . \tag{8}$$

where δ is the Dirac delta function. The first term in (8), which is proportional to the resistance R, is the familiar spectral density associated with the Nyquist noise [see (3)]. The second term, which is proportional to the inductance L, is generated by the presence of the electric dipole. We predict a very sharp peak in the spectrum at $\omega = \omega_0$ but the width and height of the peak will be determined by the resolution in the measurement of the frequency [see Fig. 2]. If $(\Delta\omega)$ is that resolution we should average the term in curly brackets of (8) in the interval $[\omega_0 - (\Delta\omega)/2, \omega_0 + (\Delta\omega)/2]$ and get for that term

$$R\left[1 + \frac{3\pi}{4}(\tau\omega_0)\frac{\omega_0}{(\Delta\omega)}\frac{c}{l}\frac{a^2}{y^2}\frac{L}{R}\right] , \tag{9}$$

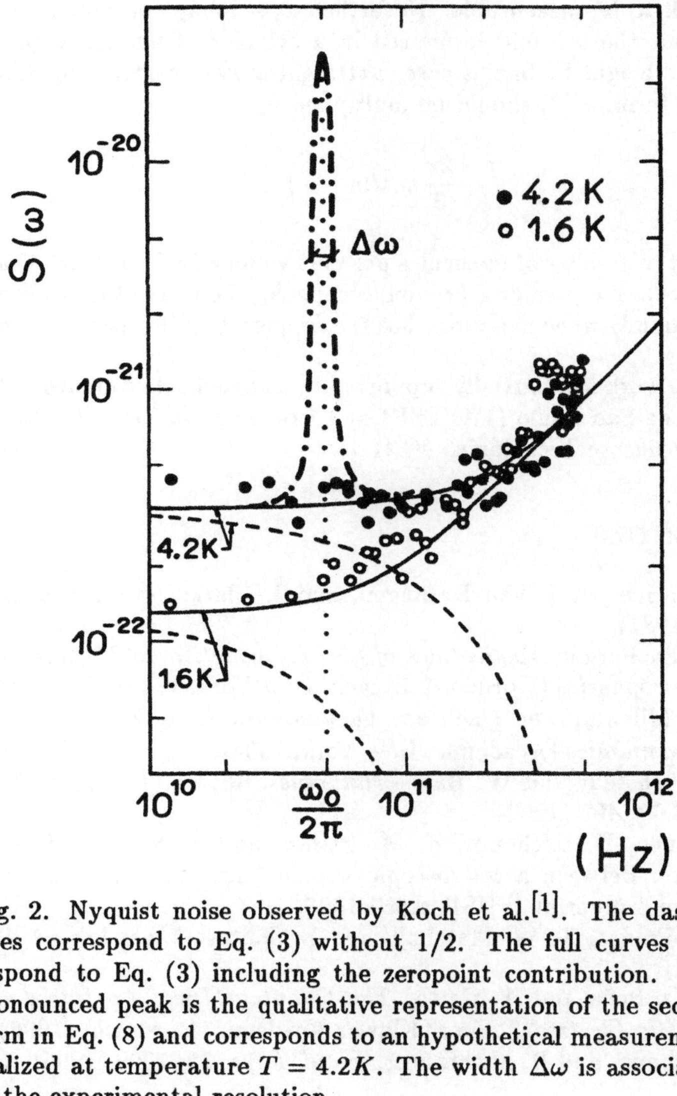

Fig. 2. Nyquist noise observed by Koch et al.[1]. The dashed lines correspond to Eq. (3) without 1/2. The full curves correspond to Eq. (3) including the zeropoint contribution. The pronounced peak is the qualitative representation of the second term in Eq. (8) and corresponds to an hypothetical measurement realized at temperature $T = 4.2K$. The width $\Delta\omega$ is associated to the experimental resolution.

if ω lies inside the interval. Even in the closest approach of the oscillator to the solenoid (i.e., $y \simeq a$) we get for the second term of (9) a value of about 10^{-13} to be compared with 1 using $(\Delta\omega)/\omega_0 \simeq 10^{-2}$, and the parameters of the Möllenstedt solenoid plus an ammonia molecule ($\omega_0 \simeq 10^{11} sec^{-1}$). This is too small to be measurable. Nevertheless, we may consider not a single molecule but the solenoid immersed in a cylinder of ammonia gas having radius l and height l. In this case, a straightforward calculation shows that the second term of (9) should be multiplied by

$$\frac{2\pi}{3} n a^2 l \ln\left(\frac{l}{a}\right) , \qquad (10)$$

where n is the number of molecules per unit volume ($2.7 \times 10^{19} cm^{-3}$ at room temperature). We predict a pronounced peak (about five times bigger than the background) when ω approaches the frequency of inversion of ammonia.

This work was partially supported by Fundação de Amparo à Pesquisa do Estado de São Paulo (FAPESP) and Conselho Nacional de Desenvolvimento Científico e Tecnológico (CNPq).

REFERENCES

1. R. H. Koch, D. J. Van Harlingen, and J. Clarke, *Phys. Rev. Lett.* **47**, 1216 (1981).
2. V. L. Ginzburg in *Applications of Electromagnetism in Theoretical Physics and Astrophysics* (Gordon & Breach, New York, 1989), Chap. 14.
3. P. W. Milonni, *The Quantum Vacuum: An Introduction to Quantum Electrodynamics* (Academic, New York, 1994).
4. G. Möllenstedt and W. Bayh, *Naturwiss.* **49**, 81 (1962). W. Bayh, *Z. Phys.* **169**, 492 (1962).
5. R. Blanco, K. Dechoum, H. M. França, and E. Santos, "Casimir Interaction Between a Microscopic Dipole Oscillator and a Macroscopic Solenoid," preprint IFUSP, April 1996.
6. H. M. França, T. W. Marshall, and E. Santos, *Phys. Rev. A* **45**, 6436 (1992).
7. L. de la Peña and A. Cetto, *The Quantum Dice: An Introduction to Stochastic Electrodynamics* (Kluwer Academic, Dordrecht, 1996).
8. K. Dechoum and H. M. França, *Found. Phys.* **25**, 1599 (1995); **26**, 1573 (1996).

COHERENT STATES AND THE MEASUREMENT PROBLEM

Lajos Diósi

KFKI Research Institute for Particle and Nuclear Physics
H-1525 Budapest 114, POB 49, Hungary

The convenience of coherent state representation is discussed from the viewpoint of what is in a broad sense called the measurement problem in quantum mechanics. Standard quantum theory in coherent state representation is intrinsically related to a number of earlier concepts conciliating quantum and classical processes. From a natural statistical interpretation, free of collapses or measurements, the usual von Neumann-Lüders collapse as well as its quantum state diffusion interpretation follow. In particular, a theory of coupled quantum and classical dynamics arises, containing the fluctuation corrections versus the fenomenological mean-field theories.

Key words: quantum measurement, wave function collapse, open quantum systems.

1. INTRODUCTION

The state of a canonical quantum system is characterized by its state vector $|\psi\rangle$ satisfying the Schrödinger evolution equation. The Schrödinger-equation, with Hamiltonian $\mathcal{H}(\hat{x},\hat{p})$, takes the equivalent form

$$\frac{d}{dt}\hat{\rho} = -i\left[\mathcal{H}(\hat{x},\hat{p}),\hat{\rho}\right] \tag{1}$$

if we use the density operator $\hat{\rho} = |\psi\rangle\langle\psi|$ to represent the system's quantum state.

In a certain sense, clarified ultimately by von Neumann [1], one may speak about the probability distribution of the coordinate operator $\hat{x}$, which is given by the diagonal elements $\langle x|\hat{\rho}|x\rangle$ of the density operator in coordinate representation. This distribution contains no information about the quantum system's momentum $\hat{p}$. *Vice versa*, in momentum representation

M. Ferrero and A. van der Merwe (eds.),
New Developments on Fundamental Problems in Quantum Physics, 115–124.

the distribution $\langle p|\hat{\rho}|p\rangle$ lacks information about the statistics of $\hat{x}$. Apparently, the proper statistical interpretation of the quantum state needs the distinguished notion of quantum measurement and assumes abrupt changes, i.e., collapses, of the wave function [1,2]. Quantum theory becomes dichotomic: the evolution of the quantum state is governed by the Schrödinger-equation (1) while in quantum measurements the wave function performs random jumps.

Why couldn't one incorporate the statistical interpretation of the wave function into a suitable quantum dynamics free of collapses? In this talk I suggest that this can perhaps be done within standard quantum mechanics merely by choosing a proper representation.

As the proper one, I suggest the well-known coherent state representation $|x,p\rangle$ [3]. Then the diagonal elements $\langle x,p|\hat{\rho}|x,p\rangle$ may, up to a well-defined precision, be interpreted as the joint probability distribution $\rho(x,p)$ of the coordinate and momentum. Opposite to the coordinate or momentum representations, the distribution $\rho(x,p)$ contains complete information about the system's quantum state and thus satisfies a closed evolution equation [5].

In Part 2, I recapitulate the coherent state representation in phase space coordinates. A most natural statistical interpretation, free of any reference to measurements or collapses (cf. Ref. [6]), will be proposed. In Part 3., we extend our results to the case of composite systems. In the subsequent parts I stress that typical theories of the emergence of classicality in quantum theory, including the von Neumann-Lüders' one, could be derived in coherent state representation.

2. QM IN COHERENT STATE REPRESENTATION

Coherent states $|x,p\rangle$, labelled by the canonical variables x and p, form a non-orthogonal overcomplet basis in the system's Hilbert-space. Coherent states are eigenstates of the particular non-Hermitian combination $\hat{x}+i\hat{p}$ of canonical operators:

$$(\hat{x}+i\hat{p})\,|x,p\rangle = (x+ip)|x,p\rangle. \tag{2}$$

The influence of the adjoint operator $\hat{x}-i\hat{p}$ brings derivative terms in:

$$(\hat{x}-i\hat{p})\,|x,p\rangle = (\partial_x - i\partial_p)\,|x,p\rangle + \frac{1}{2}(x-ip)|x,p\rangle. \tag{3}$$

To fix the normalization of the coherent states, we use the completeness relation in the following form:

$$\int |x,p\rangle\langle x,p|dxdp = I. \tag{4}$$

Then, we consider the diagonal elements of the density operator in coherent state representation:

$$\rho(x,p) = \langle x,p|\hat{\rho}|x,p\rangle. \tag{5}$$

Obviously, it is a mere technical task to rewrite the abstract Schrödinger-equation (1) into the above coherent state representation. It is, however, much less obvious that the diagonal part will satisfy a closed evolution equation [7]:

$$\frac{d}{dt}\rho(x,p) = -i\mathcal{H}\left(x + \frac{\partial_x + i\partial_p}{2}, p + \frac{\partial_p - i\partial_x}{2}\right)\rho(x,p) + c.c. \tag{6}$$

A few examples are given in the Appendix.

One might think of $\rho(x,p)$ as the joint probability distribution of the phase space coordinates x and p associated with $\hat{x}$ and $\hat{p}$ or, saying briefly, of the operators $\hat{x}$ and $\hat{p}$ themselves. The case is, however, more complicated because, for all composite variables $\mathcal{F}(x,p)$, one should specify with what Hermitian operators could they be associated. We suggest the following correspondence rule. Let any classical dynamic variable $\mathcal{F}(x,p)$ be associated with the fully symmetrized [8] Hermitian operator $\{\mathcal{F}(\hat{x},\hat{p})\}_{sym}$. Let us, furthermore, introduce a Gaussian coarse-graining of the dynamic variables over the Planckian phase space cell:

$$\bar{\mathcal{F}}(x,p) \equiv \left\langle \mathcal{F}(x + \xi/\sqrt{2},\ p + \eta/\sqrt{2})\right\rangle_{\xi,\eta}, \tag{7}$$

where ξ, η are standard Gaussian variables over which an average is understood on the r.h.s. of the above equation. After these preparations, one can state an equivalence theorem [9]. The quantum expectation value in state $\hat{\rho}$ of a coarse-grained fully symmetrized Hermitian dynamic variable coincides exactly with the classical expectation value of the corresponding fine-grained classical dynamic variable, using $\rho(x,p)$ for the phase space probability distribution, i.e.:

$$tr\left(\hat{\rho}\left\{\bar{\mathcal{F}}(\hat{x},\hat{p})\right\}_{sym}\right) = \int \mathcal{F}(x,p)\rho(x,p)dxdp \tag{8}$$

satisfies for all $\mathcal{F}(x,p)$. This is the *statistical interpretation* which emerges naturally in coherent state representation. To cut it short, we shall say that $\rho(x,p)$ expresses the coarse-grained joint probability distribution of $\hat{x}$ and $\hat{p}$.

3. QM IN HYBRID REPRESENTATION

Let us consider a composite quantum system $C \times Q$, with canonical variables $\hat{x}, \hat{p}$ and $\hat{X}, \hat{P}$, respectively. We introduce a coherent state basis $|x, p\rangle$ for the subsystem C while we keep the abstract Hilbert-space notation for the subsystem Q. This will be called ***hybrid representation*** [9]. Accordingly, by projecting the density operator $\hat{\rho}$ of the composite system $C \times Q$, we introduce the ***hybrid density*** [9] (cf. also [10,11])

$$\hat{\rho}(x,p) \equiv tr\Big((|x,p\rangle\langle x,p| \otimes I)\hat{\rho}\Big). \tag{9}$$

This mathematical object might seem strange for the first sight. It characterizes the quantum state of the original composite system while, formally, it is a phase space distribution for the subsystem C and density operator for the subsystem Q.

Similarly to the case in the previous Part, a closed evolution equation exists for the hybrid density, too [12]:

$$\frac{d}{dt}\hat{\rho}(x,p) = -i\hat{\mathcal{H}}\left(x + \frac{\partial_x + i\partial_p}{2}, p + \frac{\partial_p - i\partial_x}{2}\right)\hat{\rho}(x,p) + h.c. \tag{10}$$

where the compact notation $\mathcal{H}(\hat{x}, \hat{X}, \hat{p}, \hat{P}) = \hat{\mathcal{H}}(\hat{x}, \hat{p})$ has been applied to the composite system's Hamiltonian.

The hybrid density has a rich structure to interpret statistically. First of all,

$$\hat{\rho}_Q \equiv \int \hat{\rho}(x,p)dxdp \tag{11}$$

is the reduced density operator of subsystem Q. Alternatively,

$$\rho_C(x,p) \equiv tr\hat{\rho}(x,p) \tag{12}$$

is the coarse-grained phase-space distribution of $\hat{x}, \hat{p}$ of subsystem C, in the very sense of the statistical interpretation of the previous Part. And there is an additional possibility, common in statistics, that is, to introduce

$$\hat{\rho}_{Qxp} \equiv \frac{\hat{\rho}(x,p)}{\rho_C(x,p)} \tag{13}$$

as the *conditional quantum state* [9] of the subsystem Q when the subsystem C's coordinates take given values x, p. This interpretation makes the so-called quantum measurement and collapse simple consequences of the standard quantum theory as we shall see in the forthcoming Part.

4. APPLICATION: COLLAPSE

Consider von Neumann's dynamic model [1] of quantum measurement. $\mathcal{Q}$ will denote the measured system and $\mathcal{C}$ will denote the measuring apparatus. Both of them are quantum systems hence we apply standard quantum mechanics to the composite system $\mathcal{Q} \times \mathcal{C}$. We do so in hybrid representation. For simplicity, the measured system $\mathcal{Q}$ is a two-state system initially in the superposition: $|\psi_{\mathcal{Q}i}\rangle = \alpha|0\rangle + \beta|1\rangle$. The apparatus $\mathcal{C}$ is a canonical quantum system; let the pointer's position be $\hat{x}$. In coherent state representation, $\mathcal{C}$'s initial state $\rho_{\mathcal{C}i}(x,p)$ is concentrated at pointer position $x \approx 0$ with precision $\Delta x \approx 1$. Then the initial state of the composite system $\mathcal{Q} \times \mathcal{C}$ has the following hybrid density (9):

$$\hat{\rho}_i(x,p) = |\psi_{\mathcal{Q}i}\rangle\langle\psi_{\mathcal{Q}i}|\rho_{\mathcal{C}i}(x,p). \tag{14}$$

According to von Neumann, the interaction during the act of measurement can be modelled by the interaction Hamiltonian

$$\hat{\mathcal{H}}(x,p) = -g\delta(t)|1\rangle\langle 1\rangle p, \tag{15}$$

where, again, we use hybrid representation. The coupling constant g satisfies $g \gg 1$. To calculate the final entangled state of the composite system, we apply the hybrid equation of motion (10) with the above interaction Hamiltonian:

$$\frac{d}{dt}\hat{\rho}(x,p) = -g\delta(t)|1\rangle\langle 1\rangle\partial_x\hat{\rho}(x,p). \tag{16}$$

Then the initial state (14) evolves into a calculable final state which, after integrating out the irrelevant momentum variable p, takes the form:

$$\begin{aligned}\hat{\rho}_f(x) = \; & |\alpha|^2|0\rangle\langle 0|\rho_{\mathcal{C}i}(x) + |\beta|^2|1\rangle\langle 1|\rho_{\mathcal{C}i}(x-g) \\ & + (\alpha^\star\beta|1\rangle\langle 0| + \alpha\beta^\star|0\rangle\langle 1|)\rho_{\mathcal{C}i}(x-g/2).\end{aligned} \tag{17}$$

Let us invoke the statistical interpretation proposed in Part 3.: the coarse-grained probability distribution (12) of the pointer's position $\hat{x}$ in the final state (17) is

$$\rho_{\mathcal{C}f}(x) \equiv tr\hat{\rho}_f(x) = |\alpha|^2\rho_{\mathcal{C}i}(x) + |\beta|^2\rho_{\mathcal{C}i}(x-g), \tag{18}$$

i.e., $x \approx 0$ with probability $|\alpha|^2$ and $x \approx g$ with probability $|\beta|^2$. The two outcomes separate well since we assumed $g \gg 1$. The corresponding conditional final quantum states (13) $\hat{\rho}_{xf} \equiv \hat{\rho}_f(x)/\rho_{\mathcal{C}f}(x)$ are $|0\rangle\langle 0|$ if $x \approx 0$ and $|1\rangle\langle 1|$ if $x \approx g$, respectively. This is the von Neumann-Lüders collapse [1,2].

5. APPLICATION: MARKOVIAN OPEN SYSTEMS

Our example will be the composite quantum system $\mathcal{Q}\times\mathcal{C}$, where $\mathcal{Q}$ is a two-state atom interacting with the infinite number of electromagnetic field oscillators $\mathcal{C}$ (see, e.g., in Ref. [4]). For each eigenfrequency ω, the canonical oscillator variables $\hat{x}_\omega, \hat{p}_\omega$ are, for convenience, expressed by the usual non-Hermitian ones $\hat{a}_\omega, \hat{a}^\dagger_\omega$. If $|\psi_{\mathcal{Q}i}\rangle$ is the atomic initial state then, in hybrid representation, the composite system's initial state (9) takes the form

$$\hat{\rho}_i(a,a^\star) = |\psi_{\mathcal{Q}i}\rangle\langle\psi_{\mathcal{Q}i}|\exp(-|a|^2), \tag{19}$$

assuming vacuum state for the radiation field initially. The interaction Hamiltonian is, in interaction picture, modelled by

$$\hat{\mathcal{H}}(a,a^\star) = \sum_\omega g_\omega e^{-i(\omega-\omega_0)t}\hat{\sigma}_+ a_\omega + h.c. \equiv \Gamma\hat{\sigma}_+ + h.c., \tag{20}$$

where g_ω are coupling constants, $\hat{\sigma}_+ = |1\rangle\langle 0|$, and ω_0 is the atomic energy split. (The time dependence of $\hat{\mathcal{H}}(a,a^\star)$ is suppressed in notation.) For later convenience, we have introduced the (complex) effective field Γ to which the atom is really coupled. In interaction picture, the hybrid equation of motion (10) reads:

$$\frac{d}{dt}\hat{\rho}(a,a^\star) = -i[\hat{\mathcal{H}}(a,a^\star),\hat{\rho}(a,a^\star)] - i\sum_\omega\left(g_\omega e^{-i(\omega-\omega_0)t}\hat{\sigma}_+\frac{\partial\hat{\rho}(a,a^\star)}{\partial a^\star} - h.c.\right). \tag{21}$$

It is well-known that in Markovian approximation the atomic reduced density operator satisfies the master equation (see, e.g., in Ref.[4])

$$\frac{d}{dt}\hat{\rho}_{\mathcal{Q}} = \gamma\left(\hat{\sigma}_-\hat{\rho}_{\mathcal{Q}}\hat{\sigma}_+ - \frac{1}{2}\hat{\sigma}_+\hat{\sigma}_-\hat{\rho}_{\mathcal{Q}} - \frac{1}{2}\hat{\rho}_{\mathcal{Q}}\hat{\sigma}_+\hat{\sigma}_-\right) \equiv \mathcal{L}\hat{\rho}_{\mathcal{Q}}\ , \tag{22}$$

where $\gamma = \rho^2_{\omega_0}|\Gamma_{\omega_0}|^2$; the spectral density ρ_ω is defined by $\sum_\omega \to \int \rho_\omega d\omega/2\pi$. Consistent with this reduced dynamics, the Markovian limit of the composite system dynamics exists as well. The hybrid equation of motion (21) leads to coupled Wiener (diffusion) processes for the conditional pure state $\hat{\rho}_\Gamma = |\psi_\Gamma\rangle\langle\psi_\Gamma|$ of the atom and the effective field strength Γ, described respectively by the following Ito-Langevin-equations:

$$\frac{d}{dt}\hat{\rho}_\Gamma = \mathcal{L}\hat{\rho}_\Gamma + \sqrt{\gamma}\left(\dot{\xi}^\star_t(\hat{\sigma}_- - \langle\hat{\sigma}_-\rangle)\hat{\rho}_\Gamma + h.c.\right), \tag{23}$$

$$i\Gamma = \gamma\langle\hat{\sigma}_-\rangle + \sqrt{\gamma}\dot{\xi}_t, \tag{24}$$

where $\langle\hat{\sigma}_-\rangle = tr(\hat{\sigma}_-\hat{\rho}_\Gamma)$ and ξ_t is the standard complex white-noise [13].

Eq. (23) is the heuristic equation of the quantum state diffusion model [14], obtained originally to describe the wave function of individual quantum systems during their interactions with measuring apparatuses [15].

6. APPLICATION: COUPLING QUANTUM TO CLASSICAL

This time we make the crucial observation that the hybrid equation of motion (10) could formally be taken as the equation of motion for the composite system $\mathcal{Q}\times\mathcal{C}$ where $\mathcal{Q}$ is a quantum system with canonical coordinates $\hat{X},\hat{P}$ but the other subsystem $\mathcal{C}$ is genuine classical with coordinates x and p. So, the Eq. (10) could in principle describe the interaction of a true quantum and a true classical subsystem. The shorthand notation $\mathcal{H}(x,p,\hat{X},\hat{P})\equiv\hat{\mathcal{H}}(x,p)$ will be understood again.

To illustrate things, let $\mathcal{C}$ be a classical harmonic oscillator with Hamilton-function $(p^2+x^2)/2$, coupled linearly to a certain quantum system $\mathcal{Q}$ with Hamiltonian $\hat{H}$. The total Hamiltonian will be the following:

$$\hat{\mathcal{H}}(x,p)=\hat{H}+\frac{p^2+x^2}{2}+\hat{J}x, \tag{25}$$

where $\hat{J}$ is a certain Hermitian operator for $\mathcal{Q}$. For our purpose it will be useful to represent the state of the composite system by the classical phase space density (12) of $\mathcal{C}$ and by the conditional quantum state (13) of $\mathcal{Q}$. Using the Hamiltonian (25), the hybrid equation of motion (10) leads to a couple of equations for $\rho_{\mathcal{C}}(x,p)$ and $\hat{\rho}_{\mathcal{Q}xp}$. The first equation reads:

$$\frac{d}{dt}\rho_{\mathcal{C}}(x,p)=(x\partial_p-p\partial_x)\rho_{\mathcal{C}}(x,p)+\partial_p\langle\hat{J}\rangle_{xp}\rho_{\mathcal{C}}(x,p), \tag{26}$$

where $\langle\hat{J}\rangle_{xp}=tr(\hat{J}\hat{\rho}_{\mathcal{Q}xp})$. This is a classical Liouville-equation. The phase space distribution changes as if the Hamilton function were the quantum expectation value of the hybrid Hamiltonian (25):

$$\dot{x}=\partial_p\langle\hat{\mathcal{H}}(x,p)\rangle_{xp},\qquad \dot{p}=-\partial_x\langle\hat{\mathcal{H}}(x,p)\rangle_{xp}. \tag{27}$$

These equations are identical to the fenomenological mean-field equations widely used to calculate the evolution of a classical system under the influence of a quantized one (see, e.g., in Ref. [11]). The second equation will govern the quantized system's state under the influence of the classical subsystem. Our hybrid equation of motion (10) yields, after some elaboration, the following form:

$$\frac{d}{dt}\hat{\rho}_{\mathcal{Q}xp}+(\dot{x}\partial_x+\dot{p}\partial_p)\hat{\rho}_{\mathcal{Q}xp}+i[\hat{\mathcal{H}}(x,p),\hat{\rho}_{\mathcal{Q}xp}]$$

$$= -\frac{i}{2}(\partial_x s)[\hat{J},\hat{\rho}_{Qxp}] - \frac{i}{2}[\hat{J},\partial_x\hat{\rho}_{Qxp}] + \frac{1}{2}\partial_p\{\hat{J} - \langle\hat{J}\rangle_{xp},\hat{\rho}_{Qxp}\}$$
$$+ \frac{1}{2}(\partial_p s)\{\hat{J} - \langle\hat{J}\rangle_{xp},\hat{\rho}_{Qxp}\}, \tag{28}$$

where $s(x,p) \equiv \ln\rho_C(x,p)$. This equation differs from the corresponding equation of the standard mean-field theory by the terms on the r.h.s which are absent in the fenomenological theory. The standard theory can not describe the stochastic fluctuations induced in the classical subsystem $\mathcal{C}$ by the quantum subsystem $\mathcal{Q}$. The terms on the r.h.s. take these fluctuations into the account [16].

Moral. The classical content, if there is any, of a given quantum state displays itself in coherent state representation. I tend to believe that this is the general case for canonical quantum systems. It follows and is important to realize that the classical content is a relative and apparent feature. We can read it out if we use the proper framework — the coherent state representation, presumably. On the other hand, such classicality might be a robust feature against changes of the framework, here I mean canonical transformations on the first place. Obviously, the coherent state representation is not invariant canonically. But, due to the well-defined amount of coarse-graining used in the statistical interpretation, one hopes a satisfactory robustness of classicality whenever the dynamic variables $\mathcal{F}(x,p)$ have no structure at short scales $\Delta x \approx \Delta p \approx 1$.

It is rather promizing that the concept, presented in this talk, seems to cover all alternative concepts, including the wave function collapse, the quantum state diffusion, and the interpretation based on true classical subsystems [17]. I think that a Bohm-type interpretation, too, follows naturally (cf., maybe, Ref. [18]). Though I did not mention the decoherent history interpretation I have no doubts that hybrid quantum dynamics turn out to build up decoherent histories (we do not need the Markovian limit [19] either).

I am deeply indebted to the organizers of the Symposium for inviting me to participate and to give this talk. This work was supported by the Hungarian Scientific Research Fund under Grant OTKA T016047.

APPENDIX

Harmonic oscillator, $\mathcal{H} = (p^2 + x^2)/2$:

$$\frac{d}{dt}\rho(x,p) = (x\partial_p - p\partial_x)\rho(x,p).$$

Free particle, $\mathcal{H} = p^2/2$:

$$\frac{d}{dt}\rho(x,p) = -p\partial_x\rho(x,p) + \frac{1}{2}\partial_x\partial_p\rho(x,p).$$

Quartic oscillator, $\mathcal{H} = (p^2 + x^2)/2 + \lambda x^4$:

$$\frac{d}{dt}\rho(x,p) = (x\partial_p - p\partial_x)\rho(x,p) - \lambda\left(x + \frac{\partial_x}{2}\right)^3 \partial_p\rho(x,p)$$

$$+\frac{\lambda}{4}\left(x + \frac{\partial_x}{2}\right)\partial_p^3\rho(x,p).$$

REFERENCES AND FOOTNOTES

1. J.von Neumann, *Matematische Grundlagen der Quanten Mechanik* (Springer, Berlin, 1932).
2. G.Lüders, *Ann. Phys. (Leipzig)* **8**, 322 (1951).
3. V.Bargmann, *Commun. Pure Appl. Math.* **14**, 187 (1961). R.J.Glauber, *Phys. Rev.* **131**, 2766 (1963); see also in the recent textbook [4].
4. D.F.Walls and G.J.Milburn, *Quantum Optics* (Springer, Berlin, 1994).
5. A.Lonke, *J. Math. Phys.* **19**, 1110 (1978).
6. J.S.Bell, *Phys. World* **3**, 33 (1990).

7. This equation preserves the positivity of $\rho(x,p)$ provided certain analiticity conditions are satisfied initially. It preserves the pure state property of $\hat{\rho}$ though all forthcoming formulae will be valid for mixtures as well. The proofs will be given elsewhere. The Eq. (10) has an equivalent compact form:

$$\begin{aligned}\frac{d}{dt}\rho &= \mathcal{H}\, 2\sin\frac{\overleftarrow{\partial}_x\overrightarrow{\partial}_p - \overleftarrow{\partial}_p\overrightarrow{\partial}_x}{2}\exp\frac{\overleftarrow{\partial}_x\overrightarrow{\partial}_x + \overleftarrow{\partial}_p\overrightarrow{\partial}_p}{2}\,\rho \\ &= \{\mathcal{H},\rho\}_{Poisson} + h.o.t.\end{aligned}$$

   On the very right, one may observe the classical Liouville evolution equation to appear in the lowest order of the derivatives.

8. Full symmetrization (Weyl-ordering) is defined by the recursive rules

$$\{\hat{x}\mathcal{F}(\hat{x},\hat{p})\}_{sym} = \frac{1}{2}\left\{\hat{x}, \{\mathcal{F}(\hat{x},\hat{p})\}_{sym}\right\},$$

$$\{\hat{p}\mathcal{F}(\hat{x},\hat{p})\}_{sym} = \frac{1}{2}\left\{\hat{p}, \{\mathcal{F}(\hat{x},\hat{p})\}_{sym}\right\},$$

   while $\{\hat{x}\}_{sym} = \hat{x}$ and $\{\hat{p}\}_{sym} = \hat{p}$.

9. L.Diósi, *Quant. Semiclass. Opt.* **8**, 309 (1996). L.Diósi, "A true equation to couple classical and quantum dynamics," e-print archives quant-ph/9610028.
10. I.V.Aleksandrov, *Z. Naturf.* **36A**, 902 (1981).
11. W.Boucher and J.Traschen, *Phys. Rev. D* **37**, 3522 (1988).
12. Expansion in the derivatives yields the form

$$\frac{d}{dt}\hat{\rho} = -i[\hat{\mathcal{H}}, \hat{\rho}] + \frac{1}{2}\{\hat{\mathcal{H}}, \hat{\rho}\}_{Poisson} - \frac{1}{2}\{\hat{\rho}, \hat{\mathcal{H}}\}_{Poisson}$$

$$-\frac{i}{2}[\partial_x\hat{\mathcal{H}}, \partial_x\hat{\rho}] - \frac{i}{2}[\partial_p\hat{\mathcal{H}}, \partial_p\hat{\rho}] + h.o.t.,$$

showing the tricky combination of Liouville's classical and Schrödinger's quantum evolutions. The first line of the r.h.s. is identical to the Aleksandrov-bracket [10] which, in itself, would violate the positivity of $\hat{\rho}$ [11].
13. Proofs will be given elsewhere.
14. N.Gisin and I.C.Percival, *J. Phys. A* **25**, 5677 (1992).
15. L.Diósi, *J. Phys. A* **21**, 2885 (1988).
16. Obviously, if the quantum fluctuation of the "current" $\hat{J}$ is small enough then the last two r.h.s. terms go away. If, furthermore, the states $\rho_C(x,p)$ and $\hat{\rho}_{Qxp}$ are smooth enough functions of x, p then all r.h.s. terms can be ignored and we are left with the standard mean-field equation

$$\frac{d}{dt}\hat{\rho}_{Qxp} + i[\hat{\mathcal{H}}(x,p), \hat{\rho}_{Qxp}] = 0.$$

17. T.N.Sherry and E.C.G.Sudarshan, *Phys. Rev. D* **18**, 4580 (1978).
18. G.G. de Polavieja, *Phys. Lett.* **220A**, 303 (1996).
19. L.Diósi, N.Gisin, J.Halliwell, and I.C.Percival, *Phys. Rev. Lett.* **74**, 203 (1995).

CONTEXTUALISM, LOCALITY AND THE NO-GO THEOREMS

Arthur Fine

Department of Philosophy
Northwestern University
Evanston, Illinois 60208

The no-go theorems (I have in mind the Kochen-Specker and Bell type theorems) purport to tell us that the oldest interpretive idea for quantum mechanics fails; we cannot assign definite premeasurement values to the quantum mechanical observables, in particular to pairs of non-commuting observables. In fact, the no-go theorems are less categorical than that. They are a kind of quantum economics, a cost-benefit analysis that shows the cost involved for getting the benefit of premeasurement values. Conventionally, the cost is said to be contextualism (the Kochen-Specker theorem) or non-locality (the Bell theorem). My purpose here is to show that we do not actually need to pay quite so high a price. (It is not my purpose to urge one to buy.)

Key words: Bell theorem, Kochen-Specker theorem, hidden variables, contextualism, locality.

1. THE KOCHEN SPECKER AND THE BELL TYPE THEOREMS

A significant contrast between the Kochen-Specker and Bell theorems is this. The Kochen-Specker type theorems say that for a certain finite set of observables (not all commuting) no values can be assigned "decently" in any state of the system. The Bell type theorems say that for a certain finite set of observables (not all commuting) no values can be assigned "decently" in *certain special* states of the system. Moreover, in the recent Bell type theorems without inequalities, these special states are said to fix the assignment of values by means of locality alone, without further assumptions [1]. What constitutes decency has been spelled out in a number of different ways. Here

M. Ferrero and A. van der Merwe (eds.),
New Developments on Fundamental Problems in Quantum Physics, 125–132.

are four of them.

Where $\mathbf{v}$ is an assignment of values in state ϕ, the assignment is *decent* if the following two rules hold:

Spectrum Rule: the assigned value $\mathbf{v}(A)$ is always an eigenvalue of A that has non-zero probability in state ϕ

Product Rule: if $AB = BA$, then $\mathbf{v}(AB) = \mathbf{v}(A)\mathbf{v}(B)$.

Alternatively, an assignment is decent iff the Spectrum Rule holds and also the

Sum Rule: $\mathbf{v}(A + B) = \mathbf{v}(A) + \mathbf{v}(B)$.

again for commuting A and B. Although the Spectrum Rule is generally assumed, it is seldom explicitly mentioned. It does not have to be assumed if, as a third alternative, decency is defined as the satisfaction of the

Function Rule: for any Borel function $f, \mathbf{v}[f(A)] = f[\mathbf{v}(A)]$.

For the Spectrum Rule is a consequence of the Function Rule. Finally, a fourth alternative is to define an assignment $\mathbf{v}$ to be decent if it satisfies the following rule on joint distributions.

Joint Distribution Rule: the quantum mechanical joint distribution in state ϕ for commuting observables A and B is identical to the (always well-defined) joint distribution of the random variables $\mathbf{v}_A$ and $\mathbf{v}_B$ associated to A and B, respectively, by the assignment $\mathbf{v}$.

This requirement on joint distributions goes beyond the usual understanding in a hidden variables model, which only need require that the quantum probability distribution for each single observable is the same as the distribution of the associated random variable. Note that any specific joint probability assignment for commuting A and B is identical to the single distribution of a product of some spectral projection of A by one of B. Hence if all the single distributions are matched in a hidden variables model, we automatically get all the joint probabilities as well and without imposing the Joint Distribution Rule; i.e., without explicitly having to match individually each quantum joint distribution with the joint on the associated random variables.

The sense in which these are alternative ways of assigning values decently is that these four rules are actually equivalent provided values are assigned to sufficiently many observables (although the total number of observables may still be finite) [2].

2. GHZ: A BELL THEOREM WITHOUT INEQUALITIES

To see how the cost-benefit analysis works, we examine a version of the GHZ Bell-theorem without inequalities [1]. Consider three spin-1/2 systems (labeled 1, 2 and 3) in a triplet state ϕ. Imagine these as moving away from a common center along three straight lines in the same plane. For each system we are interested in the spin operators along the vertical (x) and horizontal (y) directions (where z is the direction of motion of the system). So, for example, σ_{x1} is the spin operator in the x-direction for system 1, σ_{y3} the y-spin operator for system 3, and so on. Consider an assignment of values $\mathbf{v}$ to the various spin operators where $x_1 = \mathbf{v}(\sigma_{x1}), y_3 = \mathbf{v}(\sigma_{y3})$ etc., then if the assignment respects the Spectrum Rule it seems that the following system of equations must hold.

$$x_1 x_2 x_3 = -1, \tag{1.1}$$
$$x_1 y_2 y_3 = +1, \tag{1.2}$$
$$y_1 x_2 y_3 = +1, \tag{1.3}$$
$$y_1 y_2 x_3 = +1, \tag{1.4}$$

since if we add σ to the x's and y's these four equations give the eigenvalues in ϕ for the respective product operators. Since $x_1^2 = y_1^2 = 1$ (Spectrum Rule again), if we multiply the first two equations, we get

$$x_2 x_3 y_2 y_3 = -1. \tag{2.1}$$

But if we multiply the second two; that is, (1.3) by (1.4), we get the GHZ contradiction

$$x_2 x_3 y_2 y_3 = +1. \tag{2.2}$$

The question we want to ask is what, in addition to the Spectrum Rule, is involved in the system (1) of equations. According to Mermin [1] the terms in these equations are justified by the EPR Criterion of Reality, "If, without in any way disturbing a system, we can predict with certainty (that is with probability equal to unity) the value of a physical quantity, then there exists an element of physical reality corresponding to this physical quantity" [3]. In accord with Bohr's flawed but persuasive retelling of EPR [4, 5] one might think of using the criterion to fix the x_i and the y_j at-a-distance, although this is not how the criterion is used in EPR. Rather, following measurement on one system in an entangled pair, distant values are assigned in EPR by means of the reduction of the wave packet, which yields an associated eigenstate for the unmeasured system. The criterion then insures that the corresponding eigenvalue (whose existence follows from the reduction not the criterion) amounts to a real physical attribute of that system, as opposed, for example, to an artifact of the measuring process. Few readers have caught this subtle use of the criterion which may be among the things Einstein disliked about Podolsky's text. For Boris Podolsky wrote EPR, following several discussion among the three "co-authors," and it appears that Einstein never saw EPR prior to publication. Immediately afterward he reacted strongly, complaining to Schrödinger that it did not come out very well and "obscured" the central things [6]. In his own numerous publications on the interpretation of quantum mechanics and on EPR entanglement Einstein never used or made mention of the "Criterion of Reality" (not even in his review [7] written in the summer of 1935 just after EPR was published). For these reasons it would be more precise historically to associate that criterion with Podolsky rather than with Einstein.

In the context of GHZ, say for Eq. (1.1), the idea for using the criterion to assign values at a distance is this. Since the product of measuring the x-spin on all three systems is -1, by measuring any two (with individual results either 1 or -1) we can determine the value of the third as that which would make the product of all three values equal -1. This *inference* determines the unmeasured value ("with certainty"), without "disturbing" that system, and hence by Podolsky's criterion there must be an "element of reality" corresponding to the unmeasured value. But since any one of the x-spins might be unmeasured, all three must be well-defined, and satisfy (1.1). Because the terms in each equation of (1) involve each of the three systems just once, a similar argument shows that all the x_i and y_j are well-defined and satisfy the given relations. Or does it?

Like Bohr's use in [4], this use of the criterion involves a sleight of hand; for the stated argument has a logical gap. Consider the situation where the x-spin on system 3 is determined by inference, as above, where we measure the x-spins on systems 1 and 2. It hardly follows that the *measured* value of the x component of spin found on system 1 in this situation would be the same as the x-spin on system 1 *inferred* in a different context where the x-spins of systems 2 and 3 were the ones being measured. Moreover, even if we were entitled to equate inferred values in one context with measured values in another, there is a further gap in assuming that the number x_1 assigned via the x-spin measurements corresponding to the terms in (1.1) would be the same x_1 that occurs in Eq. (1.2), where it is determined by different procedures involving the y component of spin for systems 2 and 3. Contrary to some claims, the locality built into Podolsky's criterion is not sufficient to yield premeasurement values for all the spin components in the GHZ situation, much less to justify Eqs. (1).

Actually, the logical situation is even worse. This whole line of argument from the criterion assumes, as above, that since the product of measuring certain spin components on all three systems is -1, by measuring any two we can determine the value of the third as that which makes the product of all three values equal -1. But this is clearly an assumption about the relation of product values to values of the factors, an assumption that is needed in addition to any locality conditions. One formulation of the principles at work here might be as follows.

Product Rule for Eigenstates: If ϕ is an eigenstate of the product observable ABC, where A, B and C are co-measurable, then

$$\mathbf{v}(ABC) = \mathbf{v}(A)\mathbf{v}(B)\mathbf{v}(C). \tag{3}$$

This rule is unavoidable if the **v**-values are the measured values from a simultaneous measurement of A, B and C. Given the Spectrum Rule and that ϕ is an eigenstate of ABC, that is just quantum mechanics. However, if the **v**-values are partly measured and partly inferred, as in GHZ and the other no-go theorems, there seems to be no necessity for the rule. Indeed there is little reason to suppose that just because a product observable has a determinate value, so do each of its factors, and even less reason to assume in *general* (i.e., whether or not the values are obtained from a simultaneous measurement) that multiplying the values of the factors (if they had values) would yield the value of the product. Thus the Product Rule for Eigenstates is a formal condition that goes well beyond ordinary quantum mechanics.

3. CONTEXTUALISM

In his pioneering discussion of the no-go theorems, John Bell [8] suggested a way to get around the Product Rule. His idea ("contextualism") was to relativize the values assigned observables to certain co-measurable observables. Thus we would make

$$\mathbf{v}(A) = \mathbf{v}(A|X) \text{ and } \mathbf{v}(B) = \mathbf{v}(B|Y), \tag{4}$$

i.e., the value of A relative to observables X that commute with A and the value of B relative to observables Y that commute with B. Then we would want products to satisfy some relation of the form $\mathbf{v}(AB) = \mathbf{v}(AB|??)$. But to what observables should the product be relative? Since observables in X may not commute with those in Y, even though A and B commute, there is no canonical answer. Thus the Product Rule may simply become inapplicable. In this way Bell saw that we can avoid the Kochen-Specker theorem (the first version of which Bell himself proved [8]), if we invoke contextualism and in this way justify indecent assignments of values. Moreover, Bell also thought that contextualism was bounded by locality which, in effect, cancels the contextualism and allows for the Bell theorem. One can show, however, that even when locality does cancel out contextualism the Bell theorem (the original, with inequalities) still involves the Product Rule as an additional assumption [9].

One way of combining contextualism with locality is to insist that

$$\mathbf{v}(A|B) = \mathbf{v}(A|C) \tag{5}$$

if observable A is defined on one system and observables B and C on a spatially separate system, even if B and C do not commute. This seems to capture the primitive intuition behind locality, which is that whatever we do to B or C should not (immediately) affect the spatially separate A. So in the GHZ situation we would have

$$\mathbf{v}(\sigma_{x1}|\sigma_{x2}, \sigma_{x3}) = \mathbf{v}(\sigma_{x1}|\sigma_{y2}, \sigma_{y_3}). \tag{6}$$

(Similarly for the other terms in system (1).) To combine contextualism and locality with the Product Rule we might modify the Product Rule for Eigenstates so that if A, B and C are defined on spatially separated systems

$$\mathbf{v}(ABC) = \mathbf{v}(A|B, C)\mathbf{v}(B|A, C)\mathbf{v}(C|A, B). \tag{7}$$

If we work through the equations in (1) we can see that this modified rule overcomes both logical gaps in the Podolsky criterion argument. Note that each term (i.e., each x_i and y_j) occurs in exactly two equations of the system (1). So we can define a term contextually by making it relative to the other pair of observables in the first of the two equations in which the term occurs, and then we can guarantee that the very same term occurs in the other equation by invoking locality as in (5). For x_1, this amounts to defining it by setting $\mathbf{v}(\sigma_{x1}|\sigma_{x2}, \sigma_{x3}) \equiv x_1$ and then invoking (6) to guarantee that x_1 in Eq. (1.1) is identical to $\mathbf{v}(\sigma_{x1}|\sigma_{y2}, \sigma_{y_3})$ in Eq. (1.2). Finally, systematic use of the modified Product Rule for Eigenstates (7) will reproduce all the equations in (1), and the GHZ contradiction.

Reconstructing the GHZ argument in this fashion brings to light the salient assumptions on which it relies. They are contextualism (4), locality (5) and the product rule (7). All three elements are necessary for GHZ. In particular, even with locality in place to limit contextualism, the GHZ contradiction could not be obtained without using the modified product rule (7). To get around the contradiction, instead of forsaking locality and assigning premeasurement values contextually, we could assign those values in accord with locality (cancelling contextualism as in (5)) and simply give up the product rule (7). Thus the cost-benefit analysis provided by the GHZ theorem does not single out locality as the cost of premeasurement values. We can have those values and have locality too, if we part with formal conditions on products.

4. CONCLUSION

The modification (7) of the product rule makes the eigenvalue of ABC the product of three values each of which is contextually assigned to its factors. The three contexts are mutually compatible, in the sense that putting together the observables in all three yields a co-measurable set consisting of A, B and C. Again, if these values are assigned by making a simultaneous measurement, (7) is what quantum mechanics requires. By contrast if only some of the assignments correspond to measured values, as in the argument that makes use of the Podolsky criterion, then no principle of the quantum theory justifies (7). To be sure for each single equation of (1) the factor variables are co-measurable. But there is no co-measurable context for any two of the equations, much less for all four simultaneously (which would involve both x and y components of spin on all the systems). A simple rule like (7) is insensitive to the incompatibilities that arise when we shift contexts from one equation to another. But it is precisely this shift that drives the GHZ contradiction. It is hard not to conclude that the product rule (7) is more heavily implicated in this contradiction than the combination of locality and contextualism. I would put it this way. After locality has done its work, GHZ (like the Kochen-Specker and the Bell theorems) still rests on arbitrary constraints on premeasurement values.

We might conclude with an alternative. Either not all observables have premeasurement values, or if they do then those values have a logic of their own that is not the functional logic of the quantum mechanical operators. In his correspondence with Einstein in 1935 over EPR, Schrödinger anticipated this latter alternative in his criticism of Einstein's idea for an ensemble interpretation. Schrödinger noted that such a treatment could not succeed in reproducing the measured values of quantum mechanics without "altering the connections with the concepts of classical mechanics" [6, 10].

In the context of our discussion here this simply means that products (and sums, etc.) of the operators that represent quantum mechanical observables only reflect the "true" relations between the values of those observables when the state is a simultaneous eigenstate of all the observables of interest. This is the special case where all values can be regarded as measured values. Another way to express this would be to say that the quantum joint distributions (for example, the correlations between observables in entangled states) are distributions for measured values only and are not to be thought of as holding generally for premeasurement values. It is well known that one can introduce hidden variables theories that thus break the Product Rule (or Sum Rule, etc.), even ones with a simple dynamics for how states and values change in time [11]. What is not so well known is the conclusion of the analysis carried out above. Namely, that this is a way to avoid the no-go theorems without sacrificing locality, or buying contextualism.

REFERENCES

1. N. D. Mermin, *Phys. Rev. Lett.* **65**, 3373 (1990).
2. A. Fine, *J. Math. Phys.* **23**, 1306 (1982).
3. A. Einstein, B. Podolsky, and N. Rosen, *Phys. Rev.* **47**, 777 (1935).
4. N. Bohr, *Phys. Rev.* **48**, 696 (1935).
5. M. Beller and A. Fine, "Bohr's response to EPR," in J. Faye and H. J. Folse, eds., *Niels Bohr and Contemporary Philosophy* (Kluwer, Dordrecht, 1994), 1-31.
6. A. Fine, *The Shaky Game: Einstein, Realism and the Quantum Theory*, 2nd edn. (University of Chicago Press, Chicago, 1996).
7. A. Einstein, J. *Franklin Institute* **221**, 313 (1936). Reprinted in A. Einstein, *Ideas and Opinions* (Crown, New York, 1954), pp. 290-323.
8. J. S. Bell, *Rev. Mod. Phys.* **38**, 447 (1966). Reprinted in J. S. Bell, *Speakable and Unspeakable in Quantum Mechanics* (Cambridge University Press, Cambridge, 1987), pp. 1-13.
9. A. Fine, "On the completeness of quantum theory," *Synthese* **29**, 257 (1974). Reprinted with addendum in P. Suppes, ed., *Logic and Probability in Quantum Mechanics* (Reidel, Dordrecht, 1976), pp. 249-281.
10. A. Fine, "Schrödinger's version of EPR, and its problems" in P. Humphreys, ed., *Patrick Suppes: Scientific Philosopher, Vol. 2* (Kluwer, Dordrecht, 1994), pp. 29-38.
11. J. C. Vink, *Phys. Rev. A* **48**, 1808 (1993).

A STATE-SPECIFIC PROOF OF THE BELL-KOCHEN-SPECKER THEOREM USING ONLY 5 PROPOSITIONS

Guillermo García Alcaine

Departamento de Física Teórica
Facultad de Ciencias Físicas, Universidad Complutense de Madrid
28040 Madrid, Spain

A set of 5 propositions admiting no set of non-contextual values in the singlet state of two spin 1/2 particles is presented. In terms of number of propositions involved, this is the most economic state-specific proof of the Bell-Kochen-Specker theorem up to date.

Key words: Bell theorem, Kochen-Specker theorem, hidden variables.

1. INTRODUCTION

This contribution is based on one of the new variants of the BKS theorem discussed in [1] (see also [2] for other recent results and [3] for a general discussion). In order to make a short self-contained exposition I have changed the notation and some of the arguments.

Let $A, B, \ldots$ be some compatible observables, $\widehat{A}, \widehat{B}, \ldots$ the commuting self-adjoint operators representing them, $a, b, \ldots (a \in \sigma\left(\widehat{A}\right), b \in \sigma\left(\widehat{B}\right), \ldots)$ a set of results in a joint measurement in some quantum pure state ψ, and $v(A)$, $v(B)$, ... the values determined by the hidden variables in a given individual system in the pure state ψ; these values must be some of the possible results in a measurement, i.e., $v(A) \in \sigma\left(\widehat{A}\right)$, $v(B) \in \sigma\left(\widehat{B}\right)$, ...

According to the spectral map theorem, if the commuting operators satisfy some functional relation, $f\left(\widehat{A}, \widehat{B}, \ldots\right) = 0$, the corresponding results in any joint measurement will fulfill the same relation, $f(a, b, \ldots) = 0$. In a deterministic hidden variables theory the values in any individual system must satisfy also $f(v(A), v(B), \ldots) = 0$. Furthermore, if a functional relation between commuting operators is not an identity but it is true when

M. Ferrero and A. van der Merwe (eds.),
New Developments on Fundamental Problems in Quantum Physics, 133–136.

acting on some pure state, $g\left(\widehat{A},\widehat{B},\ldots\right)|\psi\rangle = 0$, I will assume that the values determined by the hidden variables in any individual system in state $|\psi\rangle$ fulfill the relation $g\left(v\left(A\right), v\left(B\right), \ldots\right) = 0$ (some particular cases of this rule where discussed in [1]).

2. A PROOF OF THE BKS THEOREM IN THE SINGLET STATE OF TWO SPIN-1/2 PARTICLES

In the usual basis of eigenstates of $\widehat{\sigma}_z^{(1)} \otimes \widehat{\sigma}_z^{(2)}$, let us consider the following vectors:

$$|u_1\rangle = \frac{1}{\sqrt{2}}\begin{pmatrix}1\\-1\\0\\0\end{pmatrix}, |u_2\rangle = \frac{1}{\sqrt{2}}\begin{pmatrix}0\\0\\1\\1\end{pmatrix}, |u_3\rangle = \frac{1}{\sqrt{2}}\begin{pmatrix}1\\0\\1\\0\end{pmatrix}, |u_4\rangle = \frac{1}{\sqrt{2}}\begin{pmatrix}0\\1\\0\\-1\end{pmatrix},$$

$$|u_5\rangle = \frac{1}{2}\begin{pmatrix}1\\1\\-1\\1\end{pmatrix}, |\psi\rangle = \frac{1}{\sqrt{2}}\{|+-\rangle - |-+\rangle\} = \frac{1}{\sqrt{2}}\begin{pmatrix}0\\1\\-1\\0\end{pmatrix},$$

and the corresponding projection operators, $\widehat{P}_j = |u_j\rangle\langle u_j|$, $j = 1, \ldots, 5$. The orthogonality relations $\langle u_1|\, u_2\rangle = 0$, $\langle u_3|\, u_4\rangle = 0$, $\langle u_5|\, u_k\rangle = 0$, $k = 1, \ldots 4$, imply that $\widehat{P}_1$, $\widehat{P}_2$, $\widehat{P}_5$ commute (and represent compatible propositions), and similarly $\widehat{P}_3$, $\widehat{P}_4$, $\widehat{P}_5$. Moreover, the subspaces spanned by both triads of vectors, u_1, u_2, u_5 and u_3, u_4, u_5, contain the singlet state ψ, i.e., $\left(\widehat{P}_1 + \widehat{P}_2 + \widehat{P}_5\right)|\psi\rangle = |\psi\rangle$, $\left(\widehat{P}_3 + \widehat{P}_4 + \widehat{P}_5\right)|\psi\rangle = |\psi\rangle$.

Therefore, as said in the introduction, in any individual system in state $|\psi\rangle$ the values of the propositions P_j in a theory of hidden variables must satisfy the following relations:

$$v\left(P_1\right) + v\left(P_2\right) + v\left(P_5\right) = 1, \tag{1}$$

$$v\left(P_3\right) + v\left(P_4\right) + v\left(P_5\right) = 1, \tag{2}$$

where each $v\left(P_j\right)$ is either 0 or 1, as corresponding to values of propositions, and $v\left(P_5\right)$ is the same in both equations if the hidden variables are non-contextual.

On the other hand, I am going to show that in a hidden variables theory the values of the propositions $P_1, \ldots, P_4$ in any individual system in the singlet state satisfy the equation:

$$v\left(P_1\right) + v\left(P_2\right) + v\left(P_3\right) + v\left(P_4\right) = 1. \tag{3}$$

The propositions P_1, P_2 are not compatible with P_3, P_4, and hence can not be jointly measured in the same individual system; thus, in contradistinction with (1,2), Eq. (3) is not a translation to hidden variables of the spectral map theorem. But as a relation between values of physical observables in a hidden variables theory there is no a priori objection to it, because these theories allow to assign definite values to non-compatible observables in the same individual system.

To prove (3) I will use the following three properties: First, the value of any factorizable proposition is the product of the values of its factor propositions: $v\left(P^{(1)} \otimes P^{(2)}\right) = v\left(P^{(1)}\right) v\left(P^{(2)}\right)$ (the proposition on the total system is true, $v\left(P^{(1)} \otimes P^{(2)}\right) = 1$, if and only if both partial propositions are true, $v\left(P^{(1)}\right) = v\left(P^{(2)}\right) = 1$). Secondly, in the singlet state there is a perfect correlation between the spin component of both particles along an arbitrary common direction; if $P_{\mathbf{n}}^{(j)}$ denotes the proposition "particle j has $+1/2$ spin component in the direction $\mathbf{n}$", in any individual system in the singlet state $v\left(P_{\mathbf{n}}^{(1)}\right) = v\left(P_{-\mathbf{n}}^{(2)}\right)$ (both propositions are simultaneously true or false). Finally, in the two dimensional Hilbert space of the spin states of one of the particles, $\widehat{P}_{\mathbf{n}}^{(j)} + \widehat{P}_{-\mathbf{n}}^{(j)} = \mathbf{1} \Rightarrow v\left(P_{\mathbf{n}}^{(j)}\right) + v\left(P_{-\mathbf{n}}^{(j)}\right) = 1$.

The four propositions in the left hand side of (3) are factorizable; using successively the three properties stated above, this l.h.s. can be written down as follows:

$$
\begin{aligned}
& v\left(P_{+z}^{(1)}\right) v\left(P_{-x}^{(2)}\right) + v\left(P_{-z}^{(1)}\right) v\left(P_{+x}^{(2)}\right) + v\left(P_{+x}^{(1)}\right) v\left(P_{+z}^{(2)}\right) + v\left(P_{-x}^{(1)}\right) v\left(P_{-z}^{(2)}\right) \\
&= v\left(P_{+z}^{(1)}\right) v\left(P_{+x}^{(1)}\right) + v\left(P_{-z}^{(1)}\right) v\left(P_{-x}^{(1)}\right) + v\left(P_{+x}^{(1)}\right) v\left(P_{-z}^{(1)}\right) + v\left(P_{-x}^{(1)}\right) v\left(P_{+z}^{(1)}\right) \\
&= \left\{v\left(P_{+z}^{(1)}\right) + v\left(P_{-z}^{(1)}\right)\right\}\left\{v\left(P_{+x}^{(1)}\right) + v\left(P_{-x}^{(1)}\right)\right\} = 1, \quad q.e.d.
\end{aligned}
$$

Equations (1-3) lead to a very simple state-specific version of the BKS theorem:

The five propositions P_j in equations (1-3) admit no set of non-contextual values in the singlet state.

The proof is immediate: the sum of the left hand sides of the three equations is even (each proposition appears twice), while the sum of the right hand sides is 3. This state specific, additive, parity argument is the most economic in terms of number of propositions involved that I know off.

I shall finish this talk with a comment: the projection operators $\widehat{P}_j$ have the following quantum mean values in the singlet state:

$$
\left\langle \widehat{P}_5 \right\rangle_{\psi} = \frac{1}{2}, \left\langle \widehat{P}_k \right\rangle_{\psi} = \frac{1}{4}, k = 1, \ldots 4.
$$

These mean values satisfy equations similar to (1-3), with the replacement $v(P_j) \to \left\langle \widehat{P}_j \right\rangle_\psi$, that could be experimentally tested averaging measurements in an ensemble of systems in the quantum pure state ψ. But quantum mechanics is not a statistical description of some underlying hidden variables theory, because there are no sets of non-contextual values satisfying the relations (1-3) in the individual systems of the ensemble.

REFERENCES

1. A. Cabello, J. M. Estebaranz and G. García Alcaine, "New variants of the Bell-Kochen-Specker theorem," *Phys. Lett. A* **218**, 115-118 (1996).
2. A. Cabello, J. M. Estebaranz and G. García Alcaine, "Bell-Kochen-Specker theorem: A proof with 18 vectors," *Phys. Lett. A* **212**, 183-187 (1996).
3. A. Peres, *Quantum Theory: Concepts and Methods* (Kluwer Academic, Dordrecht, 1993).

PROPERTIES OF A CAUSAL QUANTUM THEORY IN PHASE SPACE

Gonzalo García de Polavieja

The Physical and Theoretical Chemistry Laboratory
South Parks Road
Oxford OX1, 3UB United Kingdom

Some properties of a causal theory for the coherent state representation of quantum mechanics are discussed.

Key words: causal theory, quantum theory, classical limit.

Pilot wave theories as the de Broglie-Bohm theory [1,2] and Epstein's [3] describe a system as a pilot wave solution of Schrödinger equation in position and momentum space, respectively, which drives the particles with well defined trajectories. Probability does not enter in any fundamental way but as a tool for predictive inference [4]. No empirical results distinguish these causal theories and standard quantum theory, although they have been critiqued on formal grounds. These theories present an asymmetry between position and momentum, do not have phase space classical mechanics as a limit but only a singlevalued classical theory and describe the particles of bound stationary states as standing still. We have presented an alternative causal model derived from the coherent state representation of the Schrödinger equation to circumvent these undesirable properties [5,6]. In the following we present the most basic results and discuss some of the relevant properties.

The coherent state representation of Schrödinger equation has the form [7,8,9]

$$i\hbar\frac{\partial}{\partial t}\langle q,p|\psi\rangle = \langle q,p|\hat{H}(\hat{Q},\hat{P})|\psi\rangle \tag{1}$$

for a Hamiltonian of the general form $H = \frac{1}{2m}P^2 + V(Q)$, and

$$\langle q,p|\hat{Q}|\psi\rangle = (q + i\hbar\nabla_p)\langle q,p|\psi\rangle, \tag{2a}$$

$$\langle q,p|\hat{P}||psi\rangle = -i\hbar\nabla_q\langle q,p|\psi\rangle. \tag{2b}$$

M. Ferrero and A. van der Merwe (eds.),
New Developments on Fundamental Problems in Quantum Physics, 137–139.

Inserting the polar form $\langle q,p|\psi\rangle = R(q,p,t)\exp(i/\hbar)S(q,p,t)$ for the wavefunction in (1) and separating real and imaginary parts, we obtain the following two equations and after some manipulations we find:

$$-\frac{\partial S(q,p,t)}{\partial t} = \frac{p^2}{2m} + V(q) + W(q,p,t), \tag{3}$$

$$\frac{\partial R^2(q,p,t)}{\partial t} + \nabla_q\left[\frac{p}{m}R^2(q,p,t)\right] + \nabla_p\left[\left(-\nabla_q V(q) + Z(q,p,t)\right)R^2(q,p,t)\right] = 0, \tag{4}$$

with the form of W and Z given in [5,6]. The structure of Eqs. (3) and (4) suggests a causal theory of the motion of the wave and the particle. The motion of the particle in phase space described by $(q(t),p(t))$ is the solution of the equations

$$\dot{q} = \frac{p}{m}, \tag{5}$$

$$\dot{p} = -\nabla_q V(q) + Z(q,p,t). \tag{6}$$

To obtain the trajectory $(q(t),p(t))$, the initial values $(q(0),p(0))$ have to be supplied as input. We interpret Eq. (9) as an equation for the total energy of the particle

$$E\left(q(t),p(t),t\right) = -\left.\frac{\partial S(q,p,t)}{\partial t}\right|_{(q,p)=(q,(t),p(t))}. \tag{7}$$

In the following, we present the case of the stationary state of the harmonic oscillator. The coherent state representation of the wavefunction for a harmonic oscillator with Hamiltonian $H = \frac{P^2}{2m} + \frac{1}{2}m\omega^2Q^2$ is

$$\langle q,p|\Psi_n\rangle = N_n'\left(\frac{\varepsilon}{\hbar\omega}\right)^{n/2}\exp\left(-\frac{\varepsilon}{2\hbar\omega} + in\tan^{-1}\left(-\frac{1}{m\omega}\frac{p}{q}\right) + \frac{i}{\hbar}\frac{pq}{2} - \frac{i}{\hbar}E_n t\right), \tag{8}$$

with $\varepsilon = \left(\frac{p^2}{2m} + \frac{1}{2}m\omega^2q^2\right)$ and the eigenenergy $E_n = \left(n+\frac{1}{2}\right)\hbar\omega$. The equations of motion and the energy of the particle, Eqs. (5), (6), and (7) reduce in this case to

$$\dot{q} = \frac{p}{m}, \tag{9}$$

$$\dot{p} = -m\omega^2 q, \tag{10}$$

$$E = E_n. \tag{11}$$

The equations of motion (9) and (10) coincide with the classical ones. However, the energy has the value of the eigenenergy irrespectively of the initial conditions, which has its classical relationship with position and momentum for the phase space loop $\varepsilon = E_n$ whereas the distribution $|\langle q, p|\Psi\rangle|^2$ is maximum at $\varepsilon = n\hbar\omega$. The energy difference between these two phase space loops is $\frac{\hbar\omega}{2}$. The classical limit for all members of the ensemble is approached at high energy corresponding to a phase space distribution concentrated on the $\varepsilon = n\hbar\omega$ loop and for which the $\frac{\hbar\omega}{2}$ contribution to the energy is negligible. This result gives a very different picture to that of the BB theory for which the particle is standing still on its initial condition for all eigenenergies. We have discussed elsewhere the postulates of the theory [5] and the multiparticle case [6]. Some of the important properties of the model are:

(a) It is a multivalued (phase space) theory.

(b) It has phase space classical mechanics as a limit even for nonintegrable cases [10] for stationary states for $Z = W = 0$.

(c) It is a second order and therefore the concept of quantum potential seems more justified than in previous first order approaches.

(d) It circumvents the criticism already discussed for the previous theories

(e) Extension to the relativistic case should be facilited by the work of Prugovecki [11].

We hope that the model presented will be further developed, extended and applied to important cases as the time of tunneling problem [12,13].

REFERENCES

1. D. Bohm and B.J. Hiley, *The Undivided Universe* (Routledge, 1993).
2. P.R. Holland, *The Quantum Theory of Motion* (Cambridge University Press, 1993).
3. S.T. Epstein, *Phys. Rev.* **89**, 319 (1952); **91**, 985 (1953).
4. G. García de Polavieja, "A Bayesian subquantum H-theorem," in preparation.
5. G. García de Polavieja, *Phys. Lett.* A **220**, 303 (1996).
6. G. García de Polavieja, *Found. Phys. Lett.*, to appear.
7. *Coherent States*, J.R. Klauder and B. Skagerstam, eds. (World Scientific, Singapore, 1985).
8. Go. Torres-Vega and J.H. Frederick, *J. Chem. Phys.* **93**, 8862 (1990).
9. Go. Torres-Vega and J.H. Frederick, *J. Chem. Phys.* **98**, 3103 (1993).
10. G. García de Polavieja and M.S. Child, "Quantum instability of the coherent state flux lines," submitted. See also G. García de Polavieja, *Phys. Rev. A* **53**, 2059 (1996).
11. A. Prugovecki, *Found. Phys.* **11**, 355 (1981).
12. C.R. Leavens, *Phys. Lett. A* **197**, 88 (1995).
13. R. Landauer and Th. Martin, *Rev. Mod. Phys.* **66**, 217 (1994).

FROM QUANTUM TO CLASSICAL: THE QUANTUM STATE DIFFUSION MODEL

N. Gisin

Group of Applied Physics University of Geneva
1211 Geneva 4, Switzerland

T. A. Brun*

Department of Physics,
Queen Mary and Westfield College, University of London
London E1 4NS, United Kingdom

M. Rigo

Department of Mathematics,
Royal Holloway College, University of London
Egham, Surrey TW20 0EX, United Kingdom

Stochastic Schrödinger equations provide qualitative and quantitative descriptions of the quantum-classical cross-over. We illustrate this with the regular harmonic oscillator and with a classically chaotic example.

Key words: quantum state diffusion, classical limit, chaos.

1. INTRODUCTION

The vagueness of the quantum-classical border in physics makes some physicists uncertain about the foundation of their science. Others, by contrast, feel that the accuracy of the verified predictions makes these foundations especially solid. This longtime state of affairs is changing in an interesting way. Recent progress in mesoscopic physics, and the so-called "new science of nanotechnology," open new domains to be explored by theoreticians and

* New address: Institute for Theoretical Physics, University of California, Santa Barbara, California 93106.

M. Ferrero and A. van der Merwe (eds.),
New Developments on Fundamental Problems in Quantum Physics, 141–150.

experimentalists alike. In this contribution we will exploit numerical and conceptual tools that were developed to explore this new domain of mesoscopic physics, and apply them to systems that are close to the quantum-classical border.

Central to classical mechanics are the concepts of trajectories and the phase space structures associated with them. Hence, we would like to illustrate our results with figures showing these trajectories and structures as they appear in our models. These figures, and the fact that they can be computed without any supplementary parameters, constitute the main result of this contribution.

As examples we could use several kinds of oscillators. The first, and the simplest, is the harmonic oscillator with its regular orbits; but this is boring! Next, systems that exhibit classical chaos, such as the Kicked anharmonic oscillator (KAOS) [1,2,3,4] and the double-well Duffing oscillator [5,6]. In both cases the structure of the strange attractors emerges out of quantum cloudiness when the quantum-classical border is approached. Another example worth mentioning is a nonlinear oscillator exhibiting hysteresis [7], the hysteresis curve appearing when the system becomes more classical.

The feature which characterizes all these systems is that they are open; they interact significantly with some kind of external environment. One cannot in general describe an open system by a single state vector; one requires a density matrix, essentially giving classical probabilities for the system to be in given quantum states. In the limit of Markovian systems, this density matrix evolves according to a linear master equation, as we shall see below.

For many reasons, we would like to get away from the density operator description, and these new tools allow us to do so. These tools require no additional parameters, only a simple idea: describe open quantum systems by state vectors, as for closed systems. Unlike closed systems, their evolution is not unitary—in fact, it is not linear and not deterministic (two general characteristics of open systems). The evolution is chosen such that when the pure state defined by the (normalized) state vector is averaged over the randomness, the usual density matrix description is recovered at all times. The numerical advantage of this approach stems from replacing an $n \times n$ matrix by an n component vector. The conceptual advantage lies in the possibility to have individual states in the theory, as in actual experiments, and to compute mean values as averages of individual outcomes, again as in actual experiments.

This approach shares with the decoherence approach the idea that the environment plays a crucial role in breaking the unitary (boring) evolution and reducing the state [8,9]. However, here the evolution provides classical probabilities for quantum amplitudes. Our approach also shares some similarities with the GRW spontaneous localization model [10], in that the evolution equations are stochastic, but preserve pure states. But here there are no *ad hoc* parameters.

2. QUANTUM STATE DIFFUSION AND QUANTUM JUMPS

Markovian open quantum systems are usually described by a master equation:

$$\dot{\rho} = -i[H,\rho] + \sum_m \Big(L_m \rho L_m^\dagger - \frac{1}{2}\{L_m^\dagger L_m, \rho\} \Big), \tag{1}$$

where ρ is the density matrix for the system, H its Hamiltonian, and the linear operators L_m describe the effects of the environment. In an *unraveling* of such master equations, one describes the system in terms of a normalized pure state $|\psi(t)\rangle$ which follows a stochastic "trajectory" in Hilbert space. By averaging the pure state projector $|\psi\rangle\langle\psi|$ over all possible trajectories with appropriate weights, one reproduces the density operator $\rho = M(|\psi\rangle\langle\psi|)$. This is analogous classically to replacing the Fokker-Planck equation for probability densities with a stochastic Langevin equation for single trajectories.

Unfortunately, unlike the case of classical Brownian motion, the unraveling of the master equation (1) is not unique. Thus, there is some ambiguity in how one separates classical and quantum uncertainties, related to the ambiguity in identifying density matrices with quantum ensembles. In this section we consider two well known unravelings of (1).

In quantum state diffusion (QSD), the (Itô) stochastic evolution equation for the normalized state vector $|\psi(t)\rangle$ reads:

$$\begin{aligned} |d\psi(t)\rangle = &-iH|\psi(t)\rangle dt - \frac{1}{2}\sum_j (L_j^\dagger L_j - 2\langle L_j^\dagger\rangle_\psi L_j + |\langle L_j\rangle_\psi|^2)|\psi(t)\rangle dt \\ &+ \sum_j (L_j - \langle L_j\rangle_\psi)|\psi(t)\rangle d\xi_j \end{aligned} \tag{2}$$

where the "noises" $d\xi_j$ are complex-valued Wiener processes of zero mean $M(d\xi_j) = 0$ and correlations $M(d\xi_j d\xi_k) = 0$, $M(d\xi_j^* d\xi_k) = \delta_{jk} dt$. This equation describes a continuous non-differentiable evolution similar to the familiar diffusive paths of a classical Brownian particle, but in Hilbert space instead of real space. QSD is the only continuous unraveling which satisfies the same symmetries as the master equation itself [11,3].

Our second example is the quantum jumps (QJ) unraveling, which is closely related to photon counting [12,13]. However, it can be defined for any master equation (1). The stochastic increment for the wave-function is

$$|d\psi(t)\rangle = -iH|\psi(t)\rangle dt - \frac{1}{2}\sum_j (L_j^\dagger L_j - \langle L_j^\dagger L_j\rangle_\psi)|\psi(t)\rangle dt + \sum_j \left(\frac{L_j|\psi(t)\rangle}{\sqrt{\langle L_j^\dagger L_j\rangle_\psi}} - |\psi(t)\rangle\right) dN_j. \tag{3}$$

The discrete Poissonian noises dN_j assume the values 0 or 1. Most of the time $dN_j = 0$ and the evolution is continuous and differentiable. However, whenever $dN_j = 1$ there is a "jump" to the state $L_j|\psi(t)\rangle/\sqrt{\langle L_j^\dagger L_j\rangle_t}$. The dN_j processes have mean values $M_{|\psi\rangle}(dN_j) = \langle L_j^\dagger L_j\rangle_\psi dt$ and correlations $dN_j dt = 0$ and $dNjdN_k = \delta_{jk} dN_j$. This means, essentially, that jumps occur randomly with an average rate $\langle L_j^\dagger L_j\rangle$.

It has been shown that a quantum jump description of balanced heterodyne detection produces an equation identical to the QSD equation (2) in the limit of a strong local oscillator [14,15], and that a QSD description of photon detection produces jump-like behavior [11]. Also, in the classical limit (i.e., high photon numbers), the quantum jump equation begins to exhibit diffusive behavior as well [16]. Thus, though these unravelings are distinct, they can behave similarly in certain limits.

3. HARMONIC OSCILLATOR

Let us illustrate these two unravelings for a simple example: the damped harmonic oscillator at finite temperature. $H = \omega a^\dagger a$, $L_1 = \sqrt{\bar{n}\gamma}\ a^\dagger$ and $L_2 = \sqrt{(\bar{n}+1)\gamma}\ a$ where $\bar{n}$ is the thermal equilibrium mean photon number, $\bar{n} = \langle a^\dagger a\rangle_\rho$, and γ is the inverse relaxation time. For QSD one can show that any initial states tends to a coherent state: $|\psi(t)\rangle \to |\alpha_t\rangle$, where $a|\alpha_t\rangle = \alpha_t|\alpha_t\rangle$ and $\alpha_t = (\langle q\rangle_\psi + i\langle p\rangle_\psi)/\sqrt{2}$. Furthermore, the evolution of α_t is governed by a classical equation,

$$d\alpha_t = -i\omega\alpha_t dt - \frac{\gamma}{2}\alpha_t dt + \sqrt{\bar{n}\gamma}d\xi_t. \tag{4}$$

Hence, for this example at least, the QSD equation fully describes how the environment localizes the quantum state down to a minimum Gaussian wavepacket, and how this wavepacket follows a classical trajectory [17,18,19,20,21,22].

For quantum jumps a similar analysis can be carried out, and for a system in a coherent state the evolution of α_t is governed by the equation

$$d\alpha_t = -i\omega\alpha_t dt - \frac{\gamma}{2}\alpha_t dt + \sqrt{\bar{n}\gamma}\frac{\alpha}{\sqrt{|\alpha|^2+1}}dW_t, \tag{5}$$

where dW_t is a real, non-Gaussian stochastic differential variable with mean $M(dW_t) = 0$ and $M(dW_t dW_t) = dt$, representing the difference between the mean of the jump processes and their actual values. In the classical limit of large α, this dW_t approaches a diffusion process, and Eq. (5) has behavior essentially identical to (4).

Somewhat surprisingly, states tend towards small wavepackets for quantum jumps as well as QSD, so the same "classical limit" seems to exist for both [16].

4. CHAOTIC DUFFING OSCILLATOR

A good example of a nonlinear system is the forced, damped Duffing oscillator. This has a classical equation of motion

$$\frac{d^2x}{dt^2} + 2\Gamma\frac{dx}{dt} + x^3 - x = g\cos(t), \tag{6}$$

and for some choices of Γ and g is chaotic [23].

Because the equation of motion includes explicit time-dependence, the solutions lie in a three-dimensional phase space x, p, t. It is helpful to consider a discrete surface of section of this system. Let (x_0, p_0) be the initial point of the forced, damped Duffing oscillator at time $t_0 = 0$. Then we can define a *constant phase map* in the x-p plane by the sequence of points $(x_n, p_n) = (x(t_n), p(t_n))$ at times $t_n = 2\pi n$. Figure 1 illustrates this in the chaotic regime, where we can clearly see from the surface of section the fractal structure of the strange attractor.

Quantizing the Duffing oscillator is straightforward using the QSD formalism. The Hamiltonian operator is

$$H(Q, P, t) = P^2/2m + Q^4/4 - Q^2/2 + g\cos(t)Q + \sqrt{\Gamma}(QP + PQ), \tag{7}$$

and the damping is represented by a Lindblad operator

$$L = 2\sqrt{\Gamma}a = \sqrt{2\Gamma}(Q + iP), \tag{8}$$

where we have assumed $\hbar = 1$. The last term in the Hamiltonian is an ansatz, added to give the correct equations of motion in the classical limit; it is necessary due to the simplistic model of the dissipative environment.

This system is far from classical. To go to the classical limit, we introduce a scaling factor β,

$$H_\beta(Q, P, t) = P^2/2m + Q^4/4\beta^2 - Q^2/2 + g\beta\cos(t)Q + \sqrt{\Gamma}(QP + PQ). \tag{9}$$

As we increase β, the scale of the problem (compared to $\hbar$) increases by β in x and p, without altering the dynamics. Thus, $\beta \to \infty$ is the classical

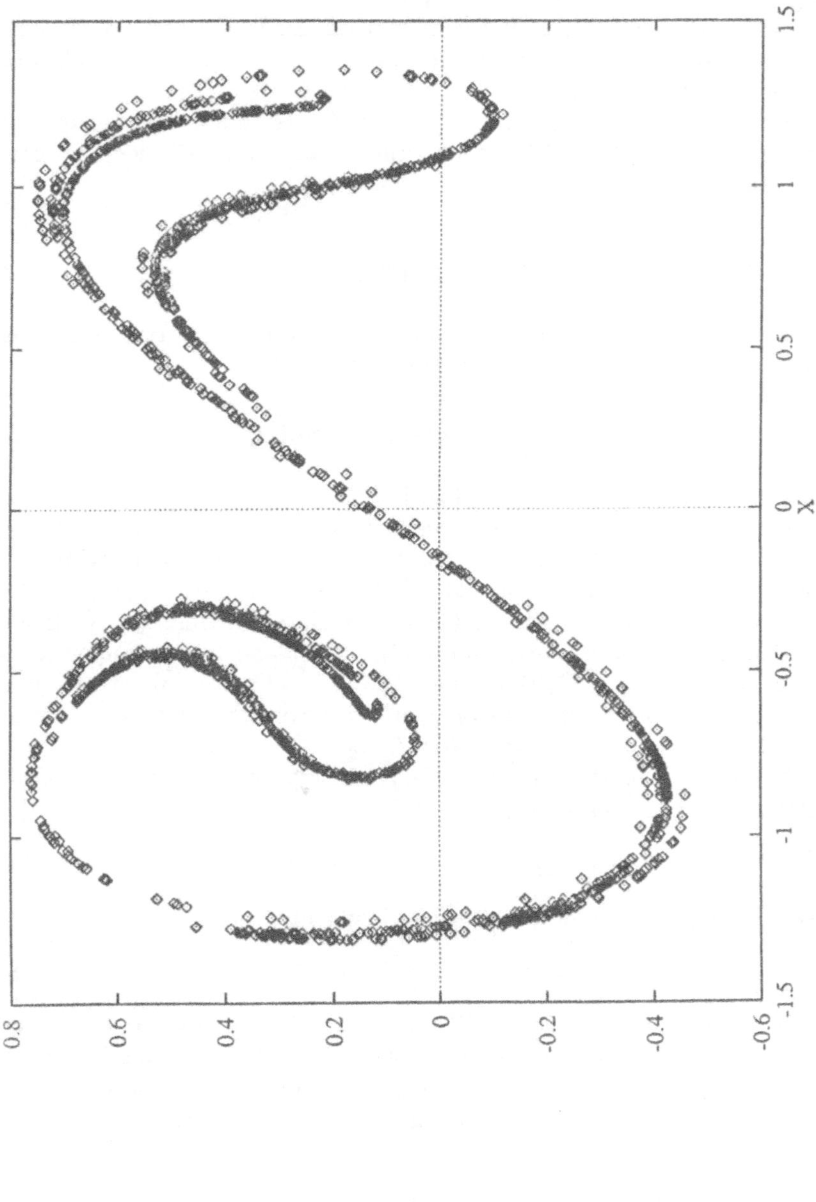

Fig. 1. The constant phase surface of section for the classical forced, damped Duffing oscillator in the chaotic regime $\Gamma = 0.125$, $g = 0.3$.

limit of this system. Classical behavior should emerge from the system in this limit. This is supported by the numerical calculations. (See Figs. 1 and 2.)

The classical unscaled problem is bounded within a small region of phase space. (See Fig. 1.) Clearly, the quantized version of this problem (with $\hbar = 1$) should be far from the classical limit. One would expect to observe little trace of the classical fractal structure. The expectation values of Q and P should be dominated by noise. We solved the forced, damped Duffing oscillator numerically by integrating the QSD equation for three different values of the parameter β.

Examining Fig. 2a, we see that the results of the numerical calculation match our expectations very well. The expectation values appear randomly distributed; they are dominated almost completely by the stochastic terms of the equation.

Approaching the classical limit of large β, more and more of the classical structure of the attractor appears. At first the broad outlines of the attractor are formed, then increasing levels of substructure. (Figs. 2b–2c.) The full fractal structure of a strange attractor is only attainable in an unattainable purely classical limit. In an actual physical system, the uncertainty principle provides a lower cutoff to the scale-invariance of the strange attractor.

It can easily be shown that the quantum jumps formalism has a "diffusive limit" similar to QSD as one approaches the classical limit [16]. Thus, exactly the same qualitative behavior is demonstrated in that case as well.

5. CONCLUSIONS

Neither QSD nor QJ can be regarded as a fundamental theory, since both divide the universe arbitrarily into a system and environment; and they represent two different and interesting unravelings of the master equation (though only QSD has the same symmetries as the master equation [11,3]). This is similar to the ambiguity which arises in consistent histories, in which it is possible to choose different sets of histories corresponding to different coarse-grained descriptions [24]. On the other hand, QSD (or QJ) makes testable predictions, and demonstrates the rise of classical physics in a way which is difficult or impossible to see with other theories.

Without entering into a sterile debate, let us note that there seems to be an interesting link between consistent histories and the stochastic models used in this contribution. It appears that the solutions of the stochastic equations provide consistent sets of histories. Hence, it seems that the two approaches are related in a way similar to the relationship in classical me-

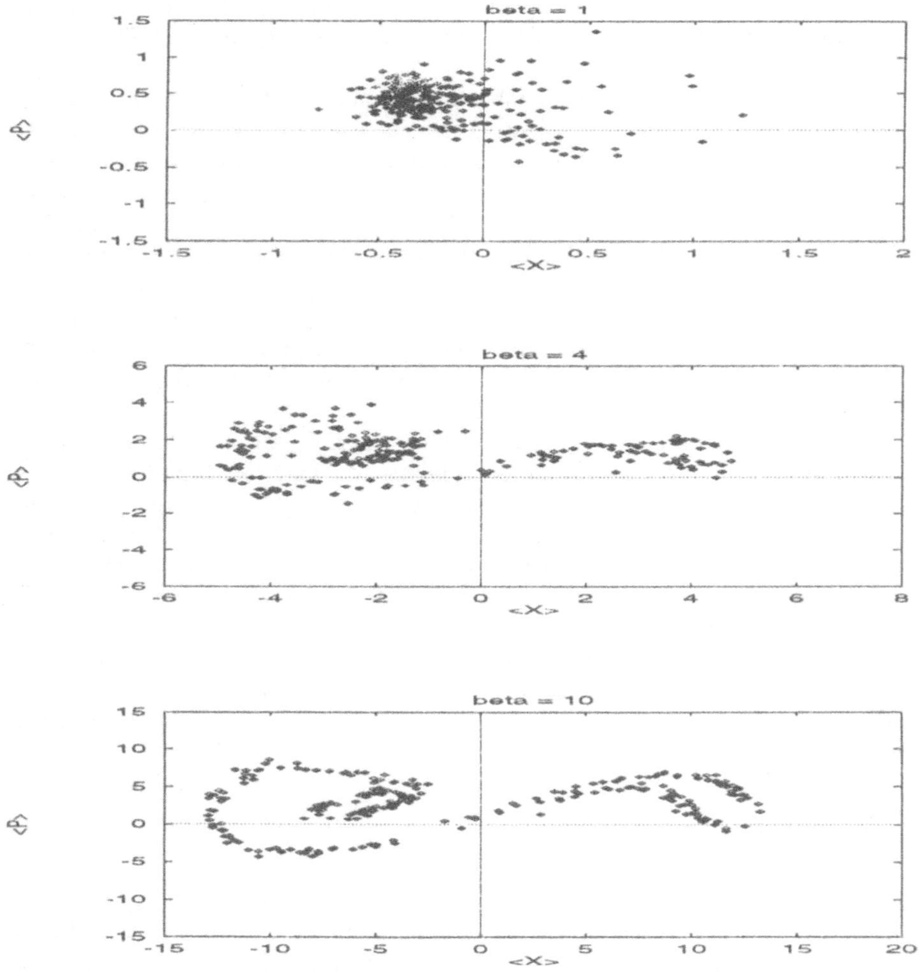

Fig. 2. The constant phase surface of section for a single QSD trajectory of the quantum forced, damped Duffing oscillator in the chaotic regime, $\Gamma = 0.125$, $g = 0.3$, for three scalings: a) $\beta = 1$, b) $\beta = 4$ and c) $\beta = 10$.

chanics between the global view given by the Maupertuis principle and the local view given by the Newton equations. The global view may be more elegant, but the local one is more useful to compute actual numbers [25,26].

Quantum mechanics is nonlocal. Classical mechanics is local. Consequently classical mechanics can not explain all quantum phenomena. Conversely, it is cumbersome to use quantum mechanics to describe classical phenomena. Not only are the computations more complex, but—and this is the main point—it is conceptually more difficult: one has to argue that nonlocality, entanglement and the principle of superposition can be set aside when crossing the "quantum → classical" border. Clearly, nonlocality, entanglement and the principle of superposition should become irrelevant in the classical limit. But why should one argue? Shouldn't it just come out of the equations? Does it come out of the equations? This contribution is about the last question. And the answer is: "it depends on which equation."

ACKNOWLEDGEMENTS

We would like to thank Lajos Diósi, Francesca Mota-Furtado, P.F. O'Mahony, Ian Percival and Rüdiger Schack for useful conversations. This research was funded in part by the UK EPSRC, the EU Human Capital and Mobility Program, and the Swiss National Science Foundation.

REFERENCES

1. T.P. Spiller and J.F. Ralph, *Phys. Lett. A* **194**, 235 (1994).
2. T.P. Spiller, J.F. Ralph, T.D. Clark, R.J. Prance, and H. Prance, *J. Low Temp. Phys.* **101**, 1037 (1995).
3. N. Gisin and M. Rigo, *J. Phys. A* **28**, 7375-7390 (1995).
4. M. Rigo and N. Gisin, *Quant. Semiclass. Opt.* **8**(1), 255 (1996).
5. T.A. Brun, *Phys. Lett. A* **206**, 167 (1995).
6. T.A. Brun, *J. Phys. A* **29**, 2077 (1995).
7. M. Rigo, G. Alber, F. Mota-Furtado, and P.F. O'Mahony, submitted to *Phys. Rev. A*.
8. E. Joos and H.D. Zeh, *Z. Phys. B* **59**, 223 (1985).
9. W.H. Zurek, *Phys. Today* **36**, October 1991.
10. G.C. Ghirardi, A. Rimini, and T. Weber, *Phys. Rev. D* **34**, 470 (1986).
11. N. Gisin and I.C. Percival, *J. Phys. A* **25**, 5677 (1992); *J. Phys. A* **26**, 2233 (1993); *J. Phys. A* **26**, 2245 (1993).
12. J. Dalibard, Y. Castin, and K. Mølmer, *Phys. Rev. Lett.* **68**, 580 (1992); see also *J. Opt. Soc. Am.* **10**, 524 (1993).
13. H.J. Carmichael, *An Open Systems Approach to Quantum Optics* (Lecture Notes in *Physics*) (Springer, Berlin, 1993).

14. H.M. Wiseman and G.J. Milburn, *Phys. Rev. A* **47**, 642 (1993).
15. P.L. Knight and B. Garraway, in *Quantum Dynamics of Simple Systems*, G.L. Oppo et al., eds. (44th Scottish University Summer School in Physics, 1996), pp. 199-237
16. T.A. Brun, N. Gisin, P.F. O'Mahony, and M. Rigo, quant-ph/9608038, *Phys. Lett. A*, in press, 1997.
17. I.C. Percival, J. Phys. A **27**, 1003 (1994).
18. J.J. Halliwell and A. Zoupas, *Phys. Rev. D* **52**, 7294 (1995).
19. R. Schack, T. Brun, and I.C. Percival, *J. Phys. A* **28**, 5401 (1995).
20. T. Steimle, G. Alber, and I.C. Percival, *J. Phys. A* **28**, L491 (1995).
21. M. Holland, S. Marksteiner, P. Marte, and P. Zoller, *Phys. Rev. Lett.* **76**, 3683 (1996).
22. M. Rigo and N. Gisin, *Quant. Semiclass. Opt.* **8**, 255-268 (1996).
23. M.C. Gutzwiller, *Chaos in Classical and Quantum Mechanics* (Springer, Berlin, 1990). J. Guckenheimer and P. Holmes, *Nonlinear Oscillations, Dynamical Systems, and Bifurcations of Vector Fields* (Springer, Berlin, 1983).
24. R. Griffiths, *J. Stat. Phys.* **36**, 219 (1984); R. Omnès, *Rev. Mod. Phys.* **64**, 339 (1992); H.F. Dowker and J.J. Halliwell, *Phys. Rev. D* **46**, 1580 (1992); M. Gell-Mann and J.B. Hartle, *Phys. Rev. D* **47**, 3345 (1993).
25. L. Diósi, N. Gisin, J.J. Halliwell, and I.C. Percival, *Phys. Rev. Lett.* **21**, 203 (1995).
26. T.A. Brun, quant-ph/9606025, submitted to *Phys. Rev. Lett.*

QUANTUM PHYSICS AND THE PROBLEM OF THE ONTOLOGICAL PRIORITY BETWEEN CONTINUOUS QUANTITY AND DISCRETE QUANTITY

Victor Gómez Pin

Departement of Philosophy
Universitat Autònoma de Barcelona
Barcelona, Spain

We have tried to display the main mathematical theories that have been confronted with the philosophical problem of the relationship between discrete quantity ad continuous quantity. What we have tried to do was to show the ontological weight of quantum physics which first at all deals with these notions and (in order to explain the duplicity of the observed phenomena) is forced to discuss their connection in such a way that it soon it becomes hard to distinguish if we remain at the level of scientific discussion or if we have penetrated into the dominion of philosophy. In fact contemporary physics re-establishes a link with Greek thought, since in some ancient physical theories the pair continuous/discrete constituted the main tool in order to try to explain the phenomena (in order to *sozein ta phainomena*, which was the main goal of the philosophy).

Key words: discrete-quantity, continuous-quantity, ontology, mathematics, physics.

An electron can revolve in an orbit around the nucleus without losing energy, provided that the orbit ... is a whole number of de Broglie wavelength in circumference.

Even if it has later been replaced by a more accurate model, Bohr's picture (in 1913) of an electron's behaviour is well adjusted to the philosophical question that we would raise.

The problem is centred on the connected words *whole* and *number*. Of course at a certain level everyone knows what these expressions mean, and we are conscious that the scientist's work can hardly be subordinated to the philosophical controversies about the basic concepts of the mathematical background of the discipline.

M. Ferrero and A. van der Merwe (eds.),
New Developments on Fundamental Problems in Quantum Physics, 151–158.

Nevertheless, contemporary physics (quantum physics as well as relativity theory) has shown the impossibility of preserving the rigid division between scientific and philosophical work; and this is for two reasons:

(a) Some of the prominent, and unanswered questions raised by quantum physics become necessary outstanding topics in the philosphy of science.

(b)The second and perhaps more important argument: nowadays when physicists have a look at the problems which would have arisen in the conceptual background of their disciplines by the very discoveries they make, they find exactly the same problems which are found in the classical content of ontology. In short present physical knowledge tends intrinsically to become a reflexion about the categories that constitute the basis of the discipline itself and perhaps the basis of human knowledge.... In this way contemporary physics reestablishes a link with the Greeks' physics, and particularly with Aristotle's physics steps: description of the *fysis* behaviour, explanation of the recorded phenomena by displaying the causes that are immediatelly operating and finally, afterwards,*meta*, research concerning the real explanation (looking after the last causes) which implies the evoked reflexion about the concepts presupposed by the previous work, and perhaps modified by the issues themselves (for example the *backward causation* as revolution in the concept of cause itself).

By reintegrating the third step the physics of today again becomes intrinsically reflexion after work, literally, *meta-ta-fysika.*

Before coming to the main point of my purpose, I'd like to recall that the giving up of the third step, the rupture of the intrinsic link between science and philosophy has a nitid bithday (even if it was announced by previous tendencies). In fact the official assumption of the divorce adopted the form of a new marriage: the repudiated philosophy as research of the last causes, and a global explanation, was replaced by a philosophy "light" christened by its very promotor as *experimental.* Let's quote the text of Newton himself, the triumphal "hypothesis non fingo" in the scolia general of third book of *Principia Mathematica*:

> *I haven't succeeded in deducing from the phenomena the grounds of gravity and hypothesis non fingo, for everything that doesn't come from phenomena is a hypothesis and hypothesis can't be admitted into experimental philosphy. In this philosphy the propositions are extracted from the phenomena and next they are generalized by way of induction.*

And Newton immediately adds that the fact that the formulas of gravity explain the phenomenon of the mouvement of celeste bodies and the phenomenon of the movement of the sea is enough for him. But the question arises: do these formulas actually explain? As René Thom claims *description*, even completed by *prevision*, doesn't mean explanation [1]. A theory which makes computations, descriptions and previsions is not a theory which *explains*.

So, experimental philosophy may replace classical natural philosophy .... Leibniz was ready to object that the new theory was in fact a renouncement of the philosophy which intrinsically *"cherche la raison et la divine sagesse qui la fournit"* [2].

Moreover Leibniz pretends that experimental philosophy doesn't actually give up hypothesis. Simply the very fruitful hypothesis, these that would display reason (causal or not) of the *phenomena* are replaced by hypothesis that merely show some kind of correlation with the *phenomena*; hypothesis in fact *asthenic* (*hypothèses faineantes* wrote Leibniz) which would imply for science a complete rupture with the demands of rationality.

In comparison with this very hard appreciation of Leibniz, the observations concerning the disturbing paradoxes in some important aspects of quantum physics look almost like friendly encouragements to improve the general presentation of the theory. In fact the first people to keep their distance with the theory were the quantum-physicists themselves, which supposed a radical difference in relation to the Leibniz-Newton controversies. In a letter of December 1926 Einstein claims that, in spite of the fact that it describes the world with a level of accuracy without precedent in science, Quantum Mechanics (as a theory which would reduce God to a die player) *"is not yet the real thing... hardly brings us closer to the secret of the Old One"* [3]. But it was Einstein himself who 20 years before was compelled to consider the beam of light as an amount of discrete entities. An attitude nothing to do with Newton self idulgence in concerning the lack of explanations (therefore the lack of meaning) in gravitational theory.

Einstein was confronted to a very contradiction between two scientific theories that he himself had helped to set up, and he could hardly be true to both of them. The choice had be made, and Einstein's option became unequivocal in 1935 with the publication of the Einstein's Podolsky's and Rosen's famous paper [4] concerning the so called E. P. R. effect, that is: a causal dependence between two phenomena which, according to the premises of quantum mechanics, should invalidate the thesis of the absolute speed of light, which was a basis of relativity theory.

It is not necessary to recall that the Einstein-Podolsky-Rosen paper has to be completed by J. S. Bell's reflexion "On the Einstein Podolsky Rosen paradox" (*Physics* **1**, 195-204 (1964)), since he demonstrated that the statistical laws of quantum theory are not compatible with both of the following assumptions: a) the univers has objectif behaviour; b) every causation is necessarily local.

As N. Cartwright suggests, contemporary physics tends intrinsically to restablish the link with the problems of philosophical tradition: the physicist, and specially the quantum physicist, without leaving his specific discipline, is sometimes confronted with questions which involvees important discussions on ontology and epistemology. But among the general issues on ontology one of the most controverssials (at least from the point of view of a mathematical contemporary perspective) concerns the word that qualifies Quantum Physics itself: *"quantity, poson* which Aristotle tried, perhaps for the first times, to determine. This not through showing the specific differences with other categories (quality particularlly), what should be impossible in Aristotle's conception of the ontological status of the first concepts, but by displaying the internal division of the concept itself. Indeed, in his work known as Categorie's Book Aristotle writes:

> *Quantity* (poson) *is either dicrete* (diorismenon) *or continuous* (suneches). *Moreover, some quantities are such that each part of the whole has a relative position to the other parts, other have within them no such relation of part to part.*
>
> *Instances of discrete quantities are numbers and speech; of continuous, lines, surfaces, solids and, besides these time* (kronos) *and place* (pou).
>
> *In the case of the parts of a number there is no common boundary at which they join. For exemple: two fives make ten, but the two fives have no common boundary, but are separate [...] A line on the other hand, is a continuous quantity for it is possible to find a common boundary at which its parts join; in the case of the line, this common boundary is the point; in the case of the plane, it is the line [...]* [5].

This distinction between two concepts of quantity has to be completed by another, drawn in *Metaphysics* **513** (1027 a 7-12): quantity (*poson*) is either plurality, multitude (*plethos*) or magnitude (*megethos*). The first has to be understood as numerable (*arithméton*) while the second has to be understood as measurable (*metreton*). Moreover while the first is divisible in parts which were discrete before the division and remain discrete afterwards, the second is divisible in parts that were continuous one another and that after the division keep separated continuity.

Finally, there is a very important aspect: while for the *plurality (megethos)* there is an absolute reference of enumeration, the unit, which is intrisically indivisible [6] the magnitude (*megethos*) has not "natural" and absolute principle, but only an arbitrary and relative one. In short: contrasting with the unit of enumeration the unit of measure appears only as a result of *"treating length as an atomic entity"* (1052 b 32-33), this is because we do not consider the fact that the *metron* has a support which is intrinsically continuous, therefore intrinsically divisible.

Aristotle claims that the ***quantity*** as ***plurality*** does not have ontological objectivity, but it results merely from the capacity that has the human spirit of making abstraction of the ontological reality. Meanwhile, for quantity as magnitude, the situation is not far from being the opposite, since the continuous is that to which we are confronted with, just when we stop in the work of making abstraction.

To summarize: while the unit of the discrete quantity is a very (atomic) unit but ontologically is a vacuum, the unit of the continuous quantity has a great ontological weight but it is in fact a false (non atomic) unit.

The Aristotelian ontological priority of the *continuum* over the discrete quantity follows directly from these premises if we assume that, given two concepts AB, A has ontological priority over B if every entity that posseses the quality B posseses necessarily the quality A. This definition of the ontological priority comes from René Thom, who explains more accurately this idea in the following sentences: ***"L'être X est ontologiquement antérieur à l'être Y six porte naturellement comme prédicat, alors que X ne peut être que difficilement être prédicat de Y... Ainsi il est lingüistiquement tout-à-fait acceptabe de parler d'une surface colorée alors qu'une couleur "superficielle" ne pourrait s'employer que très métaphoriquement"*** [7].

Before coming back to the main purpose of my talk let's recall some of René Thom's topological exemples. The main idea is that discrete entity is in fact a *continuum* which appears as discrete only by playing the role of attribute of a *continuum* of bigger dimension [8].

So if we consider a tridimensional entity, its bidimensional surfaces are discrete accidents but in fact each of these is in itself a continuous support of lines as unidimensional entities.

Moreover, the *continuum* remains locally continuous when considered as holding discrete attributes (so a body when determined by its dimension) meanwhile the entities playing the role of discrete attributes becomes locally continuous if they hold a continuous attribute.

The Aristotelian thought becomes clear when considereded from Thom contemporary perspective: the unit *arithmeton*, the foundation of discrete quantity, would never emerge, would never come to view, even merely as construction of the mind if the continuous was not there as intrinsic support.

It might be said that for Aristotle the indivisible produces only counting numbers (*arithmoi*), 1, 2, 3, and so on, which, as reduced to iteration of something intrinsically inexistent, are themselves merely abstractions of mind. On the other hand, the rational fractions (not *arithmoi*) which as results of the division of the unit interval [0 1] would intrinsically pertain to the continuous, and so explain the structure of the substance itself. The latter is in fact a tridimensional body, by essence intrinsically and infinitelly divisible. Without continuous magnitude (*sunechen megethos*) not division and not rational fractions; a fortiori not irrational portions which, as parcel of proportions theory, are not conceivable without reference to the *megethé* [9].

Therefore, the come back to Aristotle that implies René Thom's topological though is completely in contradiction with the modern mathematics originated in late nineteenth century with the works of Dedekind and Cantor, and which leads to a reduction of the continuous magnitude (*megethos*) to the discrete multitude (*plethos*).

In a work published (in 1883) under the title of *Grundlagen allgenmeinen Manigfaltigkeitslehre* [10] Georg Cantor displays his conviction that his theory of numbers carries out a revolution in both mathematical and philosophical reasoning, because it implies a modification in the concept of number itself and perhaps simply in the concept of quantity.

Of course the main point of this revolution is the legitimation of magnitudes *infinite magnas* which Leibniz repudiated [11] but a very important role is played too by the new mthematical conception of the continuum. Given an n dimensional space each point of this space is reduced to n-tuples defined from real numbers from $-\infty$ to ∞ forming a complete ordered field which in fact find legitimacy in a reflexion submited to the notion intrinsically discrete of cardinality. This last point becomes obvious if we consider the construction of real numbers as classes of equivalence of sequences of rational numbers, in such a way that every element of the complete field R becomes the limit of some denumerable, bounded and monotic sequence, either ascendent or descendent.

Well then: In 1970 René Thom published a paper with the title *"les mathématiques modernes, une erreur pédagogique et philosophique"* [12] which was a merciless condemnation of Cantor's views and proposed what René Thom himself called (some years later) "return to Aristotle". But the most explicit confrontation is this which opposes René Thom to Dedekind's views.

In his work *"was sind and was sollen die Zahlen"* Dedekind reduces the continuous line R to the completion of cuts (*schnits*) defined in rational numbers. Evoking this text Thom writes:

> *Ici je voudrais m'attaquer à un mythe profondement ancré dans la mathématique contemporaine, à savoir que le continu s'engendre (voir se definit) à partir de la générativité de l'arithmétique, celle de la suitte des entiers naturels. Je fais, bien entendu, allusions à la construction de Dedekind* [13].

But if we object to the arithmetical generativity of the *continuum*, we will be able to advance arguments backing the opposite hypothesis. Several attempts has been made before this of Professor Thom, which in fact has never been published in a very demonstrative way.

Let us simply recall that the link between contigüiity and continuity has been reversed by Leibniz when he was confronted with the classical problem of rupture of continuity, symbolized by the path from life to death. The Leibniz's "solution" consisted of displaying the contigüity as a kind of rupture of topological continuity, an interpretation which would certainly appear scandalous for contemporary mathematical set theory:

First we had only a point C, and then ... we have two points A, B. Nevertheless these points are *distincts* but not *distants*; C has become a complexe entity [14].

Of course contiguity as rupture of continuity still doesn't mean discretion. But we are in the way ... Leibniz's position shows that the problem of relationship between discrete quantity and continuous quantity is a nuclear point of the common history of philosophy and science. History that we have concentrated on the debate Aristotle-Thom/Dedekind- Cantor. In this debate there is room for a third argument to be raised: that of the Non Standard Analysis, but infortunely do not have time time to talk about it [15].

REFERENCES AND REMARKS

1. René Thom, *"Description versus explanation in science"* (IBM Conference Academia Brasileira de Ciéncias, IBM, Brasil, 1985).
2. G.W. Leibniz, *Nouveux Essais* (Gerhardt T. V., Hildesheim, N. Y. Elms, 1978), p. 39.
3. *The Born-Einstein Letters* (Walker, New York, 1971), p. 90.
4. "Can quantum mechanica description of physical reality be considered complete," *Phys. Rev.* **47**, 777-780 (1935).
5. Translated to English by E. M. Edghill, under the editorship of W. D. Ross (Clarendon Press, Oxford, 1928).
6. The evoked texts have to be completed with Met. 10.1 1052 b 22-23 ("The one is the first principle of number qua number"), et seq.
7. René Thom, *"L'anteriorité ontologique du Continu sur le Discret "* (Colloque de Cerisy, 1990, ed. Huriya Sinacoeur).
8. "Un continu ne peut être discret que s'il apparaît comme un accident d'un autre continu de dimension supérieur (une droite dans le plan par exemple)," R. Thom, idem.
9. "It is instructive to note that in the primer of the theory of proportions that constitutes Book 5 of Euclid, the definitions and theorems are all

stated in terms of megethé". [M. J. White, *The Continuous and the Discrete. Ancient Physical Theories from a Contemporary Perspective* (Clarendon Press, Oxford, 1992), p. 185].

10. G. Cantor, *Gesammelte Abhandlugen* (Springer, Berlin, 1932).
11. At less in some texts: "Je leur tegmoirai que je ne croyais pas qu'il y eût des grandeurs vritablement infinies ni véritablement infinitésimales, ce que n'était que des fictions mais des fictions utiles pour abréger et pour parler universellement." *Letter to Dagincourt*, September 11, 1916. The following sentence is in another register also interesting: "Mais comme M. le Marquis de l'Hôspital craignait que je ne trahisse la cause, ils me prièrent de n'en rien dire."
12. René Thom, in *L'Age de la Science* (Villet, Septembre 1970, Vol. III, no 3, Paris Dunod), pp. 225-242.
13. René Thom, *"L'Anériorité ontologique."*
14. As Enrico Giusti said ("Immagini del continuo nell'opera di Leibniz. L'infini chez Leibniz," Colloquio Leibniz-Gesselschaft, Roma, 1986, unedited) "cio che e privo di grandezza non e necessariamente privo di struttura." Nowadays, after Abraham Robinson's *Non Standard Analysis*, we could say that the structure C has only a lack of grandezza standard, for in the neighbour of C there exists hyperreal distances.
15. Let's mention simply that the ontological weight of N. S. A. lies in the fact that, for the first time since the problem occupied science and philosophy, the infinitely small is legitimately introduced into mathematics.

LOCALIZED AND DELOCALIZED STATES IN PYRAMIDAL MOLECULES

Isabel Gonzalo

Departamento de Optica, Facultad de Ciencias Físicas
Universidad Complutense de Madrid, 28040-Madrid, Spain
igonzalo@eucmax.sim.ucm.es

The well observed delocalized states in NH_3 due to the inversion through tunneling, have never been observed in PH_3. It is shown that PH_3 is practically free from decoherence in the interstellar medium, contrary to other heavier molecules. The observation of the inversion splitting of PH_3 in that medium is then proposed.

Key words: decoherence, inversion splitting, pyramidal molecules.

1. INTRODUCTION

A noisy environment can be the origin of the stability of localized chiral or pyramidal states in molecules [1-4] as well as of decoherence and Zeno effects [5-9]. Other causes as parity violation by weak interactions [10] and the quantum vacuum [11] has been also proposed for chiral molecules.

The inversion in pyramidal molecules leads to delocalized states whose energy splitting (Fig. 1) is well observed in NH_3 but never observed in PH_3. This one is heavier than NH_3 and must be more sensitive to the environment. Thermal radiation was shown to be no efficient [3] to induce the localization of the states, then, we consider collisions with other molecules.

2. MODEL AND ESTIMATIONS

We use the model proposed in Refs. [8,9]: The hamiltonian of the molecule in the base of pyramidal states ψ_1, ψ_2 is $H = \frac{1}{2}\begin{pmatrix} f(t) & h\nu_i \\ h\nu_i & f(t) \end{pmatrix}$ where $h\nu_i$ is the energy splitting between the delocalized eigenstates $\psi_+ = 1/\sqrt{2}\,(\psi_1 + \psi_2)$ and $\psi_- = 1/\sqrt{2}\,(\psi_1 - \psi_2)$. The perturbative noisy environ-

M. Ferrero and A. van der Merwe (eds.),
New Developments on Fundamental Problems in Quantum Physics, 159–162.

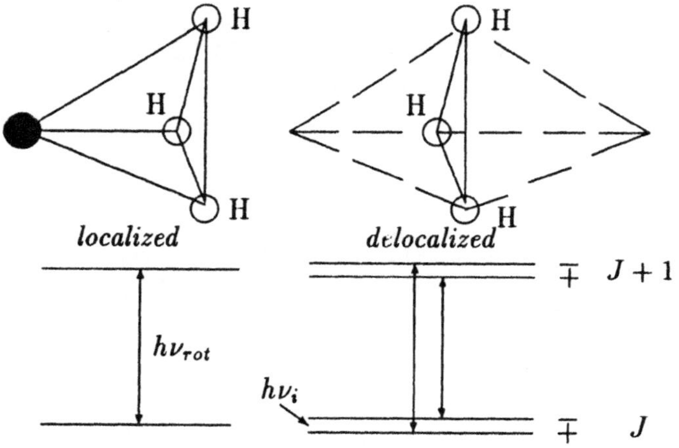

Fig. 1. Inversion frequency, ν_i, and splitting of the rotational transitions ν_{rot} in delocalized states, compared to the case of localized (pyramidal) states.

ment $f(t)$ can be characterized by the average time between collisions, t_0, able to change significantly the relative phase of ψ_1 and ψ_2.

Among the results of this model given by Berry [9], in agreement with Simonius [1] and Harris and Stodolski [10], we remark that for t_0 much lower than the inversion time $t_i \equiv \nu_i^{-1}$, decoherence occurs in $\psi_\pm$, in a time of the order of t_0, i.e., $\overline{\rho_{12}(t)} \propto e^{-t/2t_0}$. Estimation of $t_0 = \sqrt{mkT}/(2\sigma P)$ ($P = nkT$, n = number of molecules/cm^3, $\sigma = \pi b^2$, b =impact parameter for different colliding molecules [12]) is given in Table 1.

From Table 1 and decoherence criteria, it can be understood that the inversion splitting is detected in NH_3 ($\nu_i = 2.38 \times 10^{10}$ Hz) at room temperature and 1 atm ($t_0 \sim 10^{-10} - 10^{-11}$s, $t_0 \sim t_i$) disappearing at pressures greater than 1.84 atm [14] ($t_0 < t_i$), whereas the splitting is not detected in PH_3 under a resolution of 1 kHz, at room temperature and 0.04 cm Hg [15] ($t_0 \sim 10^{-8}$ s, $t_0 << t_i$). In this last case, decoherence in times $\sim t_0$ impedes the splitting to be detected, in spite of the high resolution (note that elastic collisions have no appreciable effect in broadening a microwave line).

The time t_0 is extremely large ($\sim 10^7$s) in interstellar clouds (e.g., $T = 10 - 50$ K, $n = 10^3 - 10^6$molecules cm^{-3}), then, PH_3 with estimated inversion time $t_i \sim 10^{-4}$s [13] can be considered isolated in that medium. However, the heavier molecule AsH_3 with $t_i \sim 10^7 - 10^8$s [12], cannot be considered isolated even in the interstellar medium.

Table 1. Estimated average time between collisions, t_0.

T (K)	P or n	b (Angstroms)	t_0 (s)
300	1 atm	5 (N_2)	9.6×10^{-11}
300	1 atm	10 (NH_3)	2.4×10^{-11}
300	0.04 cm Hg	10 (NH_3)	4.6×10^{-8}
300	0.04 cm Hg	5 (N_2)	1.8×10^{-7}
10	10^3 cm^{-3}	3 (H_2, He)	3.6×10^7
10	10^6 cm^{-3}	3 (H_2, He)	3.6×10^4

3. CONCLUSIONS

The molecule PH_3 is practically free from decoherence in the interstellar medium. We then propose the observation, never done, of the inversion splitting of PH_3 in that medium. The observation must be done in rotational transitions ($\nu_{rot} \sim 10^{11} - 10^{12}$ Hz) which, contrary to the inversion frequency ν_i in PH_3, are well-known frequencies and in the appropriate range for radiodetection. The unavoidable Doppler line-width [12] $\Delta\nu = (2kT\ln 2/m)^{1/2}(\nu_{rot}/c)$ for $T = 10$ K is $\Delta\nu \sim 5 \times 10^5$ Hz. Then, splittings lower than ~ 1 MHz cannot be detected.

It would be desirable to study other molecules with intermediate inversion times between those of NH_3 and PH_3. Possible candidates are $N(CH_3)_3$ and NH_2F [4]. An advantage of $N(CH_3)_3$ is its low rotational frequencies (due to high inertia momenta) improving the spectral resolution.

ACKNOWLEDGMENT

The author expresses her gratitude to V. Bujarrabal (Observatorio Astronómico Nacional, Madrid) for helpful information and to M.A. Porras for his assistance.

REFERENCES

1. M. Simonius, *Phys. Rev. Lett.* **40**, 980 (1978).
2. G. Woolley, *Isr. J. Chem.* **19**, 30 (1980).
3. E. Joos and H. D. Zeh, *Z. Phys. B* **59**, 223 (1985).
4. P. Claverie and G. Jona-Lasinio, *Phys. Rev. A* **33**, 2245 (1986).
5. B. Misra and E.C.G. Sudarshan, *J. Math. Phys.* **18**, 756 (1977).
6. E. Joos, *Phys. Rev. D* **29**, 1626 (1974).

7. W. Zureck, *Phys. Today* **36** (10) (1991).
8. Ph. Blanchard, G. Bolz, M. Cini, G.F. De Angalis, and M. Serva, *J. Stat. Phys.*, 1994.
9. M. Berry, *Ann. New York Acad. Sci., Proceedings, Fundamental Problems in Quantum Theory*, Baltimore, June 1994.
10. R. H. Harris and L. Stodolsky, *Phys. Lett. B* **78**, 313 (1978).
11. P. Pfeifer, *Ph.D. Thesis*, ETH Zürich, 1980.
12. C.H. Townes and A.L. Schawlow, *Microwave Spectroscopy* (McGraw-Hill, New York, 1955).
13. C.C. Costain and G.B.B.M. Sutherland, *J. Phys. Chem.* **56**, 321 (1952).
14. G. Birnbaum and A.A. Maryot, *Phys. Rev.* **92**, 270 (1953).
15. S.P. Belov et al., *J. Mol. Spectr.* **86**, 184 (1981).

A BIGGER CONTRADICTION BETWEEN QUANTUM THEORY AND LOCALITY FOR TWO PARTICLES WITHOUT INEQUALITIES

Lucien Hardy

Dipartimento di Fisica, Università "La Sapienza"
Piazzale Alldo Moro 2
I-00185 Rome, Italy

A previously published proof of nonlocality without inequalities for two particles is presented in a simple form. This proof goes through for only 9 % of pairs of particles. It is shown how it can be generalised so that the proof goes through for almost 50% of pairs.

Key words: Bell's theorem, quantum nonlocality, EPR argument.

Imagine the following scenario. We take a cat and a dog. They are both placed into a reasonably large black box and the lid is closed and locked. Something happens inside the box. Exactly what happens is impossible to describe, though the animals have some kind of fight. After this, the box is partitioned into two separate boxes such that the cat goes into one box (the left hand one say) and the dog goes into the other. During the partitioning, it is important that we do not see either animal. Then the two boxes are separated to a great distance. Next, two experimentalists, one at standing next to each box, simultaneously make a measurement on the animal in their box. They each have a choice of one of two measurements to make on the animal. Either they can look to see whether the animal is dead or alive, or they can look to see whether the animal has sharp or blunt teeth. They each choose which measurement to perform randomly. Which ever measurement is chosen, it must be conducted in such a way that the outcome of the other measurement is not revealed. Of course, there would be practical difficulties in looking to see whether a cat (or a dog) has sharp or blunt teeth without also finding out whether it is alive, but we will assume that experimentalists are up to the task. To further annoy animal rights activists, we also assume that we have an infinite suply of cats and dogs so that this whole proceedure

M. Ferrero and A. van der Merwe (eds.),
New Developments on Fundamental Problems in Quantum Physics, 163–170.

can be repeated an infinite (or at least a very large) number of times and any correlations between the outcomes of the measurements revealed.

Imagine that after this experiment has been performed we find that the following observations are true:

(i) If the teeth are measured on both animals, it sometimes happens that both animals are found to have sharp teeth.

(ii) If the teeth are measured on the cat and the life state is measured on the dog, it is the case that if the cat is found to have sharp teeth then the dog will certainly be found to be alive.

(iii) If the teeth are measured on the dog and the life state is measured on the cat, it is the case that if the dog is found to have sharp teeth then the cat will certainly be alive.

(iv) If the life state is measured on both animals, then it is never found that both animals are alive.

One might speculate that the reason for (ii) is that if the cat had killed the dog it would have blunted its teeth in the process (and similarly for (iii)).

If one looks at these experimental results one immeadiately spots a what looks like a contradiction. Consider a case in which both animals have sharp teeth (that this is possible follows from (i)). Since the cat has sharp teeth, it follows from (ii) that the dog is alive. Since the dog has sharp teeth it follows from (iii) that the cat is alive. This means, in these cases, that both the cat and dog are alive. However, from (iv), we see that this never happens. The natural response to this contradiction is assert that such an imagined state of affairs could never actually happen in the real world (that is outside the mind of a theoretical physicist). This is, of course, true. However, if the cats and dogs are substituted by quantum particles, then an exactly analogous state of affairs is possible and has even been realised experimentally (to within experimental error).

Consider two quantum particles each living in a two dimensional Hilbert space (such as spin half electrons). We shall continue to use nomenclature derived from the above example. Thus, particle 1 of each pair shall be called the cat, and particle 2 shall be called the dog. On each particle, we can measure to see whether the it has sharp or blunt teeth, corresponding to the eigenstates $|s\rangle_{1,2}$ and $|b\rangle_{1,2}$ respectively. Or, we can measure to see whether it is dead or alive corresponding to the eigenstates $|d\rangle_{1,2}$ and $|a\rangle_{1,2}$ respectively. Since these eigenstates simply provide different ways of spanning the same two dimensional space, we can write the blunt eigenstate as a superposition of the alive and dead eigenstates,

$$|b\rangle_i = A_i|a\rangle_i + B_i|d\rangle_i. \tag{1}$$

The state orthogonal to this must be the sharp state

$$|s\rangle_i = B_i^*|a\rangle_i - A_i^*|d\rangle_i. \tag{2}$$

These two equations can be inverted, giving

$$|a\rangle_i = A_i^*|b\rangle_i + B_i|s\rangle_i, \tag{3}$$

$$|d\rangle_i = B_i^*|b\rangle_i - A_i|s\rangle_i. \tag{4}$$

Let the two particles be prepared in the state

$$|\Psi\rangle = N(|b\rangle_1|b\rangle_2 - A_1A_2|a\rangle_1|a\rangle_2). \tag{5}$$

Note that this is a slightly unusual (though perfectly valid) way to write down a quantum state since the two terms on the RHS are not orthoganol. Now we are in a position to verify this state has the analogous properties to (i-iv) listed above.

First consider prediction (iv) that we never find both particles are alive. This means there should be no $|a\rangle_1|a\rangle_2$ term. Substitute Eq. (1) into (5):

$$|\Psi\rangle = N\big((A_1|a\rangle_1 + B_1|d\rangle_1)(A_2|a\rangle_2 + B_2|d\rangle_2) - A_1A_2|a\rangle_1|a\rangle_2\big). \tag{6}$$

We see that the last term on the RHS cancels exactly with the corresponding cross term and hence prediction (iv) is valid.

Next consider prediction (ii). Substitute (3) into (5) for $|a\rangle_1$:

$$|\Psi\rangle = N(|b\rangle_1|b\rangle_2 - A_1A_2(A_1^*|b\rangle_1 + B_1|s\rangle_1)|a\rangle_2). \tag{7}$$

If particle 1 is found to be in the state $|s\rangle_1$ (sharp teeth) then we see from this equation that particle 2 will certainly be in the state $|a\rangle_2$ (alive). This verifies (ii).

It is clear that prediction (iii) is also true for similar reasons.

Finally, consider prediction (i). Substitute (3) into (5):

$$|\Psi\rangle = N\big(|b\rangle_1|b\rangle_2 - A_1A_2(A_1^*|b\rangle_1 + B_1|s\rangle_1)(A_2^*|b\rangle_2 + B_2|s\rangle_2)\big). \tag{8}$$

The amplitude in front of the $|s\rangle_1|s\rangle_2$ term is $NA_1A_2B_1B_2$. Hence the probability that both particles have sharp teeth is $|NA_1A_2B_1B_2|^2$. This is clearly non-zero as long as the operators corresponding to the two types of measurements do not commute. It can be shown that the maximum value of this probability is about 9% (more exactly, it is $\frac{1}{\tau^5}$ where $\tau = \frac{1}{2}(1+\sqrt{5})$ is the golden mean).

The above contradiction was first presented in [1] and has been further discussed in [2] (actual experimental realisations are reported in [3]). Its resolution requires that one allows the possibility of a nonlocal influence and has been discussed in much depth in [1,2] and the reader is referred to those articles. In this article we shall be concerned with improving the extent of the contradiction. As it stands, it only goes through for nine percent of pairs, which is not very good. We will show how it can be improved to almost fifty percent.

First let us generalise to consider the possibility of many measurements at each end. To facilitate subsequent discussion we will change notation. We have end A where any of the measurements A_k (for $k = 0$ to K) can be made. Each of these measurements has the outcome A_k or $A_k^{\perp}$ with corresponding eigenstates $|A_k\rangle$ and $|A_k^{\perp}\rangle$ respectively. At the other end, B, any of the measurements B_k (for $k = 0$ to K) can be made, with eigenstates $|B_k\rangle$ and $|B_k^{\perp}\rangle$ corresponding to the outcomes B_k and $B_k^{\perp}$ respectively. The contradiction presented above has the following form with $K = 1$:

A_1 and B_1 happens with probability $p_1 \neq 0$.

A_1 implies B_0 and B_1 implies A_0.

A_0 and B_0 happens with probability $p_0 = 0$.

We can now generalise this in the following way:

A_K and B_K happens with probability $p_K \neq 0$.

A_K implies B_{K-1} and B_K implies A_{K-1}.

A_{K-1} implies B_{K-2} and B_{K-1} implies A_{K-2}.

...

A_1 implies B_0 and B_1 implies A_0.

A_0 and B_0 happens with probability $p_0 = 0$.

We see that in this way we build a ladder of inferences. We start from the proportion, p_K, of cases in which we get A_K and B_K. Then, employing the 'diagonal' inferences

A_{k+1} implies B_k and B_{k+1} implies A_k to go down the ladder rung by rung, we find that this implies that we have A_k and B_k at each rung, finally showing that we should have A_0 and B_0 for at least the proportion p_K of pairs. However, the final statement says we should never get A_0 and B_0. Hence, the proportion of pairs for which the contradiction goes through is p_K. We want to show, first of all, that it is possible to arrange for such a contradiction with quantum mechanics, and then find what the maximum proportion p_K is.

The key to doing this is to write the state of the two particles as

$$|\Psi\rangle = u_k|A_{k+1}^\perp\rangle|B_{k+1}^\perp\rangle + v_k|A_k\rangle|B_k\rangle. \tag{9}$$

We can write

$$|A_k\rangle = q_k|A_{k+1}\rangle + r_k|A_{k+1}^\perp\rangle, \tag{10}$$

$$|B_k\rangle = q_k|B_{k+1}\rangle + r_k|B_{k+1}^\perp\rangle. \tag{11}$$

From these equations, it is clear that the diagonal inferences are valid for $|\Psi\rangle$ (cf. Eq. (7)). We need to show now that it is possible to arrange that the state can be written as in (9) for $k = 0$ to $K-1$. We start with the state written in Schmidt form:

$$|\Psi\rangle = \alpha|+\rangle_A|+\rangle_B - \beta|-\rangle_A|-\rangle_B, \tag{12}$$

where $|\pm\rangle_{A,B}$ are the vectors of the Schmidt decomposition of the state. Without loss of generality we can choose both α and β to be real and positive (any complex part can be absorbed into the definition of the states $|\pm\rangle_A$). The Schmidt vectors can be written as

$$|+\rangle_A = c_k|A_k\rangle + c_k^\perp|A_k^\perp\rangle \tag{13}$$

$$|-\rangle_A = (c_k^\perp)^*|A_k\rangle - c_k^*|A_k^\perp\rangle \tag{14}$$

$$|+\rangle_B = c_k|B_k\rangle + c_k^\perp|B_k^\perp\rangle \tag{15}$$

$$|-\rangle_B = (c_k^\perp)^*|B_k\rangle - c_k^*|B_k^\perp\rangle \tag{16}$$

Substitute (13-16) into (12):

$$\begin{aligned}|\Psi\rangle =&[\alpha c_k^2 - \beta(c_k^{\perp *})^2]|A_k\rangle|B_k\rangle \\ &+ [\alpha c_k c_k^\perp + \beta c_k^{\perp *} c_k^*]|A_k\rangle|B_k^\perp\rangle \\ &+ [\alpha c_k^\perp c_k + \beta c_k^* c_k^{\perp *}]|A_k^\perp\rangle|B_k\rangle \\ &+ [\alpha c_k^{\perp 2} - \beta(c_k^*)^2]|A_k^\perp\rangle|B_k^\perp\rangle.\end{aligned} \tag{17}$$

We want to manipulate this state into the form in (9). To do this we complete the square:

$$
\begin{aligned}
|\Psi\rangle = & \left[\frac{\alpha c_k c_k^{\perp} + \beta c_k^{\perp *} c_k^*}{\sqrt{[\alpha c_k^{\perp 2} - \beta (c_k^*)^2]}} |A_k\rangle + \sqrt{\alpha c_k^{\perp 2} - \beta (c_k^*)^2} |A_k^{\perp}\rangle \right] \\
& \times \left[\frac{\alpha c_k c_k^{\perp} + \beta c_k^{\perp *} c_k^*}{\sqrt{\alpha c_k^{\perp 2} - \beta (c_k^*)^2}} |B_k\rangle + \sqrt{[\alpha c_k^{\perp 2} - \beta (c_k^*)^2]} |B_k^{\perp}\rangle \right] \qquad 18) \\
& - \left\{ \frac{(\alpha c_k c_k^{\perp} + \beta c_k^{\perp *} c_k^*)^2}{\alpha c_k^{\perp 2} - \beta (c_k^*)^2} - (\alpha c_k^2 - \beta (c_k^{\perp *})^2) \right\} |A_k\rangle |B_k\rangle.
\end{aligned}
$$

Looking at this, we see immediately that it is in the form (9) if we identify the states $|A_{k+1}^{\perp}\rangle$ and $|B_{k+1}^{\perp}\rangle$ with the expresions in the square brackets. Thus,

$$|A_{k+1}^{\perp}\rangle \propto (\alpha c_k c_k^{\perp} + \beta c_k^{\perp *} c_k^*) |A_k\rangle + (\alpha c_k^{\perp 2} - \beta (c_k^*)^2) |A_k^{\perp}\rangle, \qquad (19)$$

where we have multiplied through by $\sqrt{\alpha c_k^{\perp 2} - \beta (c_k^*)^2}$ to simplify the expression. Hence,

$$|A_{k+1}\rangle \propto (\alpha (c_k^{\perp *})^2 - \beta c_k^2) |A_k\rangle - (\alpha c_k^* c_k^{\perp *} + \beta c_k^{\perp} c_k) |A_k^{\perp}\rangle. \qquad (20)$$

We can invert equations (13) and (14) to obtain

$$|A_k\rangle = c_k^* |+\rangle_A + c_k^{\perp} |-\rangle_B, \qquad (21)$$

$$|A_k^{\perp}\rangle = c_k^{\perp *} |+\rangle_A - c_k |-\rangle_B. \qquad (22)$$

Substitution of these into (20) gives

$$|A_{k+1}\rangle \propto -\beta c_k |+\rangle_A + \alpha c_k^{\perp *} |-\rangle_A. \qquad (23)$$

Similarly,

$$|B_{k+1}\rangle \propto -\beta c_k |+\rangle_B + \alpha c_k^{\perp *} |-\rangle_B. \qquad (24)$$

Comparing (21) with (23), we see that

$$c_{k+1} \propto -\beta c_k^*, \quad c_{k+1}^{\perp} \propto \alpha c_k^{\perp *}, \qquad (25)$$

where the constant of proportionality is the same in each case and obtained by normalisation.

The probability of obtaining A_k and B_k is given by the square of the amplitude in front of the corresponding term in (27):

$$p_k = \text{Prob}(A_k B_k) = |\alpha c_k^2 - \beta(c_k^{\perp})^2|^2. \tag{26}$$

We must fix the parameters so that $p_0 = 0$. Thus, we choose $c_0 \propto \beta^{\frac{1}{2}}$ and $c_0^{\perp} \propto \alpha^{\frac{1}{2}}$. Employing (25), we obtain

$$c_k \propto (-1)^k \beta^{k+\frac{1}{2}}, \quad c_k^{\perp} \propto \alpha^{k+\frac{1}{2}}. \tag{27}$$

Substituting this into (26) and remembering normalisation, we get

$$p_k = \left(\frac{\alpha\beta^{2k+1} - \beta\alpha^{2k+1}}{\beta^{2k+1} + \alpha^{2k+1}}\right)^2. \tag{28}$$

The proportion of pairs for which the contradiction goes through is given by p_K. If $\alpha = \beta$ then $p_k = 0$ for all k so there is no contadiction. This means that this proof will only work when we have nonmaximally entangled states. Choosing $K = 1$ (as in [1,2]) we find that the maximum value of p_1 is 9.0%, realised when $\alpha\beta = 0.38$. Choosing $K = 2$ we find that the maximum value of p_2 is 17.5%, realised when $\alpha\beta = 0.43$. Assuming regular behaviour of the function, we see that as K increases, the maximum value of p_K is achieved for states that are closer to the maximally entangled case (when $\alpha\beta = 0.5$). To see what happens to p_K as $K \to \infty$ we note that (28) can be written as

$$p_k = \left(\frac{\alpha - \beta\left(\frac{\alpha}{\beta}\right)^{2k+1}}{1 + \left(\frac{\alpha}{\beta}\right)^{2k+1}}\right)^2. \tag{29}$$

Assuming $\alpha < \beta$ we see that as $K \to \infty$, $p_k \to \min(\alpha^2, \beta^2)$ (since we could have had $\beta < \alpha$). As we cannot have $\alpha = \beta$, the maximum value of p_K is (50-δ)% where δ is any small finite number and is achieved for a state that is not quite maximally entangled.

A different way of viewing the original contradiction presented at the beginning of this paper has been suggested to the author by S. Popescu. The set of statements (i) to (iv) are equivalent to the following. Cat has sharp teeth implies dog alive. Dog alive implies cat dead. Cat dead implies dog has blunt teeth. Therefore, it seems that cat has sharp teeth implies that the dog certainly has blunt teeth. But we know that sometimes both animals have sharp teeth so we have a contradiction. This contradiction is of the form $A \Rightarrow B \Rightarrow C \Rightarrow D$ but $A \not\Rightarrow D$. It is easilly seen that the generalisation presented in this paper can be understood as a generalisation of Popescu's approach where the contradiction is of the form $A \Rightarrow B \Rightarrow C \ldots \Rightarrow Y \Rightarrow Z$ but $A \not\Rightarrow Z$.

REFERENCES

1. L. Hardy, *Phys. Rev. Lett.* **68**, 2981 (1992).
2. L. Hardy, *Phys. Rev. Lett.* **71**, 1665 (1993); **73**, 2279 (1994). R. Clifton and P Niemann, *Phys. Lett. A* **166**, 177 (1992). H. P. Stapp, *Mind, Matter, and Quantum Mechanics* (Springer, New York/Berlin, 1993). T. F. Jordan, *Am. J. Phys.* **62**, 874 (1994). S. Goldstein, *Phys. Rev. Lett.* **72**, 1951 (1994). N. D. Mermin, *Am. J. Phys.* **62**, 880 (1994).
3. J. R. Torgerson, D. Branning, C. H. Monken, and L. Mandel, *Phys. Lett. A* **204**, 323 (1995). G. Di Giuseppe, F. De Martini and D. Boshi, "Experimental test of the violation of local realism in quantum mechanics without Bell inequalities," to be published.

IRREVERSIBLE QUANTUM EVOLUTION FOR UNSTABLE SYSTEMS IN LAX-PHILLIPS SCATTERING THEORY

L.P. Horwitz,[a,b] E. Eisenberg,[a] and Y. Strauss[b]

[a] *Department of Physics, Bar-Ilan University*
Ramat-Gan 52900, Israel
[b] *School of Physics and Astronomy*
Raymond and Beverly Sackler Faculty of Exact Sciences
Tel-Aviv University, Ramat-Aviv, Israel

The scattering theory of Lax and Phillips, designed primarily for hyperbolic systems, such as electromagnetic or acoustic waves, is described. This theory provides a realization of the theorem of Foias and Nagy; there is a subspace of the Hilbert space in which the unitary evolution of the system, restricted to this subspace, is realized as a semigroup. The embedding of the quantum theory into this structure, carried out by Flesia and Piron, is reviewed. We show how the density matrix for an effectively pure state can evolve to an effectively mixed state (decoherence) in this framework. Necessary conditions are given for the realization of the relation between the spectrum of the generator of the semigroup and the singularities of the S-matrix (in energy representation). It is shown that these conditions may be met in the Liouville space formulation of quantum evolution, and in the Hilbert space of relativistic quantum theory.

Key words: irreversibility, unstable systems,semigroup, quantum evolution, relativistic quantum theory.

1. INTRODUCTION

The unstable quantum system is an important example of irreversible phenomena in nature. Such systems, ranging from excited atomic states to short-lived elementary particles, are characterized by what is generally observed to be an irreversible evolution. These phenomena raise the question of explanation of such processes from first principles. Moreover, since most of the decay processes are observed experimentally to obey an exponential

M. Ferrero and A. van der Merwe (eds.),
New Developments on Fundamental Problems in Quantum Physics, 171–184.

decay law, one expects this behavior to follow from very general assumptions.

Irreversible evolution in the quantum theory has been described by the addition of non-Hermitian terms to the Hamiltonian, such that it has complex eigenvalues, and the induced evolution is non-unitary [1][2]. In this method, the non-Hermitian terms in the Hamiltonian are introduced phenomenologically, and may only indirectly be associated with some known interaction in a more fundamental Hamiltonian.

Weisskopf and Wigner [3], in a well known paper in 1930, introduced an alternative approach to the decay problem. According to their approach, the evolution takes place in a Hilbert space which is a direct sum of two subspaces: the subspace of the decaying states and that of decay products. These two subspaces are stable under the "free" evolution induced by H_0, but are combined linearly under the full evolution induced by $H = H_0 + V$. In this Hilbert space, the evolution is unitary, and hence its generator, i.e., the Hamiltonian, is self-adjoint. The decay is described as the probability flow from the subspace of the decaying states to its complement, the subspace of the decay products. They studied perturbatively, for the single-channel case, what has become known as the *survival amplitude*

$$A(t) = (\psi, e^{-iHt}\psi), \tag{1.1}$$

which is the probability amplitude for the system to remain in the discrete state until time t. In the following we will describe this approach, and pose critical problems, motivating the development of a more general theory.

The total evolution operator $U(t) = e^{-iHt}$, and the resolvent $R(z) = (z - H)^{-1}$ are related to each other by the (inverse) Laplace transform

$$U(t) = \frac{1}{2\pi i} \oint R(z) e^{-izt} dz, \tag{1.2}$$

where the integration contour is around the spectrum of H. If we project this operator into the subspace $P\mathcal{H}$, we can obtain a similar relation which expresses the reduced motion $U'(t)$ in terms of the *reduced resolvent* $R'(z) = PR(z)P$:

$$U'(t) = \frac{1}{2\pi i} \oint R'(z) e^{-izt} dz. \tag{1.3}$$

By differentiating the absolute square of (1.1) and setting $t = 0$, one sees that the initial decay rate is necessarily zero (providing that the Hamiltonian is defined on the initial state); in fact, it is easy to show that the change in $p(t)$ is $O(t^2)$. The intermediate and long time behavior follow most simply by an examination of the relation (1.3). Deforming the contour of integration which runs below the real positive spectrum of H to the negative imaginary axis, where its contribution for large times is small, the remaining contribution of the contour running above the spectrum of H can be estimated by bringing this contour continuously through the cut. When resonances exist, this contour will pass through simple poles on the way to the negative imaginary axis in the second sheet (as can be explicitly demonstrated [4] in the Lee-Friedrichs model [5]). The residues of the poles may dominate the time dependence for intermediate times, and give the approximate (due to the presence of residual contributions from the integrals along the negative imaginary axis) exponential decay behavior. For very long times, the pole contributions disappear, and the remaining integration around the branch cut results in an inverse power law asymptotic behavior [6].

It is not difficult to see that an irreversible evolution must be described by a semigroup [7] (for the reversible case this is a group induced by a unitary transformation), where we define a semigroup as follows:

Let $\{Z(t)\}$ be a family (over t) of operators on a Hilbert space; then $Z(t)$ is an element of a semigroup if

$$Z(t_1)Z(t_2) = Z(t_1 + t_2), \quad t_1, t_2 \geq 0. \tag{1.4}$$

The semigroup is said to be *strongly contractive* if $\|Z(t)\| \to 0$, for $t \to \infty$, where $\|A\|$ is the operator norm of A. On the other hand, it can be shown that the reduced motion, as described above, cannot generate a semigroup [8].

There is, furthermore, another, perhaps more fundamental, problem associated with the general method of Wigner and Weisskopf; this is that the expression (1.1) for the survival amplitude implicitly assumes the existence of a linear superposition (we restrict our discussion here to the one-channel case)

$$e^{-iHt}\psi = A(t)\psi + \chi(t), \tag{1.5}$$

where $\chi(t)$ represents the decayed system and $(\psi, \chi(t)) = 0$. In general this linear superposition does not correspond to any physical situation in our experience; a short-lived particle, for example, is seen as either the particle before the decay, or the decay products at a certain time, which can not be predicted. This linear superposition does not correspond to the object that we see experimentally in such a process.

In the framework of the theory of Weisskopf and Wigner, techniques have been developed which are capable of displaying the exact semigroup behavior of an unstable system [9]. As we have remarked above, the Lee-Friedrichs model [5] provides a simple but useful example of an unstable system for which the evolution equations are completely soluble [4]. The eigenvalue equation $Hf(z) = zf(z)$, for $f \in \mathcal{H}$ has a formal solution which does not, however, satisfy the equation. The condition to satisfy the equation coincides with the condition for a pole in the resolvent $R'(z)$, and can only be satisfied in the second Riemann sheet. The scalar product of the eigenvalue equation with a vector for which the unperturbed energy representation is analytic in a domain containing the second sheet pole can be continued to the second sheet, and at the pole position, the equation is identically satisfied. Since the "eigenvector" obtained in this way is in the space dual to a set of vectors with this restrictive analyticity requirement, corresponding to a subspace of $\mathcal{H}$, it is an element in the large space of a Gel'fand triple (rigged Hilbert space). The definition of this vector depends on the domain of analyticity chosen, and its physical interpretation is not clear, except for the fact that its (extended) unitary evolution is that of an exact exponential decay.

The problems discussed above are essentially related to the attempt to describe an unstable system in a framework more suitable to the description of reversible phenomena. In what follows we will show another approach to irreversible phenomena which attempts to solve these difficulties.

2. LAX-PHILLIPS THEORY AND THE EXACT SEMIGROUP

The theory of Lax and Phillips [10], designed for systems of hyperbolic differential equations describing the scattering of, e.g., electromagnetic or acoustic waves, and the Floquet theory [11] for periodic time dependent quantum mechanical problems are examples of such a structure. Piron [7] has shown that methods of this type are applicable to the general time dependent quantum mechanical problem. Recently, Flesia and Piron [12] have shown that scattering problems in quantum theory can be put in the form of Lax-Phillips theory (Horwitz and Piron [13] have discussed its applicability to the problem of the unstable system) by forming a direct integral of the quantum mechanical Hilbert spaces $\mathcal{H}_t$ over t in order to construct a larger space $\bar{\mathcal{H}}$ which includes t in its measure space.

Lax-Phillips theory [10] assumes the existence of a one-parameter unitary group of evolution on a Hilbert space $\bar{\mathcal{H}}$, and incoming and outgoing subspaces $\mathcal{D}_-$ and $\mathcal{D}_+$ such that

$$\begin{aligned} U(\tau)\mathcal{D}_+ &\subset \mathcal{D}_+\,, \text{ for all } \tau > 0, \\ U(\tau)\mathcal{D}_- &\subset \mathcal{D}_-\,, \text{ for all } \tau < 0, \\ \bigcap_\tau U(\tau)\mathcal{D}_\pm\}, &= \{0\}, \\ \overline{\bigcup_\tau U(\tau)\mathcal{D}_\pm} &= \bar{\mathcal{H}}, \end{aligned} \tag{2.1}$$

where τ is the evolution parameter identified with the laboratory time. It follows from a theorem of Sinai [14] that $\bar{\mathcal{H}}$ can be foliated in such a way that it can be represented as a family of (auxiliary) Hilbert spaces in the form $L^2(-\infty,+\infty;\mathcal{H}_t)$, over Lebesgue measure in t , and all the $\mathcal{H}_t$ are isomorphic (we therefore sometimes refer to these spaces simply as $\mathcal{H}$) and determined up to unitary equivalence. The scalar product in $\bar{\mathcal{H}}$ is given by

$$(f\,,g) = \int_{-\infty}^{\infty} (f_t\,,g_t)_{\mathcal{H}_t}\,dt\,. \tag{2.2}$$

Lax and Phillips show that there are unitary operators W_+^{-1}, W_-^{-1} which map the elements of $\bar{\mathcal{H}}$ into representations, called the outgoing and incoming translation representations, for which the evolution is translation in t. The subspaces $\mathcal{D}_+, \mathcal{D}_-$ correspond to the sets of functions with, in these representations, support in semi-infinite segments of the positive and negative t-axis respectively. They define the S matrix abstractly as the map from the incoming translation representation to the outgoing one, i.e., $S = W_+^{-1}W_-$. This map is defined up to unitary transformations on the auxiliary spaces $\{\mathcal{H}_t\}$, and refers to the equivalence classes for which the incoming and outgoing representations have the property that the evolution is represented by translation.

Lax and Phillips furthermore define the operator

$$\mathcal{Z}(\tau) = P_+ U(\tau) P_- \tag{2.3}$$

on $\bar{\mathcal{H}}$, where $P_\pm$ is the projection on the orthogonal complement of $\mathcal{D}_\pm$. This operator vanishes on $\mathcal{D}_\pm$ and maps the subspace

$$\mathcal{K} = \bar{\mathcal{H}} \ominus (\mathcal{D}_+ \oplus \mathcal{D}_-)\,, \tag{2.4}$$

into itself. These mappings form a semigroup [11], i.e., for $\tau_1, \tau_2 \geq 0$,

$$\mathcal{Z}(\tau_1)\mathcal{Z}(\tau_2) = \mathcal{Z}(\tau_1+\tau_2)\,, \tag{2.5}$$

and this semigroup is strongly contractive, i.e., for each $\phi \in \mathcal{K}$ and any ϵ, there exists a τ_ϕ such that

$$||\mathcal{Z}(\tau)\phi||_{\bar{\mathcal{H}}} < \epsilon \tag{2.6}$$

for $\tau > \tau_\phi$. It can be shown that $\mathcal{Z}(\tau)$ is just the unitary evolution $U(\tau)$ projected into the subspace $\mathcal{K}$. Since the states which lie in the subspaces $\mathcal{D}_\pm$, in the case of scattering, describe the incoming and outgoing waves which are not influenced by the interaction, the states which lie in $\mathcal{K}$ describe the unstable states, i.e., resonances of the scattering. From this point of view, the Lax-Phillips semigroup is analogous to the reduced motion discussed in the previous section.

This theory constitutes a constructive realization of the theorem of Foias and Nagy [15], which states that given a semigroup on a Hilbert space $\mathcal{H}$, there is a bigger Hilbert space $\bar{\mathcal{H}}$ which contains it, in which the evolution is a one-parameter unitary group, and this unitary group restricted to $\mathcal{H}$ is that semigroup.[1]

Flesia and Piron [12] have shown that the quantum theory may be embedded in a Lax-Phillips theory by considering the family of Hilbert spaces of the usual quantum theory on the parameter t as the auxiliary spaces of Lax and Phillips; the large Hilbert space $\bar{\mathcal{H}}$ is then the direct integral of these quantum mechanical spaces over all values of the time t with Lebesgue measure. The form of the theory adopted by Flesia and Piron [12] distinguishes the elements of these equivalence classes, and constructs an S-matrix which maps the auxiliary space in the incoming translation representation to the auxiliary space of the outgoing one.

In this construction, Flesia and Piron assume the form

$$\psi^{\tau}_{t+\tau} = W_t(\tau)\psi^0_t \,, \tag{2.7}$$

where, since $W_t(\tau)$ represents an evolution, it follows that

$$W_{t+\tau_1}(\tau_2)W_t(\tau_1) = W_t(\tau_1+\tau_2)\,. \tag{2.8}$$

Lax and Phillips prove that the S-matrix (in their construction) is a multiplicative operator in the spectral representation of the generator of the unitary evolution K (which is the Fourier transform of the translation representation), i.e.,

$$(S\psi)_\sigma = S(\sigma)\psi_\sigma,$$

and that the eigenvalues of the generator of the semigroup $\mathcal{Z}(\tau)$ correspond to the singularities of the analytic continuation of $S(\sigma)$. The eigenstates corresponding to these eigenvalues are analogous to the generalized eigenstates found in the framework of Weisskopf and Wigner, as discussed in Section 1. Thus, the S-matrix contains all the information about the unstable states. It can be seen [16], however, that the S matrix obtained from a model in which the evolution is given in the form (2.7) has no t-dependence, and hence its spectral representation is trivial. In this form, one therefore has no relation between the singularities of the S-matrix and the spectrum of the generator of the semigroup.

Although the generalization of Lax-Phillips theory by Flesia and Piron [12] provides a new point of view for scattering theory, we see that

[1]The theorem states that this construction is minimal. We conjecture but have not proved that the Lax-Phillips construction is minimal. We thank G. Emch for a discussion of this point.

to extend the theory further to include a description of the evolution of an unstable system, it is necessary to generalize the law of evolution to that of a nontrivial integral operator over the time.

The most general linear evolution law has the form

$$(U(\tau)\psi)_{t+\tau} = \int_{-\infty}^{+\infty} W_{t,t'}(\tau)\psi_{t'}dt'. \tag{2.9}$$

This type of evolution, which goes beyond the formulation of Flesia and Piron [12] and Floquet theory [11], corresponds to unitary evolution (with some conditions on $W_{t,t'}$) in $\bar{\mathcal{H}}$ with a nontrivial S-matrix for which the singularities of its Fourier transform are associated with the spectrum of the generator of the Lax-Phillips semigroup. As we shall show below, the form of the evolution law (2.9) has a natural realization in Liouville space as well as in the framework of relativistic quantum theory.

Let us now note, for this general evolution, some properties of the S-matrix

$$(S\psi)_t = \int S_{t,t'}\psi_{t'}dt';$$

one can show [16] in this general case that the S-matrix has the form $S_{t,t'} = S(t-t')$. This is a very important property of the S-matrix, according to which, when one goes to the spectral representation $\hat{\psi}_\sigma = \int e^{-i\sigma t}\psi_t dt$, the S-matrix takes the simple form

$$\hat{S}_{\sigma,\sigma'} = \frac{1}{2\pi}\int e^{-i\sigma t}S_{t,t'}e^{i\sigma' t'}dtdt' = \delta(\sigma-\sigma')\hat{S}(\sigma),$$

where

$$\hat{S}(\sigma) = \int e^{-i\sigma t}S(t)dt, \tag{2.10}$$

i.e., in this basis the S-matrix is diagonal, and the S-operator is multiplication on the subspaces, labeled by σ, of $\{\mathcal{H}_\sigma\}$, the set of (isomorphic) Hilbert spaces which are the Fourier dual to the set $\{\mathcal{H}_t\}$.

Lax and Phillips [10] prove that the semigroup defined by (2.3) is strongly contractive. It is shown in ref. [13] that under the free motion, the generator of the semigroup is *dissipative* due to the defect subspaces associated with the non-self-adjointness of the operator dual to the time. This non-self-adjointness follows from the fact that the spectrum of the time is partitioned to two semibounded and one bounded segment [13]. The addition of a self-adjoint interaction does not affect this conclusion. We see from this [13] the *essential mechanism of Lax-Phillips theory.* The non-self-adjointness of $P_+K_0P_-$ corresponds to the restriction of $-i\partial_t$ to a finite interval, so that, in fact the operator has imaginary eigenvalues. In the presence of interaction, these eigenvalues emerge as the actual eigenvalues of the generator of the semigroup in $\mathcal{K}$, corresponding to the singularities of $S(\sigma)$.

There are three distinct types of time operator. The incoming time operator T^{in}, provides a spectral family in terms of which the incoming representation can be constructed, and in which functions in $\mathcal{D}_-$ have definite support and functions in $\bar{\mathcal{H}}$ evolve by translation. In this representation, the norm of the evolving states in $(-\infty, 0)$ must decrease. After sufficient laboratory time τ passes, the states evolve to $\mathcal{D}_+$, and in the outgoing representation, provided by the spectral family of the outgoing time operator T^{out}, they have definite support in (ρ, ∞). The mapping of functions in the incoming representation to the outgoing representation is provided by the Lax-Phillips S-matrix, and the time operators are related by

$$T^{out} = ST^{in}S^{\dagger} . \tag{2.11}$$

The third type of time operator corresponds to the "free" representation and is related to T^{in}, T^{out} by the Lax-Phillips wave operators. The spectral family for this operator provides the "standard" representation (analogous to Dirac's choice of "standard" spectral families), which we have used above.

There is an interval, in general, when the system is in interaction, and its state is neither in $\mathcal{D}_-$ nor $\mathcal{D}_+$. The expectation value of the operator T^{in} in the state ψ^τ projected into $\mathcal{K} \oplus \mathcal{D}_+$ (corresponding to the projection P_-) can be interpreted as the interaction interval. If the system in interaction is considered as an unstable particle (a resonance), this interval is its *age* after creation at $t = 0$. The expectation value of T^{in} then moves out of $(-\infty, 0)$. The expectation value of T^{in} in the state $P_-\psi^\tau$ is

$$< T^{in} >_\tau = \int t|\ {}_{in}\langle t|P_-\psi^\tau)|^2 dt\,; \tag{2.12}$$

here $|\ {}_{in}\langle t|P_-\psi^\tau)|^2$ is the probability density for the age t at time τ, an intrinsic dynamical property of the system. The positive value that the expectation value develops corresponds to the average age. One can similarly compute the expected time after decay, the expected lifetime, and the expected value of any other observable of interest as a property of the unstable system.

The structure of the theory is somewhat similar to the Wigner-Weisskopf idea, in that a subspace is associated with the decaying system. The decay of the system is associated with the probability flow out of the subspace. As in the original Wigner-Weisskopf formulation, the process of decay may be represented as a continuous evolution from the original unstable state to the final state through a changing linear superposition. In this framework, let us choose a vector ψ in the subspace $\mathcal{K}$ to represent the state of an unstable system. Then, under the full evolution,

$$\begin{aligned}(\psi, U(\tau)\psi) &= (\psi, P_{\mathcal{K}}U(\tau)P_{\mathcal{K}}\psi) \\ &= (\psi, \mathcal{Z}(\tau)\psi),\end{aligned} \tag{2.13}$$

so that *the reduced evolution is an exact semigroup.*

Moreover, in the Lax-Phillips theory the expectation value of an observable which is decomposable in the free or outgoing representations, where $\mathcal{D}_+$ has definite support properties, necessarily reduces to the sum of the expectation values in the subspaces $\mathcal{K}\oplus\mathcal{D}_-$ and in the subspace $\mathcal{D}_+$ (the decay products), i.e.,

$$\langle A\rangle = \int dt(\psi_t, A\psi_t) = \sum_{M=D_\pm,\mathcal{K}} \int dt\,(\psi_t^M, A\psi_t^M). \tag{2.14}$$

Note that there are no cross terms. *There is, therefore, an exact superselection rule for measurements of the system by means of such decomposable operators.*

3. APPLICATIONS

3.1. Measurement According to Namiki and Machida

Recently, Machida and Namiki [17] have proposed a measurement theory based on a direct integral space of continuously many Hilbert spaces and a continuous superselection rule. As pointed out by Tasaki *et al* [18], although they had some success, their theory has a conceptual difficulty. Indeed, in their theory, while the apparatus is described by many Hilbert spaces, the system corresponds to a single Hilbert space as in the conventional theory. Thus, one needs to specify the boundary between the system and the apparatus. As discussed by von Neumann, this is impossible.

Most measurement processes are concerned with measurements of observables which are time-independent in the Schrödinger picture. Therefore, if two different Lax-Phillips states give the same expectation value for all time-independent observables, these two states are essentially indistinguishable. In this sense, we define the following:

1. A Lax-Phillips vector $\psi \in \bar{\mathcal{H}}$ is called "effectively pure" if there exists a pure state

   $$\rho_0 = \phi_0\phi_0^*, \qquad \phi_0 \in \mathcal{H},$$

   such that

   $$\langle \hat{A}\rangle_\psi = \mathrm{Tr}\rho_0 A = (\phi_0, A\phi_0), \tag{3.1}$$

where $\hat{A}$ is the "lift" of A on $\mathcal{H}$ to $\bar{\mathcal{H}}$, for every element of the algebra of bounded linear operators associated with the spectral families of the time-independent observables[2] on the original space $\mathcal{H}$.

2. A Lax-Phillips vector is called "effectively mixed" if no such (pure) ρ_0 exists.

It can be shown [16][18] that $\psi = \{\psi_t\} \in \mathcal{H}$ is effectively pure if an only if it has the form

$$\psi_t = f(t)\phi_0. \tag{3.2}$$

We now discuss the possibility of decoherence, or the evolution from effectively pure to effectively mixed states. Consider the Schrödinger evolution for a time-dependent Hamiltonian. The solution of the time-dependent Schrödinger equation can always be written formally as $\psi_t = U(t,t')\psi_{t'}$, where $U(t,t')$ satisfies the chain property $U(t,t')U(t',t'') = U(t,t'')$, and can be expressed in terms of the integral of a time-ordered product. We define $W_t(\tau) = U(t+\tau,t)$, and lift the evolution to $\bar{\mathcal{H}}$ as follows

$$\psi^{\tau}_{t+\tau} = W_t(\tau)\psi_t, \tag{3.2}$$

where $W_t(\tau)$ is given by (T implies the time-ordered product)

$$W_t(\tau) = T\left(e^{-i\int_t^{t+\tau} H(t')dt'}\right). \tag{3.3}$$

For this kind of time-evolution we obtain

$$\langle\hat{A}\rangle_\psi = \int dt\ (W_t(\tau)\psi_t, AW_t(\tau)\psi_t)_{\mathcal{H}}, \tag{3.4}$$

where we have taken the normalization as unity. For the effectively pure states we have

$$\langle\hat{A}\rangle_\psi = \int dt\ |f(t)|^2(W_t(\tau)\phi_0, AW_t(\tau)\phi_0)_{\mathcal{H}}. \tag{3.5}$$

It follows from our previous argument that the effective state corresponding to (3.5) is mixed-like if $W_t(\tau)\phi_0 \neq W_{t'}(\tau)\phi_0$ (i.e., the state ρ_ψ induced from $\psi^{\tau}_{t+\tau} = W_t(\tau)\psi_t = f(t)W_t(\tau)\phi_0$ is not pure in $\mathcal{H}$). This result is true for the generalized evolution $W_{tt'}$ of (2.9) as well (in this case, also for a closed system [16]).

[2] We wish to emphasize that what is meant is *explicit* time-dependence in the Schrödinger picture; we do not refer here to the dynamical time-dependence that may arise in the Heisenberg picture if A is not a constant of the motion.

3.2. Intrinsic Decoherence in Classical and Quantum Liouville Evolution

It has long been emphasized by Prigogine and his co-workers [19] that the natural description for the evolution of a system with many degrees of freedom is that of the evolution of the density matrix ρ, through the Liouville equation,

$$i\frac{d\rho}{dt} = [H, \rho]. \tag{3.6}$$

The density matrix ρ ($\rho \geq 0, Tr\rho = 1$) has the property that $Tr\rho^2 \leq 1$, where the equality is attained only for a pure state. In general, one considers the space of Hilbert-Schmidt operators A for which

$$Tr\, A^* A < \infty ; \tag{3.7}$$

the positive (normalized) elements of such a space correspond to the physical states, the density matrices. On this space, the commutator with the Hamiltonian H defines a linear operator $\mathcal{L}$, called the Liouvillian, for which

$$i\frac{d\rho}{d\tau} = \mathcal{L}\rho, \tag{3.8}$$

where one assumes that $\mathcal{L}$ is self-adjoint in the Liouville space.

The Hamiltonian evolution of states in classical mechanics is known by the Liouville theorem to be non-mixing, i.e., to preserve the entropy of the system [20]. The same property holds for the quantum evolution as well, and follows from the unitarity of the evolution operator. This has been an obstacle to the consistent description of irreversible processes from first principles. The usual use of techniques of coarse graining or truncation to achieve a realization of the second law does not follow from basic dynamical laws, and is fundamentally not consistent with the underlying Hamiltonian structure [21]. We shall now show that the existence of a time operator in the Liouville space provides a natural and consistent mechanism for the decoherence of physical states, i.e., that pure states become mixed during the evolution, both for quantum and classical systems [22].

In particular, for a Hamiltonian of the form of the sum of an unperturbed operator H_0 and a perturbation V, i.e., $H = H_0 + V$, the corresponding Liouvillian is

$$\mathcal{L} = \mathcal{L}_0 + \mathcal{L}_I . \tag{3.9}$$

Now suppose we consider the "time operator" T_0, conjugate to $\mathcal{L}_0$ (with spectrum $(-\infty, \infty)$; it satisfies

$$[T_0, \mathcal{L}_0] = i .$$

Then, in the spectral representation of T_0,

$$_0\langle t|\mathcal{L}|t'\rangle_0 = -i\partial_t\delta(t-t') + {}_0\langle t|\mathcal{L}_I|t'\rangle_0 , \tag{3.10}$$

where the last term is, in general, not diagonal.

The method that we have described above applies as well to the formulation of classical mechanics on a Hilbert space defined on the manifold of phase space which was introduced by Koopman [23] and used extensively in statistical mechanics [21]. Misra [24] has shown that dynamical systems which admit a Lyapunov operator necessarily have absolutely continuous spectrum in $(-\infty, \infty)$; therefore one can construct a time operator on the classical Liouville space for such systems. The structure of the theory, and the conclusions we have reached, are therefore identical to those of the quantum case.

3.3. Relativistic Quantum Mechanics

The form of relativistic quantum mechanics introduced by Stueckelberg [25], extended to the many-body case by Horwitz and Piron [26], covariantly describes the evolution of a system according to the Stueckelberg-Schrödinger equation

$$i\frac{\partial \psi_\tau}{\partial \tau} = \frac{p_\mu p^\mu}{2M}\psi_\tau \equiv K_0\psi_\tau, \tag{3.11}$$

where M is an intrinsic property of the particle ("on-shell" mass).

Since the d'Alembertian, corresponding to the operator K_0, has spectrum $(-\infty, \infty)$, there exists an operator ξ which satisfies [27]

$$[K_0, \xi] = i. \tag{3.12}$$

Note that the operator t of the relativistic theory will not serve this purpose, since its commutator with K_0 is iE/M, which only approaches i in the non-relativistic limit.

If ξ is a function of $\mathbf{x}, t$, we may construct the transformation function $\langle \xi', \beta | x\rangle$ using the defining commutation relation; a short calculation results in

$$i\frac{\partial}{\partial \xi'}\langle \xi', \beta | x\rangle = -\frac{\partial^\mu \partial_\mu}{2M}\langle \xi', \beta | x\rangle. \tag{3.13}$$

We thus see that the relativistic quantum theory provides a natural framework for the Lax-Phillips formulation of the description of an unstable system. It is interesting that the continuous spectrum of K_0 is essential to the construction; this implies that we must have both positive and negative mass-squared states in the spectrum, i.e., that the so-called tachyons, at least in the form of intermediate states, play a fundamental role in the relativistic description of unstable systems.

The general structure of the relativistic Lax-Phillips theory remains in the Galilean limit (for finite but large c) [28], providing an additional term, a pseudifferential operator in time, in the residual Hamiltonian. The experimental signature of such an additional term in the Hamiltonian would be, for example, an interference effect in time. The experiment would be of the same design as the test of such an effect in the full covariant relativistic theory [29].

Acknowledgements. One of us (LPH) wishes to thank C. Piron for many discussions, and he is grateful for his hospitality in Geneva during several visits. We wish to thank our colleagues in Brussels, I. Antoniou, B. Misra, I. Prigogine, and S. Tasaki, for helpful discussions and comments which contributed significantly to many of the results that we have presented.

REFERENCES

1. G. Gamow, *Z. Phys.* **51**, 204 (1928).
2. T.D. Lee, R. Oehme, and C.N. Yang, *Phys. Rev.* **106**, 340 (1957); T.T. Wu and C.N. Yang, *Phys. Rev. Lett.* **13**, 380 (1964).
3. V.F. Weisskopf and E.P.Wigner, *Z. Phys.* **63**, 54 (1930); **65**, 18 (1930).
4. L.P. Horwitz and J.P. Marchand, *Helv. Phys. Acta* **42**, 1039 (1969); L.P. Horwitz and J.P. Marchand, *Rocky Mtn. J. Math.* **1**, 225 (1973).
5. K.O. Friedrichs, *Comm. Pure Appl. Math.* **1**, 361 (1948); T.D. Lee, *Phys. Rev.* **95**, 1329 (1956).
6. N. Bleistein, H. Neumann, R, Handelsman, and L.P. Horwitz, *Nuovo Cimento* **41A**, 389 (1977).
7. C. Piron, *Foundations of Quantum Physics* (Benjamin/Cummings, Reading, 1976).
8. L.P. Horwitz, J.P. Marchand, and J. LaVita, *J. Math. Phys.* **12**, 2537 (1971); D. Williams, *Comm. Math. Phys.* **21**, 314 (1971).
9. L.P. Horwitz and I.M. Sigal, *Helv. Phys. Acta* **51**, 685 (1980); W. Baumgartel, *Math. Nachr.* **75**, 133 (1978). See also, G. Parravicini, V. Gorini, and E.C.G. Sudarshan, *J. Math. Phys.* **21**, 2208 (1980); A. Bohm, *The Rigged Hilbert Space and Quantum Mechanics* (Springer Lecture Notes on Physics **78**, Berlin, 1978); A. Bohm, *Quantum Mechanics: Foundations and Applications* (Springer, Berlin, 1986); A. Bohm, M. Gadella, and G.B. Mainland, *Am. J. Phys.* **57**, 1103 (1989); T. Bailey and W.C. Schieve, *Nuovo Cimento* **47A**, 231 (1978).
10. P.D. Lax and R.S. Phillips, *Scattering Theory* (Academic, New York, 1967).

11. See, for example, L.E. Reichl, *The Transition to Chaos in Conservative Classical Systems: Quantum Manifestation* (Springer, New York, 1992), and references therein. I.M. Sigal, personal communication.
12. C. Flesia and C. Piron, *Helv. Phys. Acta* **57**, 697 (1984).
13. L.P. Horwitz and C. Piron, *Helv. Phys. Acta* **66**, 694 (1993).
14. I.P. Cornfeld, S.V. Fomin, and Ya.G Sinai, *Ergodic Theory* (Springer, Berlin, 1982).
15. C. Foias and B.Sz. Nagy, *Acta Sci. Math.* **23**, 106 (1962). F. Riesz and B.Sz. Nagy, *Functional Analysis*, Appendix by B. Sz. Nagy, 2nd edn. (English translation: Dover, New York, 1990).
16. E. Eisenberg and L.P. Horwitz, in *Advances in Chemical Physics*, to be published (1996).
17. S. Machida and M. Namiki, *Prog. Theor. Phys.* **63**, 1457 (1980); **63**, 1833 (1980). M. Namiki and S. Pascazio, *Phys. Rev.* **A44**, 39 (1991).
18. S. Tasaki, E. Eisenberg, and L.P. Horwitz, *Found. Phys.* **24**, 1179 (1994).
19. C. George, *Physica* **65**, 277 (1973). I. Prigogine, C. George, F. Henin, and L. Rosenfeld, *Chemica Scripta* **4**, 5 (1973). T. Petrosky, I. Prigogine, and S. Tasaki, *Physica* **A173**, 175 (1991); T. Petrosky and I. Prigogine, Physica **A175**, 146 (1991), and references therein.
20. See, for example, J. Yvon, *Correlations and Entropy in Classical Statistical Mechanics* (Pergamon, Oxford, 1969).
21. I. Prigogine, *Non-equilibrium Statistical Mechanics* (Wiley, New York, 1961).
22. E. Eisenberg and L.P. Horwitz, *Phys. Rev. A* **52**, 70 (1995).
23. B. Koopman, *Proc. Nat. Acad. Sci.* **17**, 315 (1931).
24. B. Misra, *Proc. Nat. Aca. Sci.* **75**, 1627 (1978); B. Misra, I. Prigogine, and M. Courbage, *Proc. Nat. Acad. Sci.* **76**, 4768 (1979).
25. E.C.G. Stueckelberg, *Helv. Phys. Acta* **14**, 372, 588 (1941); **15**, 23 (1942).
26. L.P. Horwitz and C. Piron, *Helv. Phys. Acta* **46**, 316 (1973).
27. Y. Strauss and L.P. Horwitz, in preparation.
28. L.P. Horwitz, W.C. Schieve, and C. Piron, *Ann. Phys. (New York)* **137**, 306 (1981).
29. L.P. Horwitz and Y. Rabin, *Lett. Nuovo Cimento* **17**, 501 (1976).

DICHOTOMIC FUNCTIONS AND BELL'S THEOREMS

A. F. Kracklauer

Belvederer Allee 23c
99425 Weimar, Germany
kracklau@fossi.uni-weimar.de

It is shown that correlations of dichotomic functions can not conform to results from quantum mechanics. Also, it is seen that the assumptions attendant to optical tests of Bell's inequalities actually are consistent with classical physics so that in conclusion, Bell's theorems do not preclude hidden variable interpretations of quantum mechanics.

Key words: Bell's theorems, Kochen-Specker theorem, quantum mechanics.

The analysis attending Bell's theorems imposes on hidden variable formulations of quantum mechanics (QM) the requirement that they be able to correlate dichotomic functions (representing spin) so as to duplicate the result from QM:

$$P_{\mathbf{a},\mathbf{b}} = \int \rho(\lambda)\, A(\mathbf{a},\lambda) B(\mathbf{b},\lambda) d\lambda, = -\cos(\,\mathbf{a},\mathbf{b}), \tag{1}$$

where content and notation are taken from Bell [1].

Note, however, that *inter alia*, only the right side of Eq. (1) is a harmonic function. This means that the equation itself is an absurdity. To see this, consider the derivative of the QM expectation expressed as a correlation of dichotomic functions, $D_{a,b}$ which change sign at the points x_j (δ is the Dirac delta function):

$$-\frac{\partial cos(\tau)}{\partial \tau} \propto \frac{\partial}{\partial \tau}\frac{1}{T}\int_0^T D_a(x-\tau)D_b(x)dx,$$

$$sin(\tau) \propto \frac{1}{T}\int_0^T \sum_n^N \pm\delta(x_n)D_b(x)dx,$$

M. Ferrero and A. van der Merwe (eds.),
New Developments on Fundamental Problems in Quantum Physics, 185–188.

$$\propto \frac{1}{T}\sum_{n}^{N}(\pm)D_b(x_n) = \frac{integer}{T}. \tag{2}$$

(The sign of individual terms in the above series depends on the specific form of $D_{a,b}$, which, for present purposes, is not required.) A further differentiation by τ yields $cos(\tau) \equiv 0$, a false equation.

Nothing was in play here but the dichotomic nature of the functions D; it is clear that such functions simply can not be correlated to yield harmonic functions.[2] Obviously, if dichotomic functions cannot yield the QM result, then QM is doing something other than correlating them. If this be the case, then a hidden variable alternative to QM need not be required to so correlate them and the logic of Bell's theorem is broken.

Likewise, the famous result from Greenburger et al.[3] showing that the correlation of three or more dichotomic functions representing three or more particles is never; i. e., even for perfect correlations, compatible with the harmonic functions given by QM, is attributable to the same structural cause. This is most easily seen by expressing multiple correlations (for three particles, say) in terms of dichotomic sequences $A, B, C...$ (i. e., data points) as follows

$$-\cos(\tau+\theta) = \frac{1}{NM}\sum_{j,k}^{N,M} A_{j,k}B_{j,k}(\tau)C_{j,k}(\theta), \tag{3}$$

where (ignoring for the moment obvious analytic absurdities) it is clear that even for $\tau+\theta = 0$ or π; i. e., for perfect (anti) correlations, the equation can not hold because odd (even) multiples of -1 do not (do) cancel. This exposes an internal contradiction in no need of empirical verification. (Degenerate cases occurring for even multiples of dichotomic functions do not invalidate the principle. Moreover, for those with strong intuition, it is clear that "n-chotomic" functions of any rank will result in analytic problems.)

The question arises: if QM is not doing the mathematically impossible, what is it doing? For the case of polarized "photons," used virtually exclusively for testing Bell's inequalities, QM is in fact giving a classical result. The analysis of optical analogues of the EPR event artificially associates an unpolarized with an anticorrelated state by *defining* [4] the "coefficient of correlation" to be

$$E(\mathbf{a},\mathbf{b}) \equiv P_{(++)} + P_{(--)} - P_{(+-)} - P_{(-+)}, \tag{4}$$

where the $P's$ are the "quantum mechanical predictions" for detection coincidences

$$P_{(++)} = P_{(--)} = \frac{1}{2}\cos^2(\mathbf{a},\mathbf{b}),$$
$$P_{(+-)} = P_{(-+)} = \frac{1}{2}\sin^2(\mathbf{a},\mathbf{b}), \tag{5}$$

so that Eq. (4) becomes

$$E_{\text{Optical}} = \cos 2(\mathbf{a}, \mathbf{b}). \tag{6}$$

Note, however, that the P's just give the intensity of polarized light as measured with respect to an axis different from the axis of polarization according to Malus' law—a non quantum rule. Of course, the P's must be suitably interpreted to correspond ultimately to "click" probabilities in experiments performed at minimal intensity where the statistics within each polarization state may require more than traditional physics.

This is in total accord with the fact that creation/annihilation operators for photons of different states of polarization commute. As is well known, QM differs from classical physics where and only where conjugate operators do not commute. (This observation conforms with work reported elsewhere in these proceedings[5] wherein supposed QM polarization (or spin) correlations were obtained from a model involving rubber bands or "random gun" and no Planck's constant!) Thus, tests to plumb the ontological implications of QM; e. g., nonlocality, can not be made on the basis of optical experiments involving polarization states. As the creation/annihilation operators for photons of different colors also commute, these arguments extend directly to experiments exploiting parametric down conversion and so on.

QM multiparticle correlations, such as Eq. (3) or its n-particle generalizations, as noted elsewhere[6], make contact with the Bell-Kochen-Specker Theorem. Although analysis of such equations, restricted as it is to considering perfect (anti-) correlations, is less general than the full Theorem, it is more transparent because it is just tracking factors of -1. In addition, it is here very clear that dependence on variables in addition to τ and θ; e. g., by investing A, B, C with nonlocality, can not circumvent inconsistency which results only from their dichotomic values. Again, however, Eq. (3) can be rendered rigorously correct and physically sensible in any n-particle generalization if dichotomic sequences A, B, C etc. are replaced by cosines and sums converted to integrals; i. e., by considering classical polarization correlations.

In conclusion, it is seen that a demand that hidden variable theories should correlate dichotomic functions to get a harmonic result is ill founded. When this stipulation is relaxed, then the extraction of Bell's inequalities does not go through and therefore hidden variable theories duplicating QM polarization correlations; e. g., [7], are not precluded in principle. Moreover, because the mathematics describing spin correlations; i. e., their interstate structure, is isomorphic to that for classical polarization, the underlying phenomena also must be classical in essence.

In other words, it is clear that Bell's theorems establish beyond dispute that self consistent, intuitively clear hidden variable theories can not duplicate all of QM. Indeed such alternatives will be (and should be) unable to replicate those very aspects of QM that have been the source of confusion and contest from the beginning, namely the measurement projection hypothesis and an under restrictive identification of vectors in relevant Hilbert spaces as physically realizable states; e. g., pure and "cat" states and especially "photons" [8]. These two features of orthodox QM lead not only to the logical contradictions exposed by Bell's theorems but appear essentially untestable in all but certain "n-chotomic" circumstances, where a classical explanation fits anyway. These observations are not here unique. Many roads lead away from "Copenhagen," another such, for example, ends in LA [9].

REFERENCES

1. J. S. Bell, *Speakable and Unspeakable in Quantum Mechanics* (Cambridge University Press, Cambridge, 1987).
2. P. Claverie and S. Diner, *Israeli J. Chem.* **19**, 54 (1980), make the same observation in a different context. (I thank professors A. M. Cetto and L. de la Peña for this reference.)
3. D. M. Greenburger, M. A. Horne, A. Shimony, and A. Zeilinger, *Am. J. Phys.* **58**(12), 1131 (1990).
4. A. Aspect and P. Grangier, *Proceedings, International Symposium on the Foundations of Quantum Mechanics* (Tokyo, 1983), pp.214-224.
5. B. Coecke, B. D'Hooghe, and F. Valckenborgh, these preceedings.
6. M. D. Mermin, *Rev. Mod. Phys.* **65**(2), 803 (1993).
7. A. O. Barut in *Waves and Particles in Light and Matter*, A. van der Merwe and A. Garuccio, eds. (Plenum, New York, 1994), pp. 9-18.
8. T. W. Marshall, these proceedings.
9. E. J. Post, *Quantum Reprograming* (Kluwer Academic, Dordrecht, 1995).

PRODUCTION AND USES OF HYPER-ENTANGLED STATES

Paul G. Kwiat

Physics Division, P-23, MS-H803
Los Alamos National Laboratory
Los Alamos, New Mexico 87545

Entangled states are central to the new field of quantum information, including quantum dense coding, teleportation, and computation. However, only a relatively small class of entangled states has been discussed extensively, much less investigated experimentally. In particular, efforts to date have focussed on two particles entangled in a single degree of freedom, for example polarization, or energy, or momentum direction. Novel phase-matching arrangements in spontaneous parametric down-conversion allow the preparation of pairs of photons that are simultaneously entangled in polarization, momentum-direction, and energy. We shall call such a multiply-entangled state "hyper-entangled". In addition, an even more general state – a non-maximally entangled state – should be realizable, in which the amplitudes of the contributing terms are not equal.

Key words: entangled states, quantum states, Bell inequalities, nonlocality, parametric down-conversion.

1. INTRODUCTION

Entangled states are arguably the most "quantum-esque" aspect of quantum mechanics. In fact, Schrödinger described entanglement as "*the* characteristic trait of quantum mechanics, the one that enforces its entire departure from classical lines of thought." [1] Entangled states are inextricably linked to the measurement problem, and are central to the demonstration of nonlocality, e.g., in tests of Bell's inequalities. They form the basis, figuratively and literally, of experiments in quantum information, such as quantum dense coding [2] and quantum teleportation [3]. And in terms of applications, they

M. Ferrero and A. van der Merwe (eds.),
New Developments on Fundamental Problems in Quantum Physics, 189–202.

have been proposed for use in quantum cryptography [4], and are crucial in all implementations of quantum computers [5]. Nevertheless, even at the two-particle level, only a small class of entangled states has been discussed extensively, much less investigated experimentally. In particular, efforts to date have focussed on two particles entangled in a single degree of freedom, for example polarization [6], or energy [7], or momentum direction [8].[1]

A novel type-II phase-matching arrangement in spontaneous parametric down-conversion has recently permitted the demonstration of the first of these with very encouraging results [6]. All four of the Bell states were produced, and strong violations of Bell's inequalities were observed in all cases. The desired polarization-entangled states are produced directly out of the nonlinear crystal, and the source was seen to have unprecedented brightness and stability. Actually, due to the very nature of the down-conversion process, the photons are automatically also produced into an energy-entangled state. Moreover, using a modification of the phase-matching arrangement, one should be able to prepare momentum direction-entangled photon pairs. The final result will be pairs of photons that are simultaneously entangled in polarization, momentum-direction, and energy. We shall call such a multiply-entangled state "hyper-entangled". Finally, it should be possible to prepare photon pairs in a non-maximally entangled state, in which the two contributing terms do not have equal amplitudes. Such a state would be very important for enabling a loophole-free test of Bell's inequalities [11,12].

2. ENTANGLEMENT VIA PARAMETRIC DOWN-CONVERSION

An entangled state is a nonfactorizable sum of product states of two (or more) quantum systems [13]. More particularly, it is a state which cannot be factorized in *any* basis. The general form of such a state for two particles is:

$$\Psi_{1,2} = \sum_i c_i |\alpha_i\rangle \otimes |\beta_i\rangle, \tag{1}$$

where α and β are the basis vectors of particles 1 and 2, respectively. As we will see below, the sum sometimes is extended into an integral, when the relevant Hilbert space is continuous. The most familiar example of an entangled state is the Bohm singlet state, $\Psi_{1,2} = (|\uparrow_1\rangle|\downarrow_2\rangle - |\downarrow_1\rangle|\uparrow_2\rangle)/\sqrt{2}$, which represents the state of two spin-1/2 particles decaying from a spin-zero parent. The two particles are correlated — their spins are always anti-parallel — and they remain that way no matter the separation between them. In this sense entangled systems can demonstrate *nonlocal* quantum effects.

Albeit identical in principle, it is much easier in practice to work with

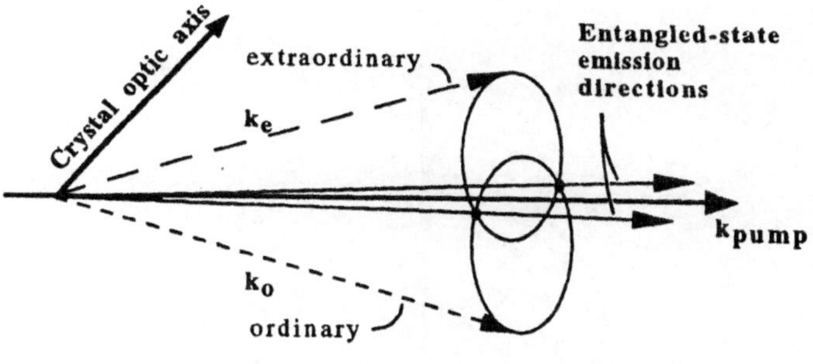

Fig. 1. Spontaneous down-conversion cones from type-II phase-matching.

photons that are correlated, because they are (now) easier to produce and their intrinsic correlation is not readily destroyed by decoherence through interaction with the environment. While initial experiments used photon pairs produced in an atomic cascade, now the source of choice is spontaneous parametric down-conversion [14]. In this process, an ultraviolet "pump" photon (typically produced in a laser) spontaneously decays inside a crystal with a $\chi^{(2)}$ nonlinearity into two highly correlated red photons of nearly equal energy. Energy and momentum are conserved in this process: $\hbar\omega_p = \hbar\omega_s + \hbar\omega_i$, $\hbar\mathbf{k}_p \approx \hbar\mathbf{k}_s + \hbar\mathbf{k}_i$, where $\hbar\omega_i$ and $\hbar\mathbf{k}_i$, $(i = p, s, i)$ are the respective energies and momenta of the parent [p] and two daughter photons (conventionally called the "signal" [s] and the "idler" [i]).

The conservation of momentum inside the crystal, which is also known as "phase-matching," is enabled by using the birefringent properties of the crystal itself to compensate for the dispersion of the material. The result is that the down-conversion photons are produced in a rainbow of colored cones, with conjugate photons (i.e., the members of a given pair) lying on opposite sides of the pump beam (see Fig. 1). For the case of "type-II" phase-matching, one of each pair is ordinary-polarized and its conjugate photon is extraordinary-polarized. For a negative uniaxial crystal such as BBO, the axis of the extraordinary cones is between the pump beam direction and the crystal optic axis, while the axis of the ordinary cones is further away from the optic axis than the pump beam (in both cases, the pump beam, crystal optic axis, and cone axis lie in a plane).

In Fig. 2 we see that there is a rich structure to the emissions from a down-conversion crystal. Various types of entangled states are produced for photons emitted in particular directions. For example, a pair of photons emitted along the directions 3 and 3′ are automatically in a polarization-entangled state. Table 1 lists the various entanglements present for the

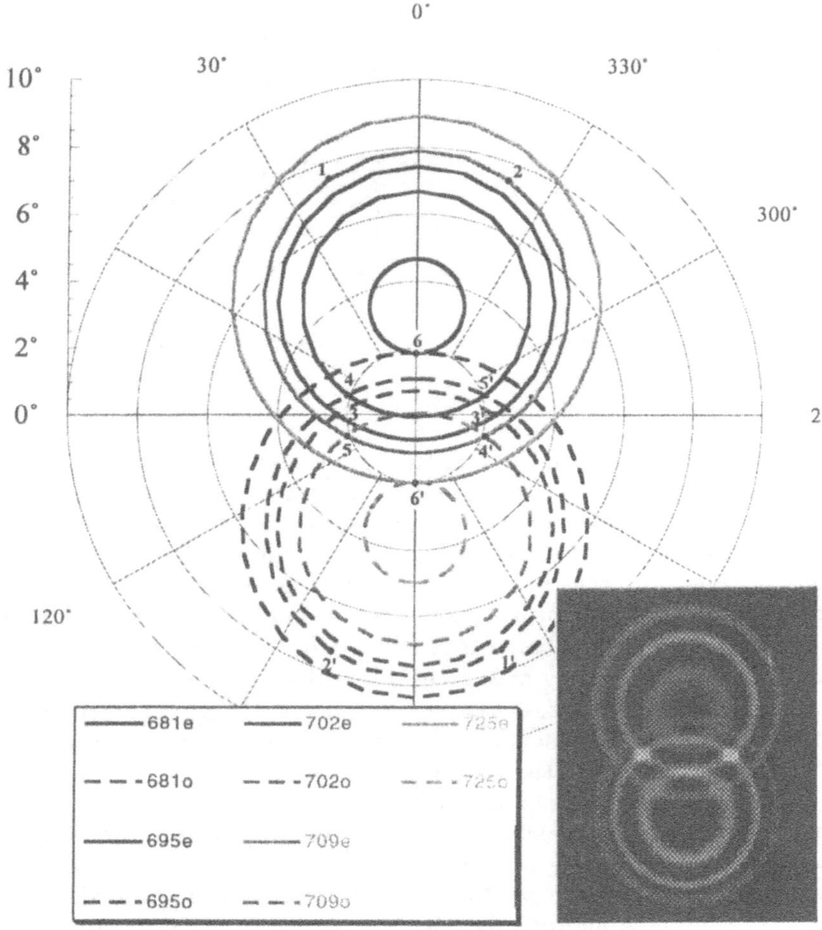

Fig. 2. Calculated type-II phase-matching curves in BBO (for a crystal cut at $\phi = 49.2°$, with a pump wavelength of 351 nm), showing various cones of down-conversion light (the pump beam would be coming directly out of the paper at the origin). The wavelengths of the five extraordinary-polarized cones (upper half, solid lines) are, from innermost to outermost: 681nm, 695 nm, 702 nm, 709 nm, and 725 nm. The ordinary-polarized conjugate cones to these (lower half, dashed lines) have their wavelengths reversed (e.g., smallest bottom cone is 725 nm). The numbered points represent sampling areas to extract various entangled states, as listed in Table 1. Inset: Photograph (courtesy of Michael Reck, Innsbruck) of three sets of conjugate cones emerging from a BBO crystal.

Table 1. The various entanglements available at the numbered points indicated in Fig. 2.

Conjugate points from Fig. 1	Energy-Time entangled	Momentum-entangled	Polarization-entangled	Non-maximally entangled
1 - 1'	✓			
1 - 1' -- 2 - 2'	✓	✓		
3 - 3'	✓		✓	
4 - 4' -- 5 - 5'	✓	✓	✓	
6 - 6'	✓		✓	✓

emission directions indicated in Fig. 2.

3. SINGLY ENTANGLED STATES

3.1. Time-Energy Entangled States

The energy-entangled states from down-conversion photons are in some sense the most universal, because they are present for *any* pair of photons. Due to the fact that there are many ways to partition the energy of the parent photon, each daughter photon has a broad spectrum, and hence a narrow wave packet in time. However, the *sum* of the two daughter photons' energies is extremely well-defined, since they must add up to the energy of the extremely monochromatic parent laser photon.[2] This correlation is represented by the following energy-entangled state:

$$|\Psi\rangle = \int_0^{E_p} dE A(E)|E\rangle_s|E_p - E\rangle_i, \tag{2}$$

where each ket describes the energy of one of the photons, s and i stand for "signal" and "idler", respectively, and $A(E)$ is essentially the spectral distribution of the collected down-conversion light.

That the photons can display nonlocal correlations arising from this entanglement was demonstrated by a number of groups [7], who implemented the experiment first proposed by Franson [15]. Each of the photons is directed into its own unbalanced Mach-Zehnder interferometer (see Fig. 3), giving it a long path (L) and a short path (S) to the detectors. Because the path length difference is much longer than the coherence length of the photons, no interference is observed in the singles rates at any of the detectors when the phase in, say, one of the long paths is changed. However, there *is* interference in the coincidence rate between detectors. The reason is that there are two processes that could lead to such a coincidence

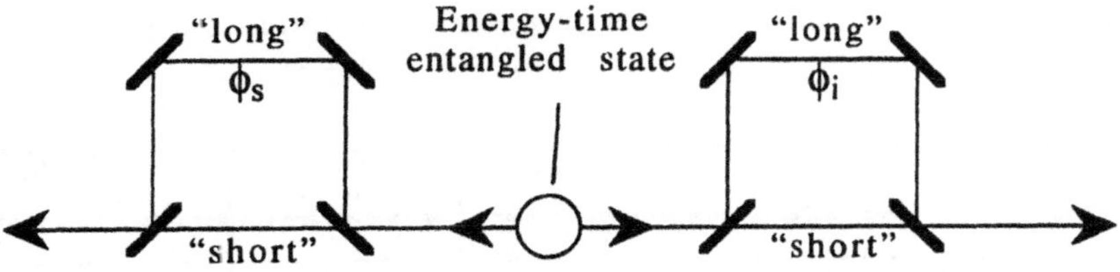

Fig. 3. Basic setup for the Franson experiment

count – both photons could have taken their respective long paths or both could have taken their respective short paths – these processes are indistinguishable, and so interfere. The relevant state is the first two terms of the following:

$$|\psi\rangle = \frac{1}{2}\left(|S_s, S_i\rangle - e^{i(\phi_s+\phi_i)}|L_s, L_i\rangle + e^{i\phi_i}|S_s, L_i\rangle + e^{i\phi_s}|L_s, S_i\rangle\right). \quad (3)$$

[Note, however, that it is first necessary to use a large path imbalance and very fast detectors to be able discard the second two terms, arising from the distinguishable (and hence *non*-interfering) events in which one photon takes its short path and the other takes the long path. We will see below how use of a doubly-entangled state can remove this requirement.] The rate of coincidences is thus dependent only on the *sum* of phases in the two interferometers. This nonlocal effect allowed experimenters to test a suitable version of Bell's inequalities [7].

3.2. Momentum Entangled States

The next most "common" entangled state from the down-conversion process is one in which the *momentum directions* of the photons are entangled (though of course the photons will still automatically possess the above-mentioned energy entanglement[2]). For simplicity we consider photons that have nearly the same energy (though this is not necessary), but which are emitted into two sets of directions (c.f. points 1-1′ – 2-2′ from Fig. 2). Concerning ourselves now only with the spatial modes, the emitted state can then be written as

$$|\psi\rangle = \left(|1_s, 1'_i\rangle + e^{i\phi}|2_s, 2'_i\rangle\right)/\sqrt{2}\ , \quad (4)$$

where the 1, 1′, 2, and 2′ indicate the directions of the photons. Such a state was employed by Rarity and Tapster to demonstrate a violation of

Bell's inequalities based on momentum entanglement [8], also depending nonlocally on separated phase elements.

3.3. Polarization Entangled States

Perhaps the simplest examples of entangled states of two photons are the polarization-entangled "Bell states" [16]:

$$
\begin{aligned}
|\psi^{\pm}\rangle &= \left(|H_s, V_i\rangle \pm |V_s, H_i\rangle\right)/\sqrt{2}, \\
|\phi^{\pm}\rangle &= \left(|H_s, H_i\rangle \pm |V_s, V_i\rangle\right)/\sqrt{2},
\end{aligned}
\tag{5}
$$

where H and V denote horizontal and vertical polarization, respectively. It is with these states that the advantage of type-II phase-matching over the more familiar type-I (in which the down-conversion photons both have the same [ordinary] polarization, and there is only a single set of cones, centered on the pump beam) becomes clear: Such states cannot be produced simply using the latter. To see how they arise in type-II phase-matching, consider points 3 and 3′ from Fig. 2. These points both lie on the degenerate set of cones (i.e., exactly half the pump frequency goes to each daughter photon). Moreover, at these points – the intersections of the two cones – a given photon belongs to *both* the extraordinary and the ordinary cones. Hence it does not have a definite polarization. Nevertheless, due to the nature of the emission process, if one of the photons is measured to have ordinary polarization (call it H), then the other is certain to have extraordinary polarization (call it V). In this way, merely by selecting out the correct directions from the down-conversion crystal we prepare the state $|\psi^+\rangle$.

By using only standard optical elements in one of the two output beams, it is possible to transform any one of the Bell states into any of the others. For example, using a polarization rotator to exchange H and V for one photon immediately changes $|\psi^+\rangle \leftrightarrow |\phi^+\rangle$ and $|\psi^-\rangle \leftrightarrow |\phi^-\rangle$. A birefringent phase-shifter in one of the beams similarly transforms $|\psi^+\rangle \leftrightarrow |\psi^-\rangle$ and $|\phi^+\rangle \leftrightarrow |\phi^-\rangle$. We were therefore able to test Bell's inequalities with all four states [6], and observe large violations in all cases (see Fig. 4), modulo the usual auxilliary assumptions to account for low detection efficiency and no rapid analyzer setting. In fact, due to an unexpected robustness in the source to larger collection irises, we were able to obtain the necessary statistics for a 102 σ violation in less than 5 minutes.

The four Bell states were also employed in a recent experiment [17] to demonstrate the possibility of "quantum dense coding", in which up to 2 bits of information may be encoded in the polarization state of a single photon [2]. A transmitting party and a receiving party each receive one member of a correlated pair of photons, initially in the state $|\psi^+\rangle$. By making one of four operations on his particle alone, the transmitter can change the entire

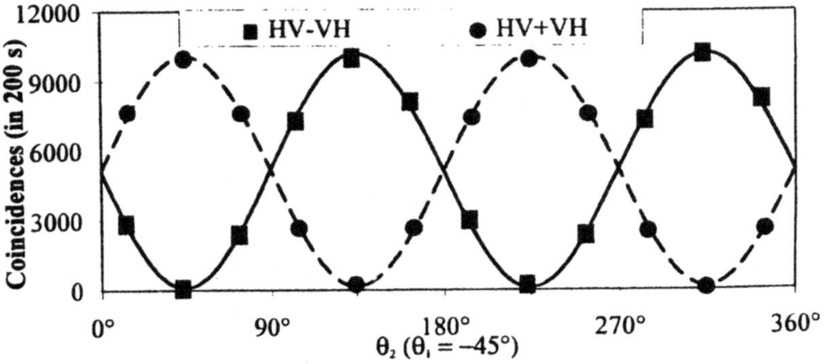

Fig. 4. High-visibility polarization correlations allowing one to observe strong violations of Bell's inequality.

joint two-particle system into one of the four Bell states. He then sends his particle to the receiver, who must make a joint measurement on the two particles to determine which of the four Bell states (which of four messages) was prepared by the transmitter. In fact, this Bell-state measurement is non-trivial, and to date it has only been possible to distinguish three of the four possibilities [17]. Below we will see how the use of a multiply-entangled state can remedy this.

4. MULTIPLY-ENTANGLED STATES

4.1. Improved Franson Experiment

As mentioned earlier, the energy-time correlations are present for all pairs of down-conversion photons, so that, in particular, they are present when the photons are also in a polarization-entangled state. The extra degrees of freedom allow one to improve some of the previous experiments. For example, one can perform a Franson-type experiment in which there is no need at all to discard any counts. Let us start out with the photons jointly in an energy-entangled state and in the polarization-entangled state $|\phi^+\rangle$, so that the photons have the same polarization. Consider now the original Franson setup (Fig. 3), but with *polarizing* beam splitters, which transmit horizontal and reflect vertical polarized light. Under this case it is clear that either the photons will both follow the short paths (term 1 of $|\phi^+\rangle$) or they will both follow the long paths (term 2 of $|\phi^+\rangle$) of their respective interferometers. In this case, *there is no non-interfering background* of "long-short" processes, for they simply do not occur. At the end we use a polarizer at 45° to "erase" the polarization information.[3] This system has a tremendous

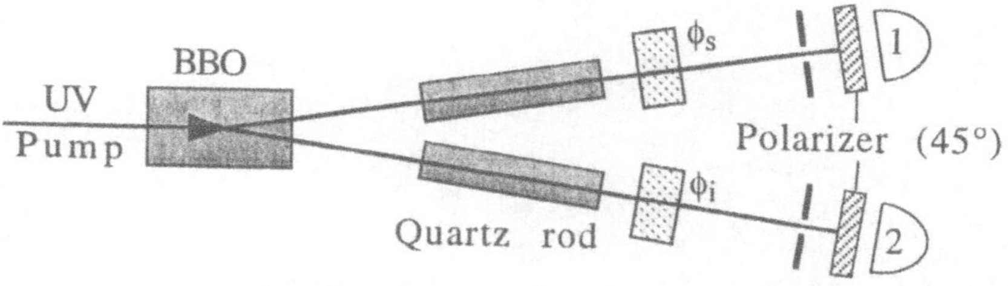

Fig. 5. Setup demonstrating the use of multiply-entangled photons (in energy-time and in polarization) to dramatically improve the performance of the Franson-type experiment.

advantage over the previous implementations in that it can be much smaller and therefore more robust.

A recent implementation [18] of this idea went one step farther, and used a birefringent quartz crystal as the entire interferometer (see Fig. 5). The analog of the short and long paths become the fast and slow modes in the quartz — the quartz is aligned with the fast axis along horizontal and the slow axis along vertical. The high-visibility correlations (again observed only in the coincident rate) implied a 17 σ violation of Bell's inequality. Moreover, the system was remarkably stable, and thus might find use in a realizable quantum cryptography setup.

4.2. Improved Bell-State Analysis

As a second example of the application of multiply-entangled photons, we consider the problem of Bell-state analysis. We wish to make a joint measurement on two photons and determine which of the four polarization Bell states (5) they are in. As alluded to in Sec. 3.3, at present there is no way to distinguish more than three of the Bell states (unless one has a nonlinear interaction that is significant at the single-photon level). This problem can be solved by embedding the polarization entangled states into a larger Hilbert space, in whose degrees of freedom the photons are also entangled. Below we briefly describe one example based on the extra time correlations (connected with the energy entanglement) of the down-converted photons. A similar scheme based instead on an additional momentum entanglement is also possible.

Consider Fig. 6, in which the two photons are directed to opposite sides of a 50-50 beam splitter. It can be shown that only for the state ψ^- will the photons exit the beam splitter in different directions – for the other three states the photons will travel off together [17] On this basis alone we can distinguish the first of the four states. Next, we pass the photons

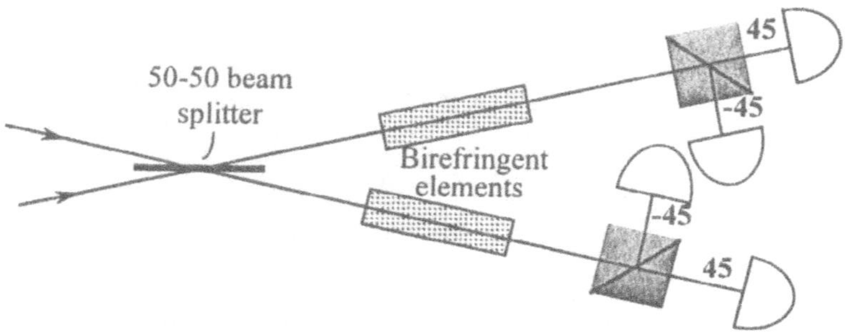

Fig. 6. Improved Bell-state analysis using doubly-entangled photons.

through very birefringent elements, so that the horizontal components are slowed relative to the vertical. In this case, the state ψ^+ (for which the photons exit with a relative delay) becomes distinguishable by timing from the states $\phi^\pm$ (for which the photons exit with no relative delay). Finally, by analyzing the pair with a polarizing beam splitter at 45°, we can distinguish the final two: for ϕ^+ both photons will go to the same detector, while for ϕ^- each photon will go to a different detector [19]. Thus, by taking advantage of the extra time-energy correlations of the down-conversion photons, one is able to solve an otherwise difficult problem.

4.3. Hyper-Entangled States

Finally, we consider the quantum state associated with photons emitted along the directions 4-4′ – 5-5′. They are apparently in a momentum-entangled state of the sort described above. They are also polarization-entangled, because each photon belongs to both an extraordinary cone and an ordinary cone. And of course they are automatically energy-entangled due to the down-conversion process. Therefore, the quantum state of these pair of photons is triply- or "hyper-entangled":

$$|\Psi\rangle = \left(\int_0^{E_p} dE A(E) |E\rangle_s |E_p - E\rangle_i \right) \otimes \frac{(|1_s, 1_i'\rangle + e^{i\phi}|2_s, 2_i'\rangle)}{\sqrt{2}} \otimes \frac{(|H_s, V_i\rangle + |V_s, H_i\rangle)}{\sqrt{2}} \tag{6}$$

All the advantages of this sort of state, and the benefits it engenders are not yet known to us, but it is our hope that, just as the Franson experiment and quantum dense-coding are improvable using doubly-entangled states, other experiments in quantum information may be simplified or enabled using such hyper-entanglements. In particular, there may be a way to allow purely optical quantum gates, requiring only linear optical elements (aside from the down-conversion crystal itself).

5. NON-MAXIMALLY ENTANGLED STATES

Until now we have concerned ourselves implicitly with entangled states in which the magnitudes of the contributing terms were equal. There is, however, a more general state in which this is not the case; such states are known as "non-maximally" or "partially"entangled, and have the form: $|\psi\rangle \propto |H_1, V_2\rangle + \epsilon|V_1, V_2\rangle$. In the limit of $\epsilon \rightarrow 1$, we recover a standard entangled state, while $\epsilon = 0$ gives a product state.

The desired state should be obtainable from type-II phase-matching by selecting the points 6 and 6′, as shown in Fig. 2. First, notice that the photons again each belong to both an extraordinary cone and an ordinary one, as in the previously discussed situation for points 3 and 3′. However, because the size of the cones is very different, so too should be the relative contributions of them. In particular, if a photon is detected along direction 6, it is more likely to have come from the *smaller* cone – the extraordinary-polarized one – because the angular density of emitted photons should be higher than for the larger ordinary- polarization cone. If the density scales simply as the reciprocal of the cone circumference, then we would obtain a value for ϵ of about 0.5.[4]

Such partially-entangled states have been shown to be useful in *gedanken* experiments demonstrating the nonlocality of quantum mechanics without inequalities [20]. More importantly for experiments, Eberhard has shown that by using a non-maximally entangled state, one can reduce the required detector efficiency for a loophole-free test of Bell's inequalities from 83% to 67% (in the limit of no background) [11]. This, in turn, is relevant because we now have single photon detectors with efficiencies of 75-80% [21], so there is hope of completing a true test of quantum nonlocality.

6. CONCLUSION

While we do not claim to understand all the possible ramifications or benefits of multiply-entangled states, it is clear that they enable a wealth of new phenomena. For example, using the accompanying energy-entanglement, it is possible to distinguish all four of the polarization-Bell states – otherwise, only three may be distinguished [17]. Such states may also simplify experimental implementations of quantum teleportation, and true loophole-free tests of quantum nonlocality.

Finally, there exists an even more general class of entangled states, even for a single degree of freedom, in which the amplitudes of the contributing terms have different magnitudes. Such states have been called "non-maximally entangled", and have been shown at least to reduce the experimental constraints in tests of Bell's inequalities. Using novel phase-matching possibilities it should be possible to produce such states, and even to produce and study non-maximal hyper-entangled states, the most general possible quantum state of two particles. Some of the above results have already been observed experimentally [6,17,18], while results for the other schemes are expected in the next months.

REFERENCES

1. E. Schrödinger, *Proc. Camb. Phil. Soc.* **31**, 555 (1935).
2. C. Bennett and S. J. Wiesner, *Phys. Rev. Lett.* **69**, 2881 (1992).
3. C. Bennett *et al.*, *Phys. Rev. Lett.* **70**, 1895 (1993).
4. A. K. Ekert, *Phys. Rev. Lett.* **67**, 661 (1991); A. K. Ekert, J. G. Rarity, P. R. Tapster, and G. M. Palma, *ibid.* **69**, 1293 (1992).
5. A. Barenco, D. Deutsch, A. Ekert, and R. Jozsa, *Phys. Rev. Lett.* **74**, 4083 (1995). T. Sleator and H. Weinfurter, *ibid.* **74**, 4087 (1995). J. I. Cirac and P. Zoller, *ibid.* **74**, 4091 (1995).
6. P. G. Kwiat *et al.*, *Phys. Rev. Lett.* **75**, 4337 (1995).
7. J. Brendel, E. Mohler, and W. Martienssen, *Europhys.Lett.* **20**, 575 (1992). P.G. Kwiat, A.E. Steinberg, and R.Y. Chiao, *Phys. Rev. A* **47**, R2472 (1993). P.R. Tapster, J.G. Rarity, and P.C.M. Owens, *Phys. Rev. Lett.* **73**, 1923 (1994).
8. J. G. Rarity and P. R. Tapster, *Phys. Rev. Lett.* **64**, 2495 (1990).
9. M. Zukowski and A. Zeilinger, *Phys. Lett. A* **155**, 69 (1991).
10. T. B. Pittman, Y. H. Shih, A. V. Sergienko, and M. H. Rubin, *Phys. Rev. A* **51**, 3495 (1995).
11. P. H. Eberhard, *Phys. Rev. A* **47**, R747 (1993).
12. P. G. Kwiat, P. H. Eberhard, A. M. Steinberg, and R. Y. Chiao, *Phys. Rev. A* **49**, 3209 (1994).
13. E. Schrödinger, *Proc. Am. Phil. Soc.* **124**, 323 (1980).
14. D. C. Burnham and D. L. Weinberg, *Phys. Rev. Lett.* **25**, 84 (1970).
15. J. D. Franson, *Phys. Rev. Lett.* **62**, 2205 (1989).
16. S. L. Braunstein, A. Mann, and M. Revzen, *Phys. Rev. Lett.* **68**, 3259 (1992).
17. K. Mattle, H. Weinfurter, P. G. Kwiat, and A. Zeilinger, *Phys. Rev. Lett.* **76**, 4656 (1996).
18. D. V. Strekalov *et al.*, *Phys. Rev. A* **54**, R1 (1996).

19. P. G. Kwiat and H. Weinfurter, in preparation.
20. L. Hardy, *Phys. Rev. Lett.* **68**, 2981 (1992). L. Hardy, *Phys. Lett. A* **167**, 17 (1992).
21. P. G. Kwiat, A. M. Steinberg, R. Y. Chiao, P. Eberhard and M. Petroff, *Phys. Rev. A* **48**, R867 (1993). P. G. Kwiat, A. M. Steinberg, R. Y. Chiao, P. H. Eberhard and M. D. Petroff, *Appl. Opt.* **33**, 1844 (1994).

NOTES

1. There has been some work looking at two particles with a mixed entanglement, e.g., where the polarization of one particle is entangled with the direction of the other [9]. Also, one group has examined the case where one variable is truly entangled, and another *emulates* an entanglement by means of a post-selection [10].
2. We are considering here only the case of a cw (continuous wave) pump, therefore assumed to be nearly monochromatic. If, on the other hand, one were to employ a very short pulsed pump, then the time correlations of the down-conversion photons would remain, but the energy correlations would be smeared out by the inherent spectral width of the pump.
3. It might seem that we *are* actually performing a post-selection on our data by using a polarizer at 45°; indeed, this only transmits half of the light. However, just as the transmitted photons display perfect 100%-visibility fringes (as one of the phases is changed), so too would the absorbed photons. If we were to use instead a polarizing beam splitter, oriented in the 45°/-45° basis, then out one port we would obtain fringes while out the other we would obtain anti-fringes.
4. It does not matter that we are looking at a non-degenerate solution — the color of the light in no way yields information about whether the photon along direction 6 (or 6′) came from the ordinary or the extraordinary cone.

Ode to Entangled States

Photons twins, at birth separated
And yet they remain so well correlated
Their colors, directions and spins synchopated
No wonder these states are so celebrated

If that one goes this way, this one goes that
If this one comes early, that one comes late
Like two random roulette wheels, yet somehow both "fixed"
To hit the same number though they're never mixed

They drove EPR to say "It's incomplete"
They've got the Bell inequalities beat
When factoring primes they allow you to cheat
Who knows what new marvel is next at our feet

Just out of reach were problems that dangled
Current attempts to solve them seemed wangled
Perhaps what's required is something new-fangled
Enter the states called *hyper-entangled*

FROM NEWTON TO SCHRÖDINGER AND BEYOND

Willis E. Lamb, Jr.

Optical Sciences Center
University of Arizona
Tucson, Arizona 85721

Newton's second law of classical mechanics describes the motion of a "particle" by use of a second-order ordinary differential equation. A slight change in the history of physics would instead have given us a partial differential equation for a complex function "$\psi(x,t)$," and a pointer function, $W(x,t)$, to describe the motion of the particle. The form of the p.d.e. can be found, essentially uniquely, from the requirement that the peak of a sufficiently narrow pointer function should *point* at all times to the Newtonian particle. In fact, this p.d.e. is identical with the non-relativistic wave equatiom of Schrödinger.

As a result, classical mechanics becomes a limiting form of quantum mechanics. Quantum mechanics becomes a straightforward generalization of classical mechanics. Although it is not usually pointed out, it is possible to give a fully classical theory of measurement based on classical mechanics. An important fringe benefit of this is that it is very easy to give a fully quantum mechanical theory of measurement. To treat the quantum theory of measurement, it is only necessary to translate the classical procedures into the quantum domain. Reference 129 describes this classical theory. Furthermore, large parts of the quantum theory were already given in the author's work of 1986 and later.

Among the advantages gained thereby are that one can forget many of the fuzzy ideas that cluttered the path from classical mechanics to quantum mechanics. These include: Bohr's correspondence principle, Heisenberg's uncertainty principle, Bohr's principles of complementarity and wave-particle duality, Dirac's and von Neumann's absurd notions of the collapse of wave packets. We might have been spared much of the agony needlessly suffered with two-slit diffraction pattern paradoxes, the Einstein-Podolsky-Rosen paradoxes, the Schrödinger cat paradoxes, hidden-variable theories, and Bell's inequalities. We could also do without discussion of the environment, or decoherence. Feynman path integrals still present a useful way

M. Ferrero and A. van der Merwe (eds.),
New Developments on Fundamental Problems in Quantum Physics, 203–205.

thinking graphically about problems in perturbation theory. However, the zig-zag path motions are highly unphysical. A satisfactory theory of the universe, if one is ever given, will have to wait for a more meaningfull extension of quantum mechanics into the fully relativistic domain.

Key words: quantum mechanics, quantum measurement, paradoxes of quantum mechanics.

It might help any person interested in my views on quantum mechanics to have the following selected items from my numbered list of publications:

REFERENCES

62. "An operational interpretation of nonrelativistic quantum mechanics," *Physics Today* **22**, 23-28 (1969); reprinted (in Polish translation) in *Postepy Fizyki* **22**, 479-493 (1971).
91. "Von Neumann's reduction of the wave function," reprinted from *The Centrality of Science and Absolute Values* (Proceedings of the Fourth International Conference on the Unity of the Sciences) (ICF Press, New York, 1975), pp. 297-303.
96. "Remarks on the interpretation of quantum mechanics," in *Science of Matter*, S. Fujita, ed. (Gordon & Breach, New York, 1979), pp. 1-8.
105. "A letter to Kip Thorne," in *Proceedings, CCNY Physics Symposium, in Celebration of Melvin Lax's Sixtieth Birthday*, H. Falk, ed. (City University of New York, 1983), pp. 40-43.
108. "Quantum chaos and the theory of measurement," in *Chaotic Behavior in Quantum Systems: Theory and Applications*, G. Casati, ed. (Plenum Press, New York, 1985) (Proceedings, NATO Advanced Research Workshop on Quantum Chaos, Como, June 1983) pp. 253-261.
109. "Schrodinger's cat," in *Paul Adrien Maurice Dirac*, B. N. Kursunoglu and E. P. Wigner, eds. (Cambridge University Press, Cambridge, 1987), pp. 249-261.
111. "Quantum theory of measurement," *Ann. N. Y. Acad. Sci.* **480**, 407-416 (1986).
113. "Theory of quantum mechanical measurements," in *Proceedings, 2nd International Symposium on the Foundations of Quantum Mechanics In the Light of New Technology*, M. Namiki et al., ed. (Japan Physical Society, Tokyo, 1987) pp. 185-192.
114. "Sequential measurements in quantum mechanics," in *Quantum Measurement and Chaos*, E. R. Pike and S. Sarkar, eds. (Plenum, New York, 1987) (Proceedings, Como NATO Advanced Research Workshop in Quantum Optics, September, 1986) pp. 183-193.

115. "Classical measurements on a quantum mechanical system," in *Proceedings, International Symposium on Spacetime Symmetries*, Y. S. Kim and W. M. Zachary, eds., reprinted from *Nuc. Phys. B (Proc. Suppl.)* **6**, 197-201 (North-Holland, Amsterdam, 1989), pp. 197-201.
118. "Quantum theory of measurement," in *Proceedings, Turin NATO Advanced Research Workshop on Noise and Chaos in Nonlinear Dynamical Systems* (March, 1989), F. Moss, L. Lugiato, and W. Schleich, eds. (Cambridge University Press, Cambridge, 1990), pp. 1-14.
122. "Computational approach to the quantum Zeno effect: Position measurements," with H. Fearn, *Phys. Rev. A* **46**, 1199-1205 (1992).
124. "Reply to 'Comment on Computational Approach to the Quantum Zeno Effect: Position Measurement'," with H. Fearn, *Phys. Rev. A* **48**, 2505-2507 (1993).
125. "Suppose Newton had invented wave mechanics," *Am. J. Phys.* **62**, 201-206 (1994).
126. "Anti-photon," *App. Phys. (Lasers and Optics) B* **60**, 77-84, (1995). (To celebrate the sixtieth birthday of Herbert Walther.)
127. "Comments on the quantum Zeno effect,"with H. Fearn, in "Quantum and semiclassical optics," *J. Eur. Opt. Soc. B* **7**, 211-214 (1995).
128. "Comments by W. E. Lamb," in *Proceedings, Workshop on Quantum Cryptography and Quantum Computing* (February 15, 1995), H. Everitt, ed. (U. S. Army Research Office), pp. 24-25.
129. "Classical theory of measurement, a big step towards the quantum theory of measurement," with H. Fearn, in *Amazing Light* (dedicated to C. H. Townes on his 80th birthday), R. Chiao, ed. (Springer, New York, 1996), pp. 373-389.
130. "Borderland between quantum and classical mechanics," dedicated to memory of George Series, to be published in *Physica Scripta*, 1997.

QUENCHING OF SPONTANEOUS EMISSION VIA QUANTUM INTERFERENCE: TIME EVOLUTION

Hwang Lee, Pavel Polynkin, and Marlan O. Scully

Department of Physics, Texas A&M University
College Station, Texas 77843
Texas Laser Laboratory, HARC, The Woodlands, Texas 77381
Max-Planck-Institut für Quantenoptik
85748 Garching, Germany

A three-level atom having two close upper levels decaying to a common third level is considered. It is shown that interference between these two spontaneous transitions drastically changes the spectrum of emitted light. Particular attention is paid to the time evolution of the spectrum. This yields insight into a novel and counterintuitive aspect of the quantum theory of radiation.

Key words: spontaneous emission, Fano interference, atomic coherence.

1. INTRODUCTION

The concept of interference between transitions from discrete quantum states to a common continuum was introduced by Fano in the early 60's [1]. Cancellation of emission into a single mode based on Fano interference was proposed in [2], and on simple quantum coherence in [3]. Fano interference physics led to one of the earliest proposals of lasing without inversion [4]. Futhermore, the elimination of a spectral line was theoretically predicted [5] and experimentally demonstrated [6]. Recently, it was found that a complete quenching of spontaneous emission from an atom driven by a strong coherent field is possible under certain conditions [7].

The goal of the present paper is to consider the simplest possible level scheme in which interference between two spontaneous emission channels plays a crucial role. We will pay particular attention to the time dependence of the energy distribution of the emitted photons. It turns out that the characteristic feature of the spectrum, a dip at line center, appears only

M. Ferrero and A. van der Merwe (eds.),
New Developments on Fundamental Problems in Quantum Physics, 207–212.

after a time relatively large compared to the inverse radiative decay rates of the individual spontaneous transitions. Such a time dependence of the spectrum is completely different from the case of two-level atomic decay. For the latter case, the energy distribution of the emitted radiation is fixed after a time of the order of the inverse decay constant.

2. MODEL ATOM AND SPONTANEOUS EMISSION SPECTRUM

Consider the three-level model atom represented in Fig. 1(a). It has two upper levels $|a_1\rangle$ and $|a_2\rangle$ coupled by the same vacuum modes to the lower level $|b\rangle$. The spacing between upper levels, ω_{12}, is supposed to be small compared to the optical frequencies, ω_{a_1b} and ω_{a_2b}.

If there were no interference between the two spontaneous transitions $|a_1\rangle \rightarrow |b\rangle$ and $|a_2\rangle \rightarrow |b\rangle$, one would expect the spontaneous emission spectrum to be simply the sum of two Lorentzians arising from the two individual transitions. This is qualitatively right for the case when ω_{12} is large compared to the radiative decay constants of the two transitions, and Fano interference plays a negligible role. Can the level scheme of the Fig. 1(a) occur in practice? One possible realization is shown in Fig. 2 [9]. A three-level atom is driven by the coherent optical field, coupling two of its levels. Then, the spontaneous decays from the components of the Autler-Townes doublet can interfere with each other just like the decays $|a_1\rangle \rightarrow |b\rangle$ and $|a_2\rangle \rightarrow |b\rangle$ in Fig. 1(a).

The interaction picture Hamiltonian describing the interaction of our three-level atom with the vacuum field in the rotating wave approximation can be written as

$$\hat{V}(t) = -\hbar \sum_k g_k \left[e^{i(\omega_{a_1b}-\nu_k)t}|a_1\rangle\langle b|\hat{a}_k + e^{i(\omega_{a_2b}-\nu_k)t}|a_2\rangle\langle b|\hat{a}_k + \text{adj.}\right], \quad (1)$$

where g_k are the coupling constants between the k-th vacuum mode and atomic transitions from $|a_{1,2}\rangle$ to $|b\rangle$ (for simplicity we suppose them to be the same for both transitions); $\hat{a}_k$ is the annihilation operator for the k-th vacuum mode with frequency ν_k; k here represents both momentum and polarization of the vacuum mode. We suppose the atom to be initially in a coherent superposition of the two upper states; therefore, the initial state vector reads:

$$\Psi(0) = [A_1(0)|a_1\rangle + A_2(0)|a_2\rangle]\,|\{0\}\rangle\,, \quad (2)$$

where $|\{0\}\rangle$ represents the absence of light quanta in all vacuum modes. The state vector at time t then has the form:

$$\Psi(t) = [A_1(t)|a_1\rangle + A_2(t)|a_2\rangle]\,|\{0\}\rangle + \sum_k B_k(t)|c\rangle|1_k\rangle\,. \quad (3)$$

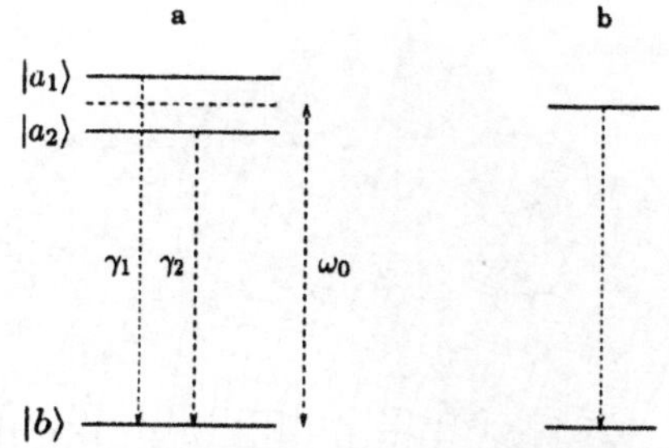

Fig. 1. (a) Level scheme of a model atom. Two upper states, $|a_1\rangle$ and $|a_2\rangle$ are coupled by the same vacuum modes to the common lower level $|b\rangle$. Decay constants γ_1 and γ_2 are the same in the calculation of spontaneous emission spectrum. (b) Two-level atom with the Lorentzian spontaneous emission spectrum, which is to be compared to the spectrum of radiation emitted by the three-level atom of the Fig. 1(a).

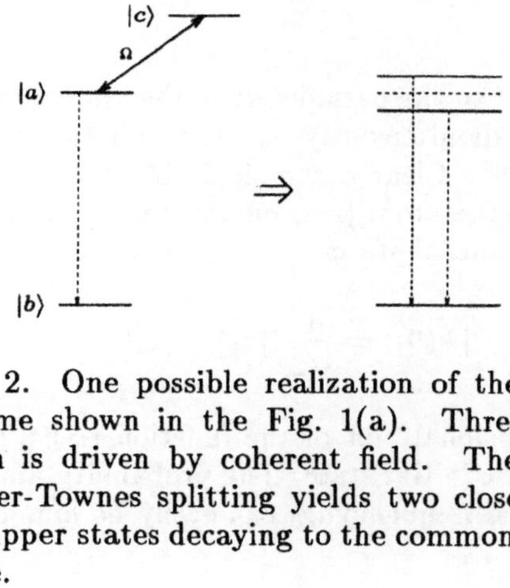

Fig. 2. One possible realization of the level scheme shown in the Fig. 1(a). Three-level atom is driven by coherent field. Then the Autler-Townes splitting yields two closely lying upper states decaying to the common lower state.

From Eqs. (1,2,3) we find:

$$B_k(t) = ig_k \left[\frac{\alpha_1 (x_1 + y_1)}{\lambda_1 + i\Delta_k} \left(e^{(\lambda_1 + i\Delta_k)t} - 1 \right) + \frac{\alpha_2 (x_2 + y_2)}{\lambda_2 + i\Delta_k} \left(e^{(\lambda_2 + i\Delta_k)t} - 1 \right) \right]. \tag{4}$$

Equation (4) relates the probability amplitude to observe a photon with the

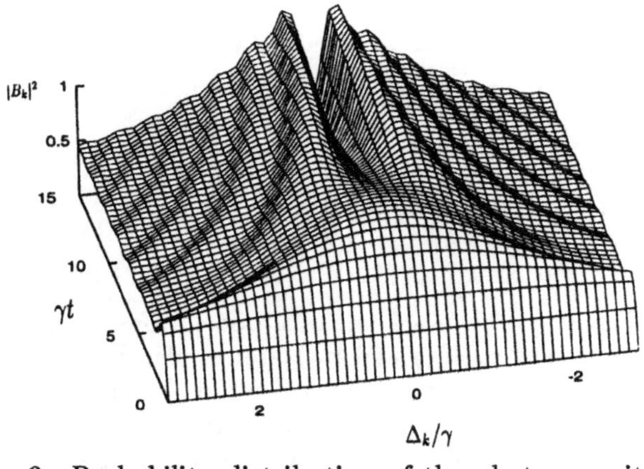

Fig. 3. Probability distribution of the photons emitted by the three-level atom of the Fig. 1(a) as a function of dimensionless time and detuning from the central frequency ω_0 (not normalized). Initial state of the atom is symmetric combination of the two upper states (5), upper level separation equals to $\gamma/2$.

wave vector k at time t to the parameters of the system and its initial state. The spectrum of the spontaneously emitted radiation up to the time t is proportional to $|B_k(t)|^2$. Clearly, the shape of the spectral curve depends on the initial state of the atom (i.e., on the constants $\alpha_{1,2}$). Consider, for example, a symmetric initial state:

$$|\Psi(0)\rangle = \frac{1}{\sqrt{2}}\left(|a_1\rangle + |a_2\rangle\right). \tag{5}$$

In Fig. 3, the 3-dimensional plot of the function $|B_k(t,\Delta_k)|^2$ is shown for the case of a symmetric initial state. The probability amplitude to observe radiation at the central frequency ω_0 can easily be found from the formula (4):

$$B_{k_0}(t) = \frac{2ig_{k_0}e^{-\gamma/2}}{\sqrt{\gamma^2-\omega_{12}^2}}\left(e^{\sqrt{\gamma^2-\omega_{12}^2}t/2} - e^{-\sqrt{\gamma^2-\omega_{12}^2}t/2}\right). \tag{6}$$

$|B_{k_0}|^2$ as a function of time is shown in the Fig. 4(a). We can see that this probability grows with t at the beginning, but after a time of the order of $5/\gamma$ it starts to decrease and goes to zero as $t \to \infty$. It is as if the atom

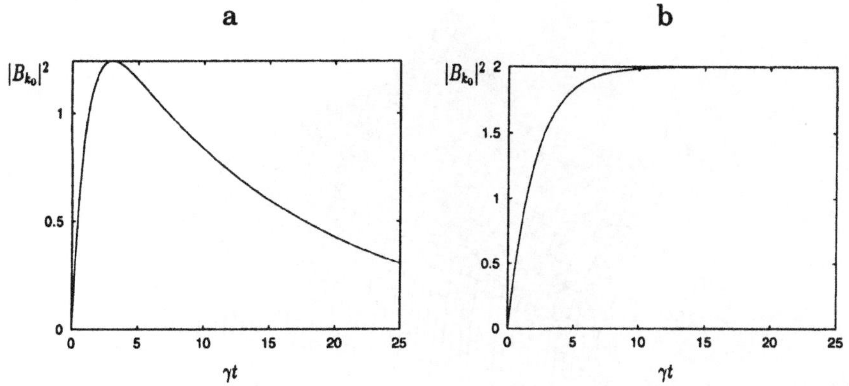

Fig. 4. a. Probability of finding a photon at the central frequency emitted by a three-level atom as a function of dimensionless time (not normalized). Initial state of the atom is symmetric combination of the two upper states, upper level separation equals to $\gamma/2$. b. The same for the two-level atom.

absorbs the photon emitted at the central frequency, and then re-emits it on the sideband! Therefore the shape of the spectrum changes with time, and only after the time $t \sim 25/\gamma$ does the spectrum approach its final form showing the Fano dip.

Consider the evolution of the joint system "atom + vacuum field". Due to interference of two spontaneous transitions this system reaches the equilibrium only after a time relatively large compared to the radiative decay times of individual levels. This behavior is completely different from that of the two-level atom decay (without Fano interference). In the Fig. 4(b) we show the probability to observe the photon on the resonance frequency emitted by the two-level atom. As expected, it grows monotonically with time, approaching its steady-state value at $t \sim 5/\gamma$. After this time, we can say that the atom has decayed, and the spectrum of the emitted radiation is fixed (it is, of course, Lorentzian, centered on the resonance frequency). The 3-dimensional plot of the spectrum as a function of time and frequency for this two-level case is shown in Fig. 5.

Acknowledgements. This work was supported by the Office of Naval Research, the Texas Advanced Research Program, and the Welch Foundation.

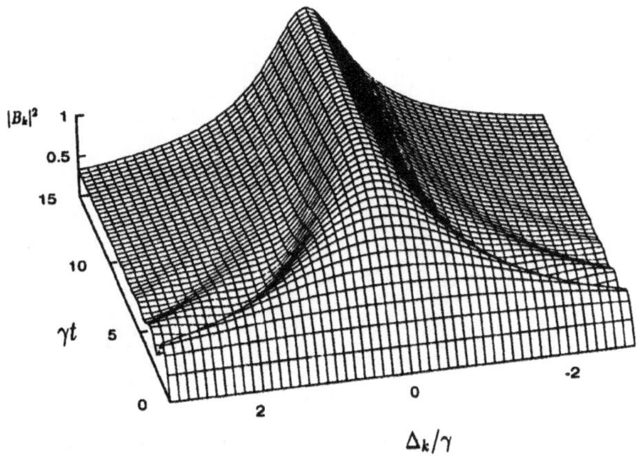

Fig. 5. Probability distribution of the photons spontaneously emitted by the two-level atom of the Fig. 1(b) as a function of dimensionless time and detuning from the resonance frequency (not normalized).

REFERENCES

1. U. Fano, *Phys. Rev.* **124**, 1866 (1961).
2. D. Agassi, *Phys. Rev. A* **30**, 2449 (1984).
3. M. Scully, S.-Y. Zhu, and A. Gavrielides, *Phys. Rev. Lett.* **62**, 2813 (1989).
4. S. E. Harris, *Phys. Rev. Lett.* **62**, 1033 (1989).
5. S.-Y. Zhu and M. O. Scully, *Phys. Rev. Lett.* **76**, 388 (1996).
6. H.-R. Xia, C.-Y. Ye, and S.-Y. Zhu, *Phys. Rev. Lett.* **77**, 1032 (1996).
7. H. Lee, P. Polynkin, and M. O. Scully, submitted to *Phys. Rev. A*.
8. V. Weisskopf and E. Wigner, *Z. Phys.* **63**, 54 (1930).
9. S.-Y. Zhu, L. Narducci, and M. O. Scully, *Phys. Rev. A* **52**, 4791 (1995).

QUANTUM MECHANICS ON DISCRETE SPACE AND TIME

Miguel Lorente

Departamento de Física, Facultad de Ciencias
Universidad de Oviedo
E-33007, Oviedo, Spain

We propose the assumption of quantum mechanics on a discrete space and time, which implies the modification of mathematical expressions for some postulates of quantum mechanics. In particular we have a Hilbert space where the vectors are complex functions of discrete variable. As a concrete example we develop a discrete analog of the one-dimensional quantum harmonic oscillator, using the dependence of the Wigner functions in terms of Kravchuk polynomials. In this model the position operator has a discrete spectrum given by one index of the Wigner functions, in the same way that the energy eigenvalues are given by the other matricial index.

A similar picture can be made for other models where the differential equation and their solutions correspond to the continuous limit of some difference operator and orthogonal polynomial of discrete variable.

Key words: spacetime, discrete variable, quantum oscillator.

1. INTRODUCTION

In the last Symposium on Fundamental Problems in Quantum Physics [1], I explored the hypothesis of a realistic interpretation of lattice theories based on some ontological model that presupposes some fundamental network previous to the concept of space and time. According to this model the structure of space-time is a consequence of the relations among these fundamental entities, and gives raise to a discrete character of the space and time variables.

Even the standard interpretation of quantum mechanics in our model is conserved the assumption of a discrete space and time introduces some drastic changes in the mathematical formulation of quantum mechanics. These consequences have been also shared by recent authors who use lattice

M. Ferrero and A. van der Merwe (eds.),
New Developments on Fundamental Problems in Quantum Physics, 213–224.

models as a mathematical tool. In particular non-commutative geometry leads in a natural way to difference operators [2]. Quantum groups and q-analisis implies a deformation of space-time groups with a q-casimir defined on a space or time lattice [3]. Lattice field theories are widely spread in an attempt to overcome infinities in perturbation theories [4]. Special functions and orthogonal polynomials on continuous variable are studied together with those of discrete variable [5]. Non-euclidean crystallography requires the use of discrete groups of hyperbolie type that has been recently developped [6]. Recent literature advocates for the application of discrete models to unify general relativity and quantum mechanics as Wheeler, Ponzano and Regge and others have proposed [7]. For this purpose some modern tools are used, such as discrete topology and partition functions defined over symplicial networks.

2. THE POSTULATES OF QUANTUM MECHANICS ON Z^4

The assumption of discrete space and time imposses some restrictions on the mathematical expressions for the postulates of Quantum Mechanics. The question is now whether this assumption keeps the analogy with standard formulation in the continuum case (in the limit both formulations coincide) and at the same time avoids the unwanted infinities. The answer is, generally speaking, in the affirmative.

In order to be more explicit, we start with the Hilbert space: this must be defined over the gaussian numbers (complex of integer components) with scalar products constructed with summation instead of integrals

$$\sum_{j=a}^{b} f_j^* g_j$$

where we use the notation $f_j \equiv f(j\varepsilon)$, ε being the fundamental length of the one-dimensional lattice, j integer numbers.

As an example, we take the following orthonormal basis

$$f_j(k_m) = \frac{1}{\sqrt{N}} \left(\frac{1 + \frac{1}{2} i\varepsilon k_m}{1 - \frac{1}{2} i\varepsilon k_m} \right)^j, \quad j = 0, 1, \cdots, N-1,$$

where

$$k_m \equiv \frac{2}{\varepsilon} \mathrm{tg} \frac{\pi m}{N}, \quad m = 0, 1, \cdots, N-1,$$

satisfying

$$\sum_{j=0}^{N-1} f_j^*(k_m) f_j(k_{m'}) = \delta_{mm'},$$

with respect to which a finite Fourier transform (equivalent to a Fourier series) can be defined:

$$F_j = \sum_{m=0}^{N-1} a_m f_j(k_m), \quad a_m = \sum_{j=0}^{N-1} f_j^*(k_m) F_j$$

Notice that although the space-time variables are discrete the functions can be continuous.

If we consider observables, the corresponding operators must be self adjoint with respect to the scalar product mentioned before, and the spectrum is always discrete. As an example, we give the eigenvalues and eigenfunctions of the position and momentum operators:

$$X: \quad j\delta_{j\ell} = \ell\delta_{j\ell}, \quad \ell \text{ fixed}; \quad j = 0, 1, \cdots, N-1,$$

$$P: \quad -i\frac{1}{\varepsilon}\Delta_j f_j(k_m) = \frac{k_m}{1 - \frac{1}{2}\varepsilon k_m} f_j(k_m),$$

where $\delta_{j\ell}$ is the Kronecker function and $f_j(k_m)$ is defined as before.

With the help of the scalar product we can defined espectation values, projection operators, density matrix, uncertainties or mean values of some operators.

Suppose now that a physical system is represented by a state vector $\psi(t)$ depending on discrete time ($t = n\tau$). If H is the operator corresponding to the hamiltonian of the system (for simplicity we take H constant in time) the Schrödinger equation for that system is

$$\frac{i}{\tau}\Delta_n \psi_n = H\tilde{\Delta}_n \psi_n,$$

the solution of which is

$$\psi_n = \left(\frac{1 - \frac{1}{2}i\tau H}{1 + \frac{1}{2}i\tau H}\right)^n \psi_0,$$

with the initial condition ψ_0.

Here Δ_n is the forward difference operator $\Delta_n \psi_n = \psi_{n+1} - \psi_n$ and $\tilde{\Delta}_n$ the mean operator $\tilde{\Delta}_n \psi_n = \frac{1}{2}(\psi_{n+1} + \psi_n)$.

As in the continuous case, we can define an unitary evolution operator

$$U_n \equiv \left(\frac{1 - \frac{1}{2}i\tau H}{1 + \frac{1}{2}i\tau H}\right)^n,$$

which is unitary (because H is self-adjoint) and satisfies the difference equation

$$\frac{i}{\tau}\Delta_n U_n = H\tilde{\Delta}_n U_n.$$

If we use the Heisenberg picture the evolution of some operators in time is given by

$$A_n = U_n^+ A_0 U_n, \tag{1}$$

where A_n are some operators depending on discrete time and U_n is the evolution operator defined before. The Heisenberg equation now reads [8]:

$$\frac{i}{\tau}\Delta A_n = \frac{1}{1-\frac{1}{2}i\tau H}\left[A_n, H_n\right]\frac{1}{1+\frac{1}{2}i\tau H}$$

the solution of which is (1).

Another scheme may be used if we take the symmetric difference operator

$$\delta_n A_n \equiv \left(A_{n+\frac{1}{2}} - A_{n-\frac{1}{2}}\right),$$

then

$$\frac{i}{\tau}\delta_n A_n = \frac{1}{\left(1+\frac{1}{4}\tau^2 H^2\right)^{\frac{1}{2}}}\left[A_n, H_n\right]\frac{1}{\left(1+\frac{1}{4}\tau^2 H^2\right)^{\frac{1}{2}}}.$$

We also have

$$\frac{i}{\tau}\left(A_{n+1} - A_{n-1}\right) = \frac{1}{1+\frac{1}{4}\tau^2 H^2} 2\left(1-\frac{\tau^2}{4}\right)\left[A_n, H_n\right]\frac{1}{1+\frac{1}{4}\tau^2 H^2}.$$

Some simplification can be achieved if we take the particular case $H^2 = 1$. One can easily prove the following equation for the operator in the Heisenberg picture:

$$\frac{i}{\tau}\Delta_n A_n = \frac{1}{\left(1+\frac{\tau^2}{4}\right)}\left(\frac{1-\frac{i}{2}\tau H}{1+\frac{i}{2}\tau H}\right)\left[A_n, H\right],$$

$$\frac{i}{\tau}\nabla_n A_n = \frac{1}{\left(1+\frac{\tau^2}{4}\right)}\left(\frac{1+\frac{i}{2}\tau H}{1-\frac{i}{2}\tau H}\right)\left[A_n, H\right],$$

$$\left(\frac{i}{\tau}\right)^2 \Delta_n \nabla_n A_n = \frac{1}{\left(1+\frac{\tau^2}{4}\right)^2}\left[\left[A_n, H\right],\ H\right],$$

$$\frac{i}{\tau}\delta_n A_n = \frac{1}{\left(1+\frac{\tau^2}{4}\right)^2}\left[A_n, H\right],$$

$$\frac{i}{\tau}\left(A_{n+1} - A_{n-1}\right) = \frac{2\left(1-\frac{\tau^2}{4}\right)}{\left(1+\frac{\tau^2}{4}\right)^2}\left[A_n, H\right].$$

In the last three equations the dependence on the Hamiltonian operator H is linear, as in the continuous case.

The realization of operators in the coordinate or position representation is given by the substitution

$$X \to j\varepsilon, \quad P \to -\frac{i}{\varepsilon}\Delta_j.$$

The substitution is not unique. We will discuss in the next section some different realization for the position and momentum operator of he harmonic oscillator.

3. QUANTUM HARMONIC OSCILLATOR OF DISCRETE VARIABLE

The quantum harmonic oscillator is described by the Schrödinger equation

$$\frac{\hbar\omega}{2}\left[-\frac{d^2}{d\xi^2}+\xi^2\right]\psi(\xi)=\lambda\psi(\xi),$$

with $\omega \equiv \sqrt{\frac{k}{M}}$, $\quad \xi = \alpha s,\quad \alpha \equiv \sqrt{\frac{M\omega}{\hbar}}$, and $\lambda = \frac{2E}{\hbar\omega}$.

For simplicity, we take $\alpha = 1$. The normalized solutions are

$$\psi_n(s) = \left(\pi^{\frac{1}{2}}2^n n!\right)^{-\frac{1}{2}} e^{-s^2/2} H_n(s), \quad n = 0, 1, 2, \cdots, \tag{2}$$

where $H_n(s)$ are the Hermite polynomials.

The $\psi_n(s)$ are eigenfunctions corresponding to the eigenvalues $\lambda = 2n+1$, and they satisfy the following recurrence relations:

$$\text{(i)}\quad 2s\psi_n(s) = \sqrt{2(n+1)}\,\psi_{n+1}(s) + \sqrt{2n}\,\psi_{n-1}(s), \tag{3}$$

$$\text{(ii)}\quad 2\frac{d}{ds}\psi_n(s) = -\sqrt{2(n+1)}\,\psi_{n+1}(s) + \sqrt{2n}\,\psi_{n-1}(s). \tag{4}$$

From these two relations, one defines the creation and annihilation operators:

$$a^\dagger\psi_n(s) \equiv \frac{1}{\sqrt{2}}\left(s - \frac{d}{ds}\right)\psi_n(s) = \sqrt{n+1}\psi_{n+1}(s), \tag{5}$$

$$a\psi_n(s) \equiv \frac{1}{\sqrt{2}}\left(s + \frac{d}{ds}\right)\psi_n(s) = \sqrt{n}\psi_{n-1}(s). \tag{6}$$

It is well known that the Hermite polynomials are the continuous limit of the Kravchuk polynomials of discrete variables $k_n(x)$ and the weight function of the Hermite polynomials is the continuous limit of the binomial distribution $\rho(x)$ which in turns is the weight function of the Kravchuk polynomial [9]. But the product of the Kravchuk polynomials time their weight function is proportional to the Wigner functions $d^j_{mm'}(\beta)$ that appear in the generalized spherical functions

$$D^j_{mm'}(\alpha,\beta,\gamma) = \exp\,(-im\alpha)\, d^j_{mm'}(\beta) \exp\,(-im'\gamma), \tag{7}$$

namely,

$$d^j_{mm'}(\beta) = (-1)^{m-m'} \frac{\sqrt{\rho(x)}}{d_n} k_n^{(p)}(x,N) \tag{8}$$

where d_n is some normalization constant, and $n = j-m$, $x = j-m'$, $p = \sin^2\,(\beta/2)$.

This connection betwen the functions of discrete and continuous variable suggests that the solution of the quantum harmonic oscillator are the continuous limit, up to a factor, of the Wigner functions. In order to prove this Ansatz we compare the recurrence relations of the two types of functions as it is done for the orthogonal polynomials of discrete variable. In our case we take the differential equation for the Wigner function [10]

$$\begin{aligned} &\pm\frac{d}{d\beta}\, d^j_{mm'}(\beta) + \frac{m' - m\cos\beta}{\sin\beta}\, d^j_{mm'}(\beta) \\ &= \sqrt{(j\mp m)(j\pm m+1)}\; d^j_{m\pm1,m'}(\beta) \end{aligned} \tag{9)(10}$$

From this we deduce two recurrence relations:

$$\begin{aligned} \text{(i)}\quad 2\,\frac{m' - m\cos\beta}{\sin\beta} d^j_{mm'}(\beta) &= \sqrt{(j-m)(j+m+1)}d^j_{m+1,m'}(\beta) \\ &\quad + \sqrt{(j+m)(j-m+1)}d^j_{m-1,m'}(\beta) \end{aligned} \tag{11}$$

$$\begin{aligned} \text{(ii)}\quad &\sqrt{(j+m')(j-m'+1)}d^j_{m,m'-1}(\beta) - \sqrt{(j-m')(j+m'+1)}d^j_{m,m'+1}(\beta) \\ &= \sqrt{(j-m)(j+m+1)}d^j_{m+1,m'} - \sqrt{(j+m)(j-m+1)}d^j_{m-1,m'} \end{aligned} \tag{12}$$

Note that the last expression has been obtained with the help of the well known property of Wigner functions

$$d^j_{mm'}(\beta) = (-1)^{m-m'} d^j_{m'm}(\beta).$$

We suppose that

$$\lim_{N \to \infty} C_n(N)\, d^j_{mm'}(\beta) = \psi_n(s), \tag{13}$$

where we take $m = j - n$, $m' = j - x$, $N = 2j$, $x = Np + \sqrt{2Npq}\, s$, and $C_n(N)$ some normalization constant to be determined.

We compare the recurrence relations i) that is to say, formulas (3) and (11). We divide the second one by $\sqrt{j}$ and substitute $d^j_{mm'}(\beta)$ by $\frac{v_n(x)}{C_n(N)}$, with $v_n(x) \equiv C_n(N)\, d^j_{j-n,j-x}(\beta)$.

The result is:

$$2\frac{j - x - (j-n)\cos\beta}{\sqrt{j}\sin\beta}\frac{v_n(x)}{C_n(N)} = \sqrt{\frac{2n(N-n+1)}{2j}}\frac{v_{n-1}(x)}{C_{n-1}(x)} + \sqrt{\frac{2(N-n)(n+1)}{2j}}\frac{v_{n+1}(x)}{C_{n+1}(x)}$$

or

$$2\left(s + \frac{(2p-1)\,n}{\sqrt{2Npq}}\right)\frac{v_n(x)}{C_n(N)} = \sqrt{2(n+1)}\sqrt{1-\frac{n}{N}}\frac{v_{n+1}(x)}{C_{n+1}(N)} + \sqrt{2n}\sqrt{1-\frac{n-1}{N}}\frac{v_{n-1}(x)}{C_n(N)}.$$

In the limit $N \to \infty$, this expression goes to the recurrence relation (3) provided $\frac{C_n(N)}{C_{n+1}(N)} = \frac{C_n(N)}{C_{n-1}(N)} = 1$, or $C_n(N) = \text{const} = 1$.

The recurrence relations (ii) formulas (4) and (12) can be compared by the same method. We substitute

$$v_n(x) \equiv d^j_{j-n,\ j-x}(\beta) \tag{14}$$

in (12) and divide both sides by $\sqrt{N}$.

The result is

$$\sqrt{\frac{(N-x)}{N}(x+1)}v_n(x+1) - \sqrt{x\frac{(N-x+1)}{N}}v_n(x-1)$$

$$= \sqrt{n\frac{(N-n+1)}{N}}v_{n-1}(x) - \sqrt{\frac{(N-n)}{N}(n+1)}v_{n+1}(x).$$

Substituting $x = Np + \sqrt{2Npq}s$ and extracting $\sqrt{2Np} \equiv \frac{1}{h}$ on the left side, we obtain

$$\frac{1}{h}\left\{\sqrt{\left(1-\frac{x}{N}\right)\left(1+\sqrt{\frac{2q}{Np}}+\frac{1}{Np}\right)}\, v_n(x+1)\right.$$
$$\left. - \sqrt{\left(1+\sqrt{\frac{2q}{Np}}s\right)\left(1-\frac{x-1}{N}\right)}\, v_n(x-1)\right\}$$
$$= \sqrt{2n\left(1-\frac{n-1}{N}\right)}\, v_{n-1}(x) - \sqrt{2(n+1)\left(1-\frac{n}{N}\right)}\, v_{n+1}(x).$$

In the limit $N \to \infty$, $h \to 0$ this expression becomes

$$\lim_{h \to 0} \frac{1}{h}\{\psi_n(s+h) - \psi_n(s-h)\} = -\sqrt{2(n+1)}\psi_n(s) + \sqrt{2n}\psi_{n-1}(s),$$

which coincides with (4).

We can use these results to construct creation and annihilation operators for the Wigner functions. We define

$$\begin{aligned} Ad^j_{mm'}(\beta) := \frac{1}{\sqrt{2}}&\left\{\frac{m' - m\cos\beta}{\sqrt{j}\sin\beta} d^j_{mm'}(\beta)\right. \\ &+ \sqrt{\frac{(j+m')(j-m'+1)}{4j}} d^j_{m,m'-1}(\beta) \\ &\left. - \sqrt{\frac{(j-m')(j+m'+1)}{4j}} d^j_{m,m'+1}(\beta)\right\} \\ &= \sqrt{\sqrt{\frac{(j-m)(j+m+1)}{2j}}}\, d^j_{m,m'+1}(\beta) \end{aligned} \tag{15}$$

$$\begin{aligned} A^\dagger d^j_{mm'}(\beta) := \frac{1}{\sqrt{2}}&\left\{\frac{m' - m\cos\beta}{\sqrt{j}\sin\beta} d^j_{mm'}(\beta)\right. \\ &- \sqrt{\frac{(j+m')(j-m'+1)}{4j}} d^j_{m,m'-1}(\beta) \\ &\left. + \sqrt{\frac{(j-m')(j+m'+1)}{4j}} d^j_{m,m'+1}(\beta)\right\} \\ &= \sqrt{\frac{(j+m)(j-m+1)}{2j}} d^j_{m-1,m'}(\beta). \end{aligned} \tag{16}$$

Using the limit of the recurrence relations, we obtain

$$\lim_{N \to \infty} A d^j_{mm'}(\beta) = \frac{1}{\sqrt{2}}\left(s + \frac{d}{ds}\right)\psi_n(s) \equiv a\psi_n(s), \tag{17}$$

$$\lim_{N \to \infty} A^\dagger d^j_{mm'}(\beta) = \frac{1}{\sqrt{2}}\left(s - \frac{d}{ds}\right)\psi_n(s) \equiv a^\dagger\psi_n(s). \tag{18}$$

Relations (15) and (16) suggest that the creation and annihilation operators are connected with the raising and lowering operators for the spherical harmonics Y_{jm}. In fact, multiplying (15) and (16) by $Y_{jm'}$ and adding for m', we find

$$A\sum_{m'} d^j_{mm'}(\beta)\, Y_{jm'} = \sqrt{\frac{(j-m)(j+m+1)}{2j}}\sum_{m'} d^j_{m+1,m'}(\beta)\, Y_{jm'}$$

or

$$AY_{jm} = \frac{1}{\sqrt{2j}}\sqrt{(j-m)(j+m+1)}Y_{j,m+1} = \frac{1}{\sqrt{2j}}J_+Y_{jm}. \tag{19}$$

Similarly,

$$A^\dagger\sum_{m'} d^j_{mm'}(\beta)\, Y_{jm'} = \sqrt{\frac{(j+m)(j-m+1)}{2j}}\sum_{m'} d^j_{m-1,m'}(\beta)\, Y_{jm'}$$

or

$$A^\dagger Y_{jm} = \frac{1}{\sqrt{2j}}\sqrt{(j+m)(j-m+1)}Y_{j,m-1} = \frac{1}{\sqrt{2j}}J_-Y_{jm}. \tag{20}$$

In order to make more transparent the connection between the creation and annihilation operators with the raising and lowering operators of the spherical harmonics, we take the commutation and anticommutation relations of the former operators:

$$\begin{aligned}(AA^\dagger - A^\dagger A)\, Y_{jm} &= \frac{1}{2j}(J_+J_- - J_-J_+)Y_{jm}\\ &= \frac{1}{2j}2J_zY_{jm} = \frac{m}{j}Y_{jm} = \left(1 - \frac{n}{j}\right)Y_{jm}.\end{aligned}$$

Substituting $Y_{jm} = \sum_{m'} d^j_{mm'}(\beta)\, Y_{jm'}$, we get

$$[A, A^\dagger]\, d^j_{mm'}(\beta) = \left(1 - \frac{n}{j}\right)d^j_{mm'}(\beta), \tag{21}$$

which in the limit $j \rightarrow \infty$ tends to

$$\left[a, a^{\dagger}\right] \psi_n(s) = \psi_n(s) .$$

Similarly,

$$\begin{aligned}\left(AA^{\dagger} + A^{\dagger}A\right) Y_{jm} &= \frac{1}{2j}\left(J_+J_- + J_-J_+\right) Y_{jm} = \frac{1}{j}\left(J^2 - J_z^2\right) Y_{jm} \\ &= \frac{1}{j}\left(j(j+1) - m^2\right) Y_{jm} = \left\{(2n+1) - \frac{n^2}{j}\right\} Y_{jm}\end{aligned}$$

or

$$\left(AA^{\dagger} + A^{\dagger}A\right) d^j_{mm'}(\beta) = \left\{(2n+1) - \frac{n^2}{j}\right\} d^j_{mm'}(\beta), \qquad (22)$$

which in the limit $j \rightarrow \infty$ becomes

$$\left(aa^{\dagger} + a^{\dagger}a\right) \psi_n(s) = (2n+1)\, \psi_n(s) .$$

If we multiply both sides by $\hbar\omega/2$ we obtain the eigenvalue equation for the hamiltonian.

The interpretation of this model can be taken from the quantum harmonic oscillator of continuous variable. The energy levels are equally distant by the amount $\hbar\omega$ and are labelled by $n = 0, 1, 2, \cdots \infty$. In the quantum harmonic oscillator of discrete variable we have also the discrete eigenvalues of the hamiltonian connected with the index $m = j - n$ of the Wigner function $d^j_{mm'}(\beta)$. These values are equally separeted but finite $(m = -j, \cdots + j)$.

Similarly the eigenvalue of the position operator $A + A^{+}$ are also discrete and connected to the index $m' = j - x$ of the Wigner functions but finite $(m' = -j, \cdots, +j)$.

The integer numbers $x = 0, 1, \cdots 2j$ are related to the quantity $x = \alpha s$ where s is the continuous variable and $\alpha = \sqrt{M\omega/\hbar}$. Since x is a pure number and s has the dimension of a length, the spacing of the one-dimensional lattice is equal to $1/\alpha = \sqrt{\hbar/M\omega}$. Therefore the Planck's constant $\hbar$ play an role with respect to discrete space similar to the role with respect to discrete energy values.

4. CONCLUDING REMARKS

The analysis we have made for the quantum harmonic oscillator of discrete variable can be applied to other cases, where the functions involved are orthogonal polynomials of continuous variable the limit of which are some orthogonal polynomial of discrete variables; we give some examples:

1. The function $f_j(k_m)$ described in Sec. 2 are polynomials of discrete variable the continuous limit of which are the exponential function. We have developed a new scheme for the Klein-Gordon and Dirac field equation that can be extended to lattice gauge theories [11].

2. The solution of the Schrödinger equation for the hidrogen atom is given in term of the orthogonal Laguerre polynomials and the spherical harmonics. The radical equation can be translated into the difference equation for the Meixner polynomial of discrete variable.

3. The quantification of the electromagnetic fiels leads to the D'Alambert equation the solution of which are given in terms of the Bessel spherical functions that are related to the trigonometric functions. These functions suggest the connection with the orthogonal polynomials of discrete variable, that are solutions of difference equations of the hypergeometric type. General speaking a parallel study of discrete and continuous model can be made similar to that made by the Russian school of mathematicians with respect to orthogonal polynomials.

ACKNOWLEDGEMENTS

I would like to thank the organizers for the invitation to this meeting.

I want to thank also to Prof. André Ronveaux for very illuminating suggestions about the connection between orthogonal polynomials of discrete and continuous variable and Professors R. Alvarez-Estrada, A. Galindo, and A. Fz. Rañada for very interesting comments about this model. This work has been supported partially by D.G.I.C.Y.T. (grant PB94-1318).

REFERENCES

1. M. Lorente, "A realistic interpretation of lattice gauge theories," in *Fundamental Problems in Quantum Physics*, M. Ferrero and A. van der Merwe, eds. (Kluwer Academic, Dordrecht, 1995), pp. 177-186.
2. A. Connes, *Geometrie non-commutative* (Intereditions, Paris, 1990).
3. A. Ballesteros, F.J. Herranz, M.A. Olmo, and M. Santander, "Deformation of space-time symmetries and fundamental scales," in *Problems in Quantum Physics*, M. Ferrero and A. van der Merwe, eds. (Kluwer Academic, Dordrecht, 1995), pp. 29-35.
4. A. I. Montvay and G. Münster, *Quantum Fields on a Lattice* (Cambridge University Press, Cambridge, 1994).
5. A.F. Nikiforov and V.B. Uvarov, *Special Functions of Mathematical Physics* (Birkhäuser, Boston, 1988).
6. M. Lorente and P. Kramer, "Non-Euclidean crystallography," *Symmetries in Science VII*, B. Gruber, ed. (Plenum, New York, 1995).

7. A. Kheyfets, N.J. Lafave, and W.A. Miler, "Nule-strut calculus," *Phys. Rev. D* **41**, 3628, 3637 (1990).
8. M. Lorente, "The method of finite differences for some operator field equations," *Lett. Math. Phys.* **13**, 229-236 (1987).
9. A.F. Nikiforov, S.K. Suslov, and V.B. Uvarov, *Classical Orthogonal Polynomials of a Discrete Variable* (Springer, Berlin, 1991), p. 48.
10. See Ref.9, p. 233.
11. M. Lorente, "A new scheme for the Klein-Gordon and Dirac fields on the lattice with axial anomaly," *J. Group Theory in Physics* **1**, 105-121 (1993).

SYMPLECTIC TOMOGRAPHY OF SCHRÖDINGER CAT STATES OF A TRAPPED ION

O.V. Man'ko

Lebedev Physical Institute
Moscow, Russia

The marginal distribution of squeezed, rotated, and shifted quadrature for two types of nonclassical states of a trapped ion – squeezed correlated states and squeezed even and odd coherent states (squeezed Schrödinger cat states) is studied

Key words: symplectic tomography, Paul trap, Schrödinger cat states.

Recently, in [1] it was shown that the steady state of a trapped ion irradiated by a bichromatic laser field is a superposition of two coherent states which is the even/odd coherent state introduced in [2] (Schrödinger cat states). The theory of an ion in a Paul trap was developed in [3-5] where the trapped ion was described by the model of quantum oscillator with a periodically varying frequency. In [6], a procedure was formulated to obtain the Wigner function of the quantum system in terms of the marginal distribution of rotated quadrature, which may be measured by a balanced homodyne detector and the scheme called optical tomography has been used experimentally [7]. In [8], a symplectic-tomography procedure was suggested in which the quantum states were measured by measuring the marginal distribution for squeezed, rotated, and shifted quadrature. In [9], a new equation in quantum mechanics was introduced describing time evolution of this marginal distribution which has completely classical form but contains all the information about the quantum system. The aim of this work is to consider two important types of nonclassical states of a trapped ion (squeezed and correlated states of the ion in a Paul trap [3–5] and even and odd coherent states of the ion irradiated by bichromatic laser field [1]) within the framework of the symplectic-tomography procedure and using new quantum evolution equation.

An ion in a Paul trap is described by the model of parametric oscil-

M. Ferrero and A. van der Merwe (eds.),
New Developments on Fundamental Problems in Quantum Physics, 225–229.

lator. For the trapped ion, the time dependence of the frequency is taken to be periodic [3]:

$$\omega^2(t) = 1 + \kappa^2 \sin^2 \Omega t\,.$$

It is easy to show that packet solutions to the Schrödinger equation may be introduced and interpreted as coherent states $(\hbar = 1)$:

$$\Psi_\alpha(x,t) = \pi^{-1/4}[\varepsilon(t)]^{-1/2} \exp\left(\frac{i\dot{\varepsilon}(t)x^2}{2\varepsilon(t)}\right) \exp\left\{-\frac{|\alpha|^2}{2} - \frac{\alpha^2\varepsilon^*(t)}{2\varepsilon(t)} + \frac{\sqrt{2}\alpha x}{\varepsilon}\right\}, \tag{1}$$

where the classical complex trajectory satisfies the equation

$$\ddot{\varepsilon}(t) + \omega^2(t)\varepsilon(t) = 0$$

with initial conditions

$$\varepsilon(0) = 1\,, \qquad \dot{\varepsilon}(0) = i\,,$$

where α is a complex number.

Other normalized solutions to the Schrödinger equation are a squeezed even coherent state and a squeezed odd coherent state [2] (squeezed Schrödinger cat states)

$$\Psi_\alpha^{(+)}(x,t) = 2N^{(+)}\Psi_0(x,t) \exp\left\{-\frac{|\alpha|^2}{2} - \frac{\varepsilon^*(t)\alpha^2}{2\varepsilon(t)}\right\}.\cosh\frac{\sqrt{2}\alpha x}{\varepsilon(t)}\,, \qquad N^{(+)} = \frac{\exp(|\alpha|^2/2)}{2\sqrt{\cosh|\alpha|^2}}\,, \tag{2}$$

$$\Psi_\alpha^{(-)}(x,t) = 2N^{(-)}\Psi_0(x,t) \exp\left\{-\frac{|\alpha|^2}{2} - \frac{\varepsilon^*(t)\alpha^2}{2\varepsilon(t)}\right\}.\sinh\frac{\sqrt{2}\alpha x}{\varepsilon(t)}\,, \qquad N^{(-)} = \frac{\exp(|\alpha|^2/2)}{2\sqrt{\sinh|\alpha|^2}}\,. \tag{3}$$

In [8], it was shown that for the generic linear combination of quadratures which is a measurable observable

$$\widehat{X} = \mu\hat{q} + \nu\hat{p} + \delta\,,$$

where $\hat{q}$ and $\hat{p}$ are the position and momentum, respectively, the marginal distribution $w(X, \mu, \nu, \delta)$ (normalized with respect to the X variable) depending upon three extra real parameters μ, ν, δ is related to the state of the quantum system expressed in terms of its Wigner function $W(q,p)$ as follows:

$$w(X,\mu,\nu,\delta)=\int \exp\left[-ik(X-\mu q-\nu p-\delta)\right]W(q,p)\frac{dkdqdp}{(2\pi)^2}\,. \tag{4}$$

The physical meaning of the parameters μ, ν, δ is that they describe ensemble of shifted, rotated, and scaled reference frames in which the position X is measured. This formula can be inverted and the Wigner function of state can be expressed in terms of the marginal distribution [8]. In [9], it was shown that for Hamiltonian systems the marginal distributions satisfy the quantum time-evolution equation, which for a trapped ion takes the form

$$\dot{w}-\mu\frac{\partial}{\partial\nu}w+\omega^2(t)\nu\frac{\partial}{\partial\mu}w=0\,. \tag{5}$$

Calculating the integral (4) one can show that for generic Gaussian packets of a trapped ion (also for particular case (1)) the marginal distribution is

$$w(X,\mu,\nu,\delta,t)=\frac{1}{\sqrt{2\pi\sigma_X(t)}}\exp\left\{-\frac{(X-\bar{X})^2}{2\sigma_X(t)}\right\}, \tag{6}$$

where the dispersion of the symplectic observable X and the mean value of the observable depend on the time and the parameters as follows:

$$\sigma_X(t)=\frac{1}{2}\left(\mu^2\mid\varepsilon\mid^2+\nu^2\mid\dot{\varepsilon}\mid^2+2\mu\nu\sqrt{\mid\varepsilon\dot{\varepsilon}\mid^2+1}\right), \tag{7}$$

$$\overline{X}=\frac{\alpha}{\sqrt{2}}(\mu\varepsilon^*+\nu\dot{\varepsilon}^*)+\frac{\alpha^*}{\sqrt{2}}(\mu\varepsilon+\nu\dot{\varepsilon})+\delta\,. \tag{8}$$

One can check that the normalized marginal distribution (6) with parameters (7) and (8) satisfy the evolution equation (5).

Now we will discuss the marginal distribution for nonclassical states of the parametric oscillator, namely, even and odd coherent states [2]. The Wigner function for even and odd coherent states is

$$\begin{aligned}W_\pm &= 4\mid N^\pm\mid^2\exp\left\{-p^2\mid\varepsilon\mid^2-\mid\dot{\varepsilon}\mid^2q^2+(\dot{\varepsilon}\varepsilon^*+\varepsilon\dot{\varepsilon}^*)pq\right\}\\ &\cdot\left\{e^{-2|\alpha|^2}\cosh\left(2\sqrt{2}\left[p\,\mathrm{Im}\,(\alpha\varepsilon^*)-q\mathrm{Im}\,(\alpha\dot{\varepsilon}^*)\right]\right)\right.\\ &\left.\pm\cos\left(2\sqrt{2}\left[q\,\mathrm{Re}\,(\alpha\dot{\varepsilon}^*)-p\mathrm{Re}\,(\alpha\varepsilon^*)\right]\right)\right\}.\end{aligned} \tag{9}$$

The marginal distribution of a trapped ion in even/odd coherent state is

$$\begin{aligned}w_\pm &= \mid N^\pm\mid^2\pi^{-1/2}\left[\mid\dot{\varepsilon}\mid^2\nu^2+\mid\varepsilon\mid^2\mu^2+2\mu\nu\mathrm{Re}\,(\dot{\varepsilon}\varepsilon^*)\right]^{-1/2}\\ &\cdot\{w_1+w_2\pm w_3\pm w_4\}\,,\end{aligned} \tag{10}$$

where

$$w_1 = \exp\left[-\frac{\{X-\delta+2\sqrt{2}\mathrm{Re}\,(\alpha[\varepsilon^*\mu+\nu\dot{\varepsilon}^*])\}^2}{|\,\dot{\varepsilon}\,|^2\,\nu^2+|\,\varepsilon\,|^2\,\mu^2+2\mu\nu\mathrm{Re}\,(\dot{\varepsilon}\varepsilon^*)}\right],$$

$$w_2 = \exp\left[-\frac{\{X-\delta-2\sqrt{2}\mathrm{Re}\,(\alpha[\varepsilon^*\mu+\nu\dot{\varepsilon}^*])\}^2}{|\,\dot{\varepsilon}\,|^2\,\nu^2+|\,\varepsilon\,|^2\,\mu^2+2\mu\nu\mathrm{Re}\,(\dot{\varepsilon}\varepsilon^*)}\right],$$

$$w_3 = \exp\left[-2\,|\,\alpha\,|^2-\frac{\{X-\delta+2i\sqrt{2}\mathrm{Im}\,(\alpha[\varepsilon^*\mu+\nu\dot{\varepsilon}^*])\}^2}{|\,\dot{\varepsilon}\,|^2\,\nu^2+|\,\varepsilon\,|^2\,\mu^2+2\mu\nu\mathrm{Re}\,(\dot{\varepsilon}\varepsilon^*)}\right],$$

$$w_4 = \exp\left[-2\,|\,\alpha\,|^2-\frac{\{X-\delta-2i\sqrt{2}\mathrm{Im}\,(\alpha[\varepsilon^*\mu+\nu\dot{\varepsilon}^*])\}^2}{|\,\dot{\varepsilon}\,|^2\,\nu^2+|\,\varepsilon\,|^2\,\mu^2+2\mu\nu\mathrm{Re}\,(\dot{\varepsilon}\varepsilon^*)}\right].$$

In the present work, we calculated the marginal distribution of a symplectic observable (which is a generic linear quadrature) for nonclassical states of a trapped ion modeled by a parametric quantum oscillator. Measurements of the marginal distribution give the possibility of measuring the quantum states. In the case of the particular choice of the parameters $\mu = \cos\varphi$; $\nu = \sin\varphi$, the measurement is reduced to finding the marginal distribution for homodyne output and reconstructing the Wigner function by means of the Radon transform of the optical-tomography scheme [6,7]. The distribution found for generic linear quadrature satisfies a new classical-like equation of quantum dynamics introduced in the symplectic-tomography scheme.

ACKNOWLEDGEMENTS

This work was partially supported by the Russian Foundation for Basic Research (Project No. 96–02–18623) and by the RF State Program "Optics. Laser Physics." O.V.M. would like to express her sincere appreciation to the Organizing Committee of the Second International Symposium on Fundamental Problems in Quantum Physics for the invitation and Professor M. Ferrero and Dr. S.F. Huelga for their hospitality.

REFERENCES

1. R.L. de Matos Filho and W. Vogel, "Even and odd coherent states of the motion of a trapped ion," *Phys. Rev. Lett.* **76**, 608–611 (1996).
2. V.V. Dodonov, I.A. Malkin, and V.I. Man'ko, "Even and odd coherent states and excitations of a singular oscillator," *Physica* **72**, 597–615 (1974).
3. R.J. Glauber, "The quantum mechanics of a particle in time–dependent quadrupole fields," in *Recent Developments in Quantum Optics, Proceed-*

ings of the International Conference on Quantum Optics (Hyderabad, 1991), R. Inguva, ed. (Plenum, New York, 1993), pp. 1–13.
4. G. Schrade, V.I. Man'ko, W. Schleich, and R.J. Glauber, "Wigner functions in the Paul trap," *Quantum Semiclass. Opt.* **7**, 307–325 (1995).
5. O.V. Man'ko, "Symplectic tomography of nonclassical states of a trapped ion," Preprint IC/96/39, ICTP, Trieste (1996); Los Alamos Report No. quant–ph/9604018; *J. Russ. Laser Research* (Plenum, New York) **17**, 439–448 (1996).
6. K. Vogel and H. Risken, "Determination of quasiprobability distributions in terms of probability distributions for the rotated quadrature phase," *Phys. Rev.* A **40**, 2847–2849 (1989).
7. D.T. Smithey, M. Beck, M.G. Raymer, and A. Faridani, "Measurement of the Wigner distribution and the density matrix of a light mode using optical homodyne tomography: application to squeezed states and vacuum," *Phys. Rev. Lett.* **70**, 1244–1247 (1993).
8. S. Mancini, V.I. Man'ko, and P. Tombesi, "Wigner function and probability distribution for shifted and squeezed quadratures," *Quantum Semiclass. Opt.* **7**, 615–623 (1995).
9. S. Mancini, V.I. Man'ko and P. Tombesi, "Symplectic tomography as a classical approach to quantum systems," *Phys. Lett.* A, **213**, 1–6 (1996).

MAGICAL PHOTON OR REAL ZEROPOINT?

Trevor W. Marshall

Department of Mathematics, University of Manchester
Manchester M13 9PL, United Kingdom

We show that the phenomenon known as "parametric down conversion" in optically nonlinear crystals, has a completely classical description once due account has been taken of Planck's zeropoint electromagnetic field. Certain properties of the radiation emitted by this system appear to be nonlocal when described in terms of photons, but are entirely local in the zeropoint-field description. It now becomes clear that these new phenomena follow a long established pattern. From Young's two-slit to the atomic-cascade-polarization ("Aspect") experiment, they may all be described in terms of either a nonlocal (that is, magical) photon concept, or by a local (that is, realist) field concept, with the zeropoint part of that field taken into account where relevant.

Key words: vacuum, fluctuations, interference, coincidence counts.

1. INTRODUCTION

The concept of light as particles was born in 1704 with Newton's *Opticks*, superseded in the early 1800s by the waves of Young and Fresnel, and resurrected a century later as Einstein's *Lichtquanten*,which since about 1925 have come to be known as photons. Yet the interference and diffraction experiments of Young and Fresnel, together with their wave description, remain as valid as ever. Einstein's photon-particles fare no better than Newton's, since they contradict the principles of locality and causality, as enunciated by Einstein, Podolsky and Rosen in 1935 and by Einstein in 1948.

Young's two-slit experiment demonstrated the wave character of light through the interference fringes behind a pair of slits illuminated by a single source. The wave explanation drew an analogy with surface water waves;

M. Ferrero and A. van der Merwe (eds.),
New Developments on Fundamental Problems in Quantum Physics, 231–245.

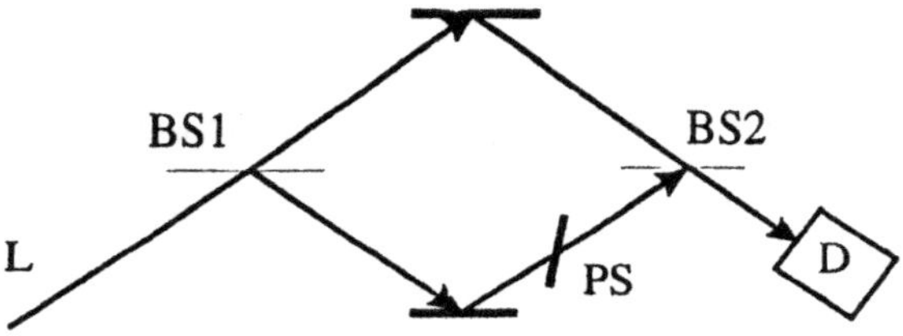

Fig. 1. Young's two-slit experiment. The prism and the slit in the first screen produce a monochromatic light source, and the interference fringes on the right hand screen demonstrate the wave nature of light. No satisfactory explanation of this experiment using a corpuscular description has ever been given.

the waves from a source in one tank spread through two holes into a second tank, and a trough from one wave group cancels a peak, or reinforces a trough, from the other, depending on their relative phases. G. I. Taylor subsequently conducted low intensity experiments ('single photons'), which implied that a photon passing through one slit would have to 'know' whether the second slit was open or not. Thus the particle-like photon conflicts with Einstein's locality principle, which states that no physical action can travel faster than light. The *wave-particle duality* idea of Louis de Broglie suffers the same fate, since the pilot waves of that theory also propagate with speeds greater than light.

Modern versions of the two-slit experiment (Fig. 1) use a laser (L) as a directed source of monochromatic light, and beam splitters (BS) to divide or combine light beams. We call the two waves incident on a beam splitter d_1 and d_2 (as we shall see shortly, there are *always* two incident waves!). Then the transmitted and reflected waves are

$$e_1 = Td_1 + iRd_2, \quad e_2 = Td_2 + iRd_1, \tag{1}$$

where R and T are real reflection and transmission coefficients, satisfying

$$R^2 + T^2 = 1. \tag{2}$$

The emergent waves are diverted by mirrors and recombined in a second beam splitter. If the path-length difference between the twin laser beams is varied by the orientation of the phase shifter (PS), we find fringes in the counting rate observed at the detector (D). The *fringe visibility*, defined as

$$\text{F.V.} = \frac{N_{\max} - N_{\min}}{N_{\max} + N_{\min}}, \tag{3}$$

where N_{max} and N_{min} are the maximum and minimum counting rates at D, is given by

$$\text{F.V.} = \frac{2R_1T_1R_2T_2}{R_1^2T_2^2 + T_1^2R_2^2} \exp\left[-\frac{|p_u - p_l|}{p_L}\right], \tag{4}$$

where we have put suffixes on R and T to indicate which of the beam splitters they refer to, p_u and p_l are the upper and lower path lengths of the laser beams, and p_L is its coherence length, typically several cm. Note that, by a suitable choice of the parameters in Eq. (4), F.V. may be made very close to one.

The explanation of this experiment in terms of the nineteenth-century wave description is straightforward, but again the photon "explanation" really explains nothing. Indeed, its inadequacy can be shown dramatically by replacing the split-laser source with a pair of independent but carefully tuned lasers. An experiment along these lines was performed by Pfleegor and Mandel in 1967 [1], and the two waves interfered, just as predicted in the wave theory, even at "single-photon" intensities.

The zeropoint field (ZPF) was conceived by Max Planck in 1911 as a wave alternative to Einstein's *Lichtquant* explanation of the photoelectric effect and black body spectrum. The ZPF has been used in more recent times to explain a range of phenomena including the Lamb shift, Casimir effect, van der Waals forces and paramagnetism [2]. The anticorrelation phenomenon [3] which occurs in the two outgoing channels of a beam splitter has also been explained with the ZPF, thus indicating the possibility that all particle-like properties may be explained using this extension of the wave theory [4,5]. The essential idea is to replace the "vacuum" by a space filled with electromagnetic waves; each mode of the field contains an intensity equivalent to half a photon. In situations like that of Fig. 1, where the first beam splitter appears to have only one wave incident upon it, we must always take account of the zeropoint wave incident in the other channel. Such an intervention makes no difference if the first incident wave is intense, the so-called "classical" situation, but becomes crucial in the case of weak ("single-photon") signals, and this is the situation in which anticorrelation occurs.

So, why has so little attention been paid to the ZPF? When Einstein,in 1905, first proposed his *Lichtquanten*, hardly anyone took them seriously, yet nowadays it is quite difficult to find an article in optics which is not couched in photon language, and it is Max Planck's alternative idea which remains in the margin. It was not always so. In 1913 [6], when the scientific leaders of the day, that is Max Planck and Walther Nernst, wanted to nominate Einstein for the chair of theoretical physics in Berlin, they said that his strange ideas about the nature of light should not stand in his way. They continued to develop the ZPF idea during the next four years, with Einstein himself also participating [7]. Yet the early successes of the new quantum mechanics in the period 1917-1930 caused us to forget the conceptual difficulties of photons. Only relatively recently have we come to realise the full strangeness these mythical objects would have to possess; a recent article [8] has, quite appropriately, described the photon description of certain phenomena (see below) as "mind boggling". And yet...everyone has to write articles as though they really existed! Clearly we must reckon, in 1996 as in 1913, with the inertia of the scientific community.

2. THE PHENOMENON OF INDUCED EMISSION

Nonlinear optics is an experimental discipline which has developed in the last thirty years; it is a side product of laser technology. Induced emission is a nonlinear extension of such long established optical phenomena as refraction and double refraction. Thus refraction - the slowing up of light waves in a material medium - results from the *polarization* of the medium by an incoming light wave; if the electric vector of this wave is $\mathbf{E}$, there is a displacement of the electrons in the medium, which takes up the configuration of an array of dipoles. We represent this effect by the linear relation

$$\mathbf{P} = \alpha \mathbf{E}, \tag{5}$$

and α is called the *polarizability* of the medium; then the relation between the velocity of light in the medium, c , and that *in vacuo* is

$$c^2 = c_{\mathbf{vac}}^2/(1 + 4\pi\alpha). \tag{6}$$

However, at optical frequencies there is a retardation effect, that is Eq. (5) should be replaced by

$$\mathbf{P}(t) = \int_0^\infty \alpha(t')\mathbf{E}(t - t')dt'. \tag{7}$$

This results in c being a function of ω, that is

$$c^2(\omega) = c_{\mathbf{vac}}^2/[1 + 4\pi\tilde{\alpha}(\omega)], \tag{8}$$

where $\tilde{\alpha}(\omega)$ is the Fourier transform of $\alpha(t)$; waves of different frequency travel at different speeds.

The above is a good description of refraction in fluids and disordered solids, but crystals exhibit double refraction. This is because α is not a scalar but a second-rank tensor, that is

$$P_i = \sum_j \alpha_{ij} E_j, \tag{9}$$

so the polarization is not, in general, parallel to the electric vector. This results in waves with two perpendicular directions of polarization, the 'ordinary' and 'extraordinary' waves, being propagated with two different speeds, even though they have the same frequency. These two directions of polarization are the eigenvectors of α.

Now if the incoming wave is intense enough - as is the case for a laser input - we have to take account of *nonlinear polarizability*, that is we replace Eq. (5) by

$$\mathbf{P} = \alpha\mathbf{E} + \beta\mathbf{E}^2, \tag{10}$$

or rather a generalized form of this relation, in which β, like α, has a time dependence as well as a tensorial nature.

The phenomena of refraction and double refraction now become richer and more complex, and one new phenomenon which emerges is induced emission. It is revealed by using *two* laser inputs, L_0 and L_1 (see Fig. 2), which we represent by plane waves

$$\mathbf{E}_0(\mathbf{r},t) = \mathbf{v}_0 \cos\left[\omega_0\left(t - \frac{\mathbf{n}_0.\mathbf{r}}{c_0(\omega_0)}\right)\right], \tag{11}$$

$$\mathbf{E}_1(\mathbf{r},t) = \mathbf{v}_1 \cos\left[\omega_1\left(t - \frac{\mathbf{n}_1.\mathbf{r}}{c_1(\omega_1)}\right)\right], \tag{12}$$

incident on a nonlinear crystal (NLC).

Then the part of $\mathbf{P}$ which comes from $(\mathbf{E}_0 + \mathbf{E}_1)^2$ contains terms representing oscillations of the crystal at the frequency $(\omega_0 - \omega_1)$. For certain combinations of ω_1 and $\mathbf{n}_1$ the dipole array associated with this frequency produces a strong coherent radiation in the direction $(\omega_0\mathbf{n}_0/c_0 - \omega_1\mathbf{n}_1/c_1)$; this is induced emission. More precisely (see Fig. 2), consider what happens when we hold ω_0 and θ_i fixed, and vary the frequency ω_1 between 0 and ω_0. Then induced emission occurs, in the direction θ_s, when ω_1 passes through the frequency ω_i, and the frequency of the emitted wave is $\omega_s = \omega_0 - \omega_i$. The (approximately) plane wave (ω_s, θ_s) is called the *signal wave*, and its conjugate (ω_i, θ_i) is called the *idler wave*; their roles may be reversed, that is if the incident angle between L_0 and L_1 is θ_s, then induced emission occurs at $\omega = \omega_s$ in the direction θ_i.

These two waves are related by the *phase matching conditions*

$$\left|\frac{\omega_0}{c_0} - \frac{\omega_i \cos\theta_i}{c_i} - \frac{\omega_s \cos\theta_s}{c_s}\right| \lesssim \frac{\sin\theta_i}{\delta_1}, \tag{13}$$

$$\left|\frac{\omega_s \sin\theta_s}{c_s} - \frac{\omega_i \sin\theta_i}{c_i}\right| \lesssim \frac{1}{\delta_0} + \frac{\cos\theta_i}{\delta_1}, \tag{14}$$

where c_0, c_i, c_s are the wave velocities, inside the crystal, of light at the corresponding frequencies. For example, if L_0 is violet (wavelength ~400nm),

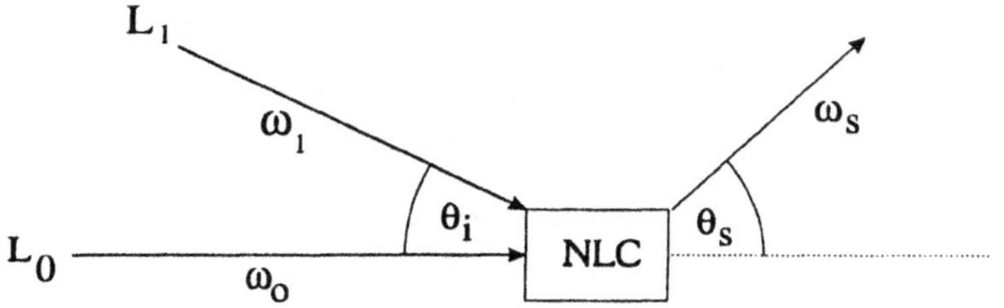

Fig. 2. Induced emission. Two lasers, L_0 with frequency ω_0 and L_1 with frequency ω_1, are incident on a nonlinear crystal. The angle, θ_i, between them is kept fixed while the frequency ω_1 is varied continuously from 0 to ω_0. Induced emission is observed, at a frequency $(\omega_0 - \omega_1)$ and in a direction θ_s, when the frequency ω_1 passes through the "resonance" at ω_i.

and if NLC has a refractive index for violet light which is 2 percent less than for red light, then, with θ_i about 11 degrees, the "resonance" will occur in the red part of the spectrum (wavelength of ω_i ~800nm), and the signal (ω_s) will also have around the same wavelength. These relations also give us the linewidth of the resonance; it is related to the widths of the laser beams, δ_0 and δ_1; the wider the beams, the sharper is the resonance.

This resonance actually has a fine structure, because the wave velocities c_0, c_i and c_s are different for the ordinary and extraordinary polarizations, and hence there will be eight different "spectral lines" corresponding to the possible combinations of polarizations of ω_0, ω_i and ω_s. Of course the relative intensities of these lines will depend not only on the wave vectors of L_0 and L_1, but also on their polarization vectors, which must be expressed as linear combinations of the ordinary and extraordinary polarizations. For example, if L_0 and L_1 are both polarized in their ordinary directions, there will be only two lines. The intensity of a given spectral line also depends on the structure of the tensor β, and the ability of the apparatus to actually separate these lines depends on the relative size of the line separations and the linewidths, which, as we saw above, are related to the laser beamwidths.

We can do a nonlinear-optics version of Young's experiment, by splitting the laser L_0, in the manner of Fig. 1, and then letting the twin lasers fall on two separate nonlinear crystals (see Fig. 3) which have been carefully aligned so that the second laser, L_1, falls on both of them. Because the twins produced at BS1 are mutually coherent, the same is true of the induced signals, s_1 and s_2. The phase shifter (PS) in Fig. 3 plays the same role as in Fig. 1; as its orientation is varied we observe fringes in the counting rate at the detector D. The phenomenon in Fig. 3 may be called *induced coherence.* Now the fringe visibility, for the setup of Fig. 3, is

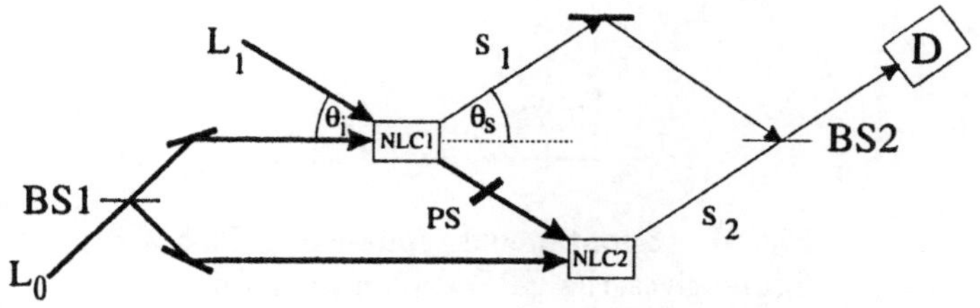

Fig. 3. Induced coherence. Two mutually coherent signals, produced by a beam splitter BS1 acting on a laser L_0, are incident on the nonlinear crystals NLC1 and NLC2. A second laser L_1 is incident on both crystals, and its angle of incidence is adjusted so that the two pairs of laser inputs produce induced emissions, s_1 and s_2, in the two crystals. The mutual coherence of s_1 and s_2 is demonstrated by the interference fringes produced at the detector D when the optical-path length separating NLC1 and NLC2 is varied with the phase shifter PS.

$$\text{F.V.} = \frac{2R_1T_1R_2T_2}{R_1^2T_2^2 + T_1^2R_2^2} \exp\left[-\frac{|p_u - p_l|}{p_L}\right] \exp\left[-\frac{|q_u - q_l|}{p_s}\right], \tag{15}$$

where q_u and q_l are the upper and lower optical path lengths from NLC1 to BS2, p_s is the coherence length of s_1 and s_2 (typically about 0.5mm), and the other symbols have the same meaning as in Eq. (4) The similarity of Eq. (15) to Eq. (4) confirms our classification of the phenomenon of Fig. 3 as another form of the two-slit experiment.

3. ENTER THE ZEROPOINT FIELD

The phenomena described in the previous section are rather simple to understand in a nineteenth-century, that is undulatory, description of light. What is somewhat surprising is that, if we reduce the intensity of the laser L_1 (see Fig. 2) to zero, the signal of frequency ω_s in the direction θ_s is reduced in intensity, but it does not disappear altogether. This phenomenon has been known since the 1960s, and was initially called *spontaneous scattering.* Nowadays it is more commonly known as *parametric down conversion*, because the modern "understanding" (which is actually a very profound error!) is that a photon of the laser is converted, by the crystal, into two photons of frequencies ω_i and ω_s. We propose to revert to the terminology of the 1960s and call it "spontaneous emission". In so doing we are saying that (see Fig. 4) *spontaneous emission is the same as induced emission,*

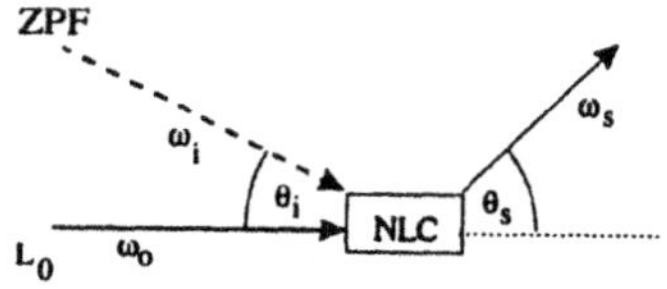

Fig. 4. Spontaneous emission. This figure gives the ZPF explanation for a quantum phenomenon, generally called "parametric down conversion". It is the same as a classical phenomenon - induced emission, see Fig. 2 - with the second laser replaced by a component of the ZPF. Since all components of the ZPF induce such emission, we actually see a rainbow of emitted signals.

with the zeropoint field taking the place of the second laser; the emission of light at frequency ω_s results from the joint action, on the crystal, of the laser L_0 and a component, of frequency ω_i, of the zeropoint field. Since *all* components of the ZPF participate, what we actually see is a rainbow of emitted radiation; any given colour may be emphasized by reinforcing the ZPF component with a secondary laser of the same frequency and direction. In this way we make the transition from the quantum phenomenon of Fig. 4 to the classical one of Fig. 2.

Recently Wang, Zou and Mandel [9] discovered that, also in the induced coherence experiment of Fig. 3, the interference fringes at D remain unchanged when the laser L_1 is removed. This is a rather striking phenomenon, which its discoverers have called "induced coherence without induced emission", and which some commentators have subsequently qualified as "mind boggling" [8]. However, once we have understood the transition from the situation of Fig. 2 to that of Fig. 4, we can make a similar transition, replacing the laser L in Fig. 3 by its corresponding ZPF component - see Fig. 5. Then the appropriate name for the phenomenon discovered by Wang, Zou and Mandel is *spontaneous coherence*; it is related to spontaneous emission in the same way as is induced coherence to induced emission, and the ZPF is the vital link in understanding this set of relations.

Wang, Zou and Mandel also demonstrated that the signals s_1 and s_2 become incoherent, so that the fringes at D disappear, if the path between NLC1 and NLC2 is interrupted by a mirror. These authors considered this to be the most surprising and nonlocal behaviour by the participating photons, and it was this feature which was subsequently described as "mind-

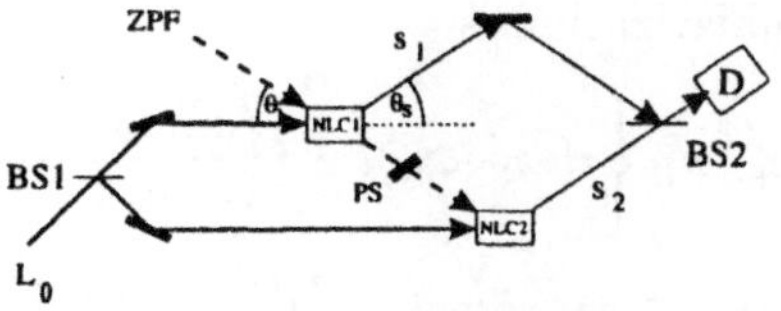

Fig. 5. Spontaneous coherence. This is the phenomenon whose photon description is mind boggling. Its ZPF explanation is simple. All we have to do is compare the present figure with Fig. 3; the laser input L_1 has again been replaced by a component of the ZPF, We have to select the same component from both rainbows in order to demonstrate the coherence effect.

boggling". In terms of the description we are offering there is no mystery; the coherence disappears because the two idler components of the zeropoint field are now completely different; the first is reflected away by the upper face of the mirror, and its place is taken by a different one which reflects off the lower face. In the photon description, there are no incident idler components; *only idler photons emitted by one or other of the two crystals.* So the "explanation" offered, in that description, for the coherence is that the alignment causes the two idler photons to be indistinguishable. Then, by the Feynman rules for combining probability amplitudes, we must add the amplitudes for these two photons before squaring to obtain a probability. This is a typically *post hoc* and nonlocal quantum mechanical "explanation". To be prepared to accept it you must indeed first boggle your mind; the *Shorter Oxford Dictionary* tells us that to boggle is to deceive!

We now consider, in more detail, the spontaneous emission process of Fig. 4. The total field incident on the crystal is

$$E(\mathbf{r},t) = E_0(\mathbf{r},t) + i\sum_{\mathbf{k}}(\pi\hbar\omega_{\mathbf{k}}/L^3)^{1/2}[z_{\mathbf{k}}e^{i(\omega_{\mathbf{k}}t-\mathbf{k}.\mathbf{r})} - \text{c.c.}], \qquad (16)$$

where $E_0(\mathbf{r},t)$ is the pump field, defined in Eq. (11) L^3 is the volume of a large cube in which the normal modes are defined, and $z_{\mathbf{k}}$ are a set of independent complex random variables, all having zero mean and unit variance, representing the mode amplitudes of the ZPF. (For more on this representation of the ZPF see Refs. [5,10,11].) For simplicity the vector character of the field has been suppressed; at this point we shall not concern ourselves with its polarization properties. Then the current associated with the part

βE^2 of the crystal's polarization is

$$j(\mathbf{r},t) = \beta v_0 \sum_{\mathbf{k}} \left[z_{\mathbf{k}}^* \exp\left\{ i(\omega_0 - \omega_{\mathbf{k}})t - i\left(\frac{\omega_0 \mathbf{n}_0.\mathbf{r}}{c_0} - \mathbf{k}.\mathbf{r}\right)\right\} + \text{c.c.}\right] \tag{17}$$
$$(\omega_0 - \omega_{\mathbf{k}})(\pi\hbar\omega_{\mathbf{k}}/L^3)^{1/2} + \text{N.R.C.},$$

where N.R.C. designates nonradiating components, associated with the frequencies $(\omega_0 + \omega_{\mathbf{k}})$. The spontaneous-emission field in the direction $\mathbf{n}$ is what this current radiates, namely [12]

$$E_s(\mathbf{r},t;\mathbf{n}) = \frac{i\beta v_0}{r} \sum_{\mathbf{k}} \frac{(\omega_0 - \omega_{\mathbf{k}})^2}{\tilde{c}_{\mathbf{k}}^2} (\pi\hbar\omega_{\mathbf{k}}/L^3)^{1/2}$$
$$z_{\mathbf{k}}^* g_{\mathbf{k}}(\mathbf{n}) \exp\left[i(\omega_0 - \omega_{\mathbf{k}})\left(t - \frac{\mathbf{n}.\mathbf{r}}{\tilde{c}_{\mathbf{k}}}\right)\right] - \text{c.c.}, \tag{18}$$

where $\tilde{c}_{\mathbf{k}}$ is the velocity, in the crystal, of a wave whose frequency is $(\omega_0 - \omega_{\mathbf{k}})$, and

$$g_{\mathbf{k}}(\mathbf{n}) = \int_V \exp\left[\frac{i\omega_0 \mathbf{n}_0.\mathbf{r}'}{c_0} - i\mathbf{k}.\mathbf{r}' - \frac{i(\omega_0 - \omega_{\mathbf{k}})\mathbf{n}.\mathbf{r}'}{\tilde{c}_{\mathbf{k}}}\right] d^3\mathbf{r}', \tag{19}$$

V being the irradiated volume of the crystal. Note that we are following the simplification of Hong and Mandel [13], by supposing that the nonlinear crystal is embedded in a linear medium of the same dielectric constant; this means we do not have to worry about reflection at the interface. The function $g_{\mathbf{k}}(\mathbf{n})$ is strongly peaked for those values of $\mathbf{k}$ and $\mathbf{n}$ satisfying the phase matching relations (13) and (14) above, but with δ_1 now, effectively, infinite. This is the origin of those relations, and it is what justifies Fig. 4; any given component of the spontaneous-emission rainbow is associated with a narrow range of components of the ZPF.

4. THE ROLE OF THE RADIATIVE REACTION

Now let us extend the analysis of Fig. 4, which we have just made, to Fig. 5. The signals s_1 and s_2 may be represented by the radiated fields

$$E_{s_1}(\mathbf{r},t;\mathbf{n}) = \frac{i\beta T_1 v_0 e^{i\omega_0 p_1/c_0}}{|\mathbf{r} - \mathbf{r}_1|} \sum_{\mathbf{k}} \frac{(\omega_0 - \omega_{\mathbf{k}})^2}{\tilde{c}_{\mathbf{k}}^2} (\pi\hbar\omega_{\mathbf{k}}/L^3)^{1/2}$$
$$z_{\mathbf{k}}^* e^{i\mathbf{k}.\mathbf{r}_1} g_{\mathbf{k}}(\mathbf{n}) \exp\left[i(\omega_0 - \omega_{\mathbf{k}})\left(t - \frac{\mathbf{n}.(\mathbf{r} - \mathbf{r}_1)}{\tilde{c}_{\mathbf{k}}}\right)\right] - \text{c.c.}, \tag{20}$$

$$E_{s_2}(\mathbf{r},t;\mathbf{n}) = \frac{\beta R_1 v_0 e^{i\omega_0 p_2/c_0}}{|\mathbf{r} - \mathbf{r}_2|} \sum_{\mathbf{k}} \frac{(\omega_0 - \omega_{\mathbf{k}})^2}{\tilde{c}_{\mathbf{k}}^2} (\pi\hbar\omega_{\mathbf{k}}/L^3)^{1/2}$$
$$z_{\mathbf{k}}^* e^{i\mathbf{k}.\mathbf{r}_2} g_{\mathbf{k}}(\mathbf{n}) \exp\left[i(\omega_0 - \omega_{\mathbf{k}})\left(t - \frac{\mathbf{n}.(\mathbf{r} - \mathbf{r}_2)}{\tilde{c}_{\mathbf{k}}}\right)\right] - \text{c.c.}, \tag{21}$$

where p_1 and p_2 are the optical path lengths from BS1 to NLC1 and NLC2, and $\mathbf{r}_1$ and $\mathbf{r}_2$ are the position vectors of two reference points in NLC1 and NLC2; if these are suitably chosen, the coefficient $g_{\mathbf{k}}(\mathbf{n})$ may be taken as identical for both of these fields. Now these two fields are mixed by BS2, in accordance with one of Eqs. (1). Of course, the wide spectrum of frequencies in these fields implies an equally wide spectrum of phase differences between the components of s_1 and s_2, so no fringes are observed unless the signals are first filtered to produce a narrow range of frequencies at D. This is what was done in the experiment of Wang, Zou and Mandel; the frequency filter effectively, through the phase matching relations (13) and (14), selects a pair of directions, $\mathbf{n}$ for the signal and $\mathbf{k}$ for the idler. Then spontaneous coherence is demonstrated, by fringes at D, when NLC1 and NLC2 are aligned so that $(\mathbf{r}_1 - \mathbf{r}_2)$ is parallel to $\mathbf{k}$. The fringe visibility has been calculated as [14]

$$\text{F.V.} = \frac{2R_1T_1R_2T_2}{R_1^2T_2^2 + T_1^2R_2^2} \exp\left[-\frac{\mid q_u - q_l \mid}{p_s}\right], \tag{22}$$

where the symbols have the same meaning as in Eq. (15), which gave the fringe visibility for induced coherence. The only differences between these two fringe visibilities are, first that we have now simplified by supposing that $(p_u - p_l)$ is small compared with p_L, and second that p_s is determined by the combination of the laser beamwidth and the filter, instead of by two laser beamwidths.

The result we have just obtained agrees exactly with the fringe visibility calculated by Wang, Zou and Mandel (WZM), using the usual quantum-optical representation for the fields, in which all the mode amplitudes are represented by operators. There is, however, a significant, though experimentally untestable, difference between the detection rates on which the calculation of F.V. is based; *the detection rates given by the above procedure are all precisely one half of the WZM rates.* This is because the vacuum mean value of the random variable $z_{\mathbf{k}}z_{\mathbf{k}}^*$ is one half that of the corresponding operator product $\hat{z}_{\mathbf{k}}\hat{z}_{\mathbf{k}}^\dagger$.

The above calculation is obviously incomplete, because the *total* field at D is, of course, the sum of the radiated and the incident fields. To be sure, the filtration procedure removes all the components associated with the pump and idler fields, but there are components of the ZPF, not taken account of in Fig. 5, which do enter the detector. A fuller picture of the situation is provided by Fig. 6, where we see that two new, independent parts of the ZPF, denoted s_{01} and s_{02}, should be taken into account. If the detection rate at D were simply proportional to the intensity, as has been the case in all semiclassical radiation theories previously considered [15], then the fringe visibility would be drastically reduced by the intervention of these extra components. But *a semiclassical radiation theory with a real*

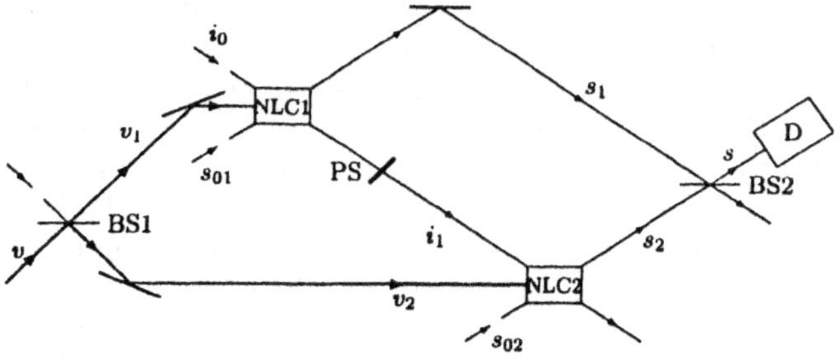

Fig. 6. This is a fuller version of Fig. 5, in which we show all the parts of the ZPF, denoted by interrupted lines, contributing to the signal s entering the detector D.

ZPF, must, of necessity have a modified theory of detection; since the ZPF on its own gives a zero (or very small) dark rate, we must assume that detectors measure only the field intensity above the ZPF background threshold. The above calculation, though it did not explicitly display this threshold subtraction, effectively calculates the fields to first-order perturbation theory, that is

$$s_1 = s_{01} + \beta s'_{01}, \quad s_2 = s_{02} + \beta s'_{02}. \tag{23}$$

Then the threshold subtraction at D removes all zeroth order terms in β, while the first-order terms average to zero. Hence only the second-order terms survive, which is equivalent to the above calculation. But we should have gone to *second*-order perturbation theory, that is

$$s_1 = s_{01} + \beta s'_{01} + \beta^2 s''_{01}, \quad s_2 = s_{02} + \beta s'_{02} + \beta^2 s''_{02}. \tag{24}$$

The above calculation included only the intensities coming from $s'_{01}s'^{*}_{01}$, $s'_{02}s'^{*}_{02}$, $s'_{01}s'^{*}_{02}$ and its complex conjugate, whereas a full calculation includes also $s_{01}s''^{*}_{01}$, $s_{02}s''^{*}_{02}$ and their conjugates. (Note that $s_{01}s''^{*}_{02}$ etc. all average to zero.) A remarkable simplification then occurs [14]; these second-order contributions simply duplicate the first-order ones, and *we end up with all of the detection rates multiplied by two.* This is precisely the same as the WZM result.

Is this just an algorithmic curiosity? - a complicated derivation of a relatively simple result from quantum optics? I say, very emphatically, NO - it is very much more! The first-order contributions are direct vacuum-field effects, whereas the second-order contributions represent the action, on the crystal, of its own radiation field; they are the continuous-medium equivalent of the classical *radiative reaction*, on a point particle, first calculated by

Lorentz. The role of this force has remained obscure throughout the whole development of quantum electrodynamics (Ref. [10], p. 142), and it has been argued, from within that discipline [16], that a spontaneous transition from an upper to a lower atomic energy level occurs under the influence of two complementary electromagnetic fields - ZPF and radiative reaction - which contribute equally to the transition rate. A similar description applies, within stochastic electrodynamics [5,11], when we consider the case of a charged linear oscillator acted on by the ZPF. Some of the most interesting features of the real physical process are hidden in the very simplicity of the quantum optical calculation.

Indeed, though the overall effect of the second-order terms is merely to duplicate that of the first-order ones, the detailed breakdown of these terms is not without interest. The term s''_{02} in Eq. (24) above differs from s''_{01} in having not only a part representing the action of the radiative field of NLC2 on NLC2, but also another part representing the action of the radiative field of NLC1 on NLC2. The fact that no such interactive part is present in s''_{01} is evidence of the causal nature of our description. This feature of our description also confirms another result obtained within stochastic electrodynamics [17], with respect to the internal oscillations of a linear crystal embedded in the ZPF. There also it is necessary to classify as "radiative reaction" the action, on a given atom in the crystal, of the fields radiated by all the other atoms. In other words, in an extended system, whether two crystals or a lattice of atoms in a single crystal, the concept of radiative reaction field should include all the retarded-field interactions between the component parts of the system.

5. CONCLUSION

Throughout most of the twentieth century two alternative descriptions have been available for the quantum properties of light - Photons (since 1905) and Zeropoint Field (since 1911). The latter has been strong in its explanatory, but underdeveloped in its algorithmic power, while the former, by contrast, has aquired a formidable algorithmic power, especially in the hands of Dirac [18] and Feynman [19], but is rather weak on explanation. More recently, however, it has become clear [5,10,11] that there is a close resemblance between the formalism used to describe the ZPF - called *stochastic electrodynamics* - and that of Dirac and Feynman - called *quantum electrodynamics*. This means we can now look forward to a return to the nineteenth-century situation, when physical theories were both powerful and understandable. The twenty-first century will return to a classical description of the electromagnetic field. Just as the nineteenth century rejected Newton's light particles, so will the twenty-first reject those of Einstein, Dirac and Feynman. But Einstein will have the consolation of being proved correct in insisting that scientific explanations must conform to his locality principle.

We remark that it is a straightforward matter to extend the ZPF description of nonlinear crystals to cover the cases of "two-photon interference" discussed in Reference [8]. As regards the celebrated set of experiments using atomic cascades, known popularly as "The Aspect Experiment" [20,21], we have already shown [22] that their description, along the lines of ZPF, does not suffer from the nonlocality exhibited by all photon descriptions. Although the treatment of processes involving single atoms is necessarily less complete than of those using nonlinear crystals - atoms are less classical than crystals - our work nevertheless indicates that, also with atoms, a deeper understanding of how the ZPF produces spontaneous emission, and of how this relates to induced emission, is what we have to seek in order to obtain a proper explanation of these experiments.

No competent magician *believes* in magic.

A competent magician is really a scientist *pretending* to be a magician.

A *competent* magician is one who knows how the rabbit got into the hat!

ACKNOWLEDGEMENT

I am grateful to M. O. Scully for drawing my attention to his work [23], where a similar attitude is taken to the zeropoint field's role in parametric down conversion.

REFERENCES

1. R. L. Pfleegor and L. Mandel, *Phys. Rev.* **159**, 1084 (1967); *J. Opt. Soc. Am.* **58**, 946 (1968).
2. T. H. Boyer, *Scientific American*, August 1985.
3. P. Grangier, G. Roger and A. Aspect, *Europhys. Lett.* **1**, 173 (1986).
4. T. W. Marshall and E. Santos, *Found. Phys.* **18**, 185 (1988).
5. T. W. Marshall, "Is there a bridge connecting stochastic and quantum-electrodynamics?," in *Fundamental Problems in Quantum Physics*, M. Ferrero and A. van der Merwe, eds. (Kluwer, Dordrecht, 1995), pp. 187-198.
6. A. Pais, *Subtle is the Lord* (Oxford University Press, Oxford, 1982), p. 384.
7. A. Einstein and O. Stern. *Ann. Phys.* **40**, 551 (1913).
8. D. M. Greenberger, M. A. Horne, and A. Zeilinger, *Phys. Today* **46**(8),

22 (1993).
9. L. J. Wang, X. Y. Zou, and L. Mandel, *Phys. Rev. A* **44**, 4614 (1991).
10. P. Milonni, *The Quantum Vacuum* (Academic, San Diego,1993).
11. L. de la Peña and A. M. Cetto, *The Quantum Dice - an Introduction to Stochastic Electrodynamics* (Kluwer Academic, Dordrecht, 1996).
12. J. D. Jackson, *Classical electrodynamics* (Wiley, New York, 1962), p. 270.
13. C. K. Hong and L. Mandel, *Phys. Rev. A* **31**, 2409 (1985).
14. A. Casado, T. W. Marshall, and E. Santos, *J. Opt. Soc. Am.*, article accepted for publication.
15. L. Mandel, *Prog. in Optics* **13**, 27 (1976).
16. J. Dalibard, J. Dupont-Roc, and C. Cohen-Tannoudji, *J. Phys. (Paris)* **43**, 1617 (1982).
17. R. Blanco, H. M. França and E. Santos, *Phys. Rev. A* **43**, 693 (1991).
18. P. A. M. Dirac, *Principles of Quantum Mechanics* (Oxford University Press, Oxford, 1930).
19. R. P. Feynman, *Quantum Electrodynamics* (Benjamin, New York, 1961).
20. S. J. Freedman and J. F. Clauser, *Phys. Rev. Lett.* **28**, 938 (1972).
21. A. Aspect, P. Grangier, and G. Roger, *Phys. Rev. Lett.* **49**, 91 (1982).
22. T. W. Marshall and E. Santos, *Phys. Rev. A* **39**, 6271 (1989).
23. M. O. Scully and U. W. Rathe, *Optics Communs.* **110**, 373 (1994).

LINEAR AND NONLINEAR OPTICAL RESPONSE OF A DILUTE BOSE GAS IN THE CONDENSATION REGIME

Jesús Martínez-Linares,[1] Pál Bogár,[1] and G. S. Agarwal[1,2]

1. Max-Planck-Institut für Quantenoptik
Hans-Kopfermann-Str. 1, D–85748 Garching, Germany
2. Physical Research Laboratory, Ahmedabad-380009, India

Analytical expressions for the linear and third-order optical susceptibilities of a Bose condensate are derived. The transition to the condensate state is marked by the remarkable changes in the peak heights and line widths. For example, at the critical point, the linewidth as a function of temperature exhibits an inflection point in the linear response and a break-point in the third-order response.

Key words: Bose-Einstein condensation, coherent matter waves, optical susceptibility.

1. INTRODUCTION

A significant implication of the quantum statistics of ideal gases is the condensation of a finite fraction of particles obeying Bose-Einstein statistics into the lowest quantum state of momentum at finite temperature [1]. Bose-Einstein condensation (BEC) occurs at such low temperatures that the wavefunction of the atoms overlap so that the individual particles become indistinguishable. The phenomenon of BEC has long been studied, for example in the context of superfluidity [2], and has recently been demonstrated experimentally in dilute atomic gases [3], opening the door to potential sources of coherent matter waves. The standard text books emphasize that the cause of condensation lies in the symmetry of the wavefunction and not in any inter particle interaction. In addition, text books emphasize the nature of the transition point as the temperature is lowered. The question arises — What is a good signature of this transition? Given the success of optical spectroscopy in physics, the optical response of a Bose gas should be one of the prime candidates for studying the transition from normal states

M. Ferrero and A. van der Merwe (eds.),
New Developments on Fundamental Problems in Quantum Physics, 247–254.

to condensate state [4].

In the present paper we calculate the first- and third-order optical susceptibility of a dilute Bose gas. We search for measurable manifestation of BEC in the optical response as the system is being cooled below the critical temperature of the condensation, T_{cr}. We demonstrate that, the temperature dependence of the linewidth of the optical response of the gas exhibits an inflection point at T_{cr}. The *inflection point* is a consequence of the first appearance of condensate particles and identifies the critical point of BEC unambiguously. T_{cr} is marked in the third-order susceptibility even more clearly: the linewidth as a function of temperature exhibits a break-point at the critical temperature in this case. In both the linear and third-order susceptibilities, therefore, the linewidth of the imaginary part of the optical susceptibility could in principle be used to diagnose the level of condensation in the gas sample.

2. OPTICAL RESPONSE

Let us consider an ideal gas of N non-interacting two-level atoms in a volume V at a fixed density of $n = N/V$. In order to account for quantum statistical effects we work in second quantization formalism in the occupation number representation. Thus, we introduce the field operators $\Psi_{e(g)}$ defined as

$$\Psi_{e(g)}(\boldsymbol{r}) = \frac{1}{\sqrt{V}} \sum_{\boldsymbol{k}} a_{\boldsymbol{k},e(g)} e^{i\boldsymbol{k}\boldsymbol{r}}, \tag{1}$$

where e and g indicate the internal states of the atoms, $\boldsymbol{k}$ their momenta, and $a_{\boldsymbol{k},e(g)}$ is the annihilation operators in Fock representation. They satisfy the bosonic commutation relations such as $[\Psi_e(\boldsymbol{r}), \Psi_e^\dagger(\boldsymbol{r}')] = \delta(\boldsymbol{r} - \boldsymbol{r}')$. In order to probe the momentum distribution of the atoms, we shine the gas with a laser field of amplitude E_o and wave vector $\boldsymbol{k}_l$. The optical transition frequency ω_0 between the upper level e and the lower level g is nearly resonant with the laser frequency ω_l. The interaction between the field and the dipole moment $\boldsymbol{d}$ of the atoms is governed by the second quantization Hamiltonian

$$H_I = -\boldsymbol{d}^* \boldsymbol{E}_0 e^{-i\omega_l t} \sum_{\boldsymbol{k}} a^\dagger_{\boldsymbol{k}+\boldsymbol{k}_l,e} a_{\boldsymbol{k},g} + H.c., \tag{2}$$

expressing that transitions in the internal state of the atoms are accompanied by a corresponding recoil in their center of mass.

We assume that, since in the range of temperatures involved $\hbar\omega_0/k_BT \gg 1$, all the atoms are initially in the lower internal state g. In their momenta $\boldsymbol{k}$, they follow Bose-Einstein statistics, $N_{\boldsymbol{k},g} = (z^{-1}e^{\beta E_{\boldsymbol{k}}} - 1)^{-1}$, where z is the fugacity, $\beta = 1/k_BT$, and $E_{\boldsymbol{k}} = \hbar^2 k^2/2m$. Applying standard time-dependent perturbation theory in the first order we find that the linear susceptibility is given by

$$\chi_{ij}^{(1)} = \left(\frac{d_i d_j^*}{\hbar V}\right) \sum_{\boldsymbol{k}} \frac{N_{\boldsymbol{k},g}}{(\Omega_{\boldsymbol{k}+\boldsymbol{k}_l,e} - \Omega_{\boldsymbol{k},g} - \omega_l) - i\Gamma}, \tag{3}$$

[5] where the relation, $E_{\boldsymbol{k},g} = \hbar\Omega_{\boldsymbol{k},g}$, has been used. Here, the homogeneous atomic linewidth, Γ, accounts for decay. The zero momentum term in Eq. (3), as is well known from the standard theory of ideal Bose gases, diverges for $z \to 1$. Therefore, we split off the term corresponding to $k = 0$ and replace the rest of the sum by an integral over k. In this way we obtain

$$\sum_{\boldsymbol{k}} N_{\boldsymbol{k},g}\mathcal{F}(\boldsymbol{k}) \longrightarrow \frac{V}{(2\pi)^3}\int N_{\boldsymbol{k},g}\mathcal{F}(\boldsymbol{k})d^3k + f_{cond}N\mathcal{F}(0), \tag{4}$$

where f_{cond} denotes the fraction of the atoms in the condensate state. We introduce the dimensionless temperature $\theta = T/T_{cr}$, normalized by the critical temperature given by

$$T_{cr} = \frac{2\pi\hbar^2}{mk_B}\left(\frac{n}{g_{3/2}(1)}\right)^{2/3}, \tag{5}$$

where the polylogarithm function is defined as $g_n(z) = \sum_{l=1}^{\infty} z^l/l^n$. In terms of the normalized temperature, f_{cond} can be written in the form [1]

$$f_{cond} = \begin{cases} 1 - \theta^{3/2}, & \text{if } \theta \leq 1 \\ 0, & \text{otherwise} \end{cases} \tag{6}$$

After performing the integral we arrive at the final form of the normalized first-order susceptibility, $\bar{\chi}_{ij}^{(1)} = \chi_{ij}^{(1)}/\chi_0^{(1)}$, where $\chi_0^{(1)} = nd_id_j^*/\hbar\Gamma$, given by

$$\bar{\chi}_{ij}^{(1)} = \left(\frac{i\sqrt{\pi}\theta}{g_{3/2}(1)A_{cr}}\right)\mathcal{W} - \left.\frac{1-\theta^{3/2}}{\bar{\Delta}+i}\right|_{\theta\leq 1}. \tag{7}$$

Here we have introduced

$$\mathcal{W}(\bar{\Delta}) = \sum_{l=1}^{\infty}\frac{z^l}{l}\, w\!\left(\sqrt{l}\,\frac{i+\bar{\Delta}}{A}\right), \tag{8}$$

where the error function is $w(\zeta) = e^{-\zeta^2}(1 - erf(-i\zeta))$ [6]. We introduce the Doppler-parameter A, as the ratio of the Doppler-linewidth to the homogeneous atomic linewidth that explicitly reads as

$$A = \frac{\omega_l}{\Gamma}\sqrt{\frac{2k_BT}{mc^2}}. \tag{9}$$

This can be rearranged in the form $A = A_{cr}\sqrt{\theta}$, by introducing

$$A_{cr} = 2\sqrt{\pi}\left(\frac{n}{g_{3/2}(1)}\right)^{1/3} \bar{\omega}_l \lambda_C, \tag{10}$$

where the Compton wavelength is given by $\lambda_C = \hbar/mc$, and the normalized laser frequency is defined as $\bar{\omega}_l = \omega_l/\Gamma$. Correspondingly, A_{cr} is the value of the normalized Doppler-linewidth at the critical temperature, $\theta = 1$. Also in Eq. (7), the normalized detuning parameter, $\bar{\Delta} = \Delta/\Gamma$, is given by $\Delta = \delta - \omega_R$, where the detuning reads as $\delta = \omega_l - \omega_0$, and ω_R denotes the recoil frequency, $\omega_R = \hbar\omega_l^2/2mc^2$. Since its effect is not relevant, ω_R will be neglected in the discussions of the results.

Typical results for the real and imaginary parts of the linear susceptibility for various temperatures are depicted in the left two panels of Fig. 1. In this example, where $A_{cr} = 9.5$ is selected, a narrowing of the lines is apparent as the temperature is decreasing. In particular, the imaginary part of the susceptibility exhibits a broad line at a high temperature that corresponds to the usual Doppler-broadening behaviour. When the temperature is decreasing the Doppler-linewidth is decreasing with $\sqrt{T}$ (see Eq. (9)) as it can be seen in the high temperature regime of the linewidth depicted in the upper panel of Fig. 2. However, the solid lines in Fig. 1 also tell us that at the critical temperature ($\theta = 1$) the graph of the linewidth reaches an inflection point. This, together with the two-term character of Eq. (7), suggests that a second physical mechanism becomes significant at this point. Below T_{cr}, the gas consists of two components, condensated and non-condensated particles, that both contribute to the line. Using Eq. (6) in Eq. (7), the first term becomes proportional to $(1 - f_{cond})^{2/3}$ and the second to f_{cond} for $\theta \leq 1$. Therefore, the first term accounting for the noncondensated particles becomes smaller (the linewidth corresponding to this term is depicted by the dashed line in the upper panel of Fig. 2), while the second one becomes larger as the temperature is decreasing. Since the condensated particles contribute to the line with a Lorentzian of temperature-independent linewidth (see Eq. (7)) the total linewidth becomes less sensitive to the temperature change as the level of condensation, f_{cond}, becomes higher. This is the reason for the inflection point in the temperature dependence of the linewidth. It can be seen by comparing the solid lines in the upper panel of Fig. 2 that the inflection point is more pronounced and, hence, the critical point of the system is identified more clearly if $A_{cr} \gg 1$. In this case, the rate of the narrowing of the line can tell us whether we have reached the condensation regime or not. Hence, the imaginary part of the optical susceptibility could in this regime of large A_{cr} be used for diagnostics of the BEC in dilute atomic gases.

In order to check how sensitive is this result to the details of the interaction, we carry the perturbation theory further in order to calculate

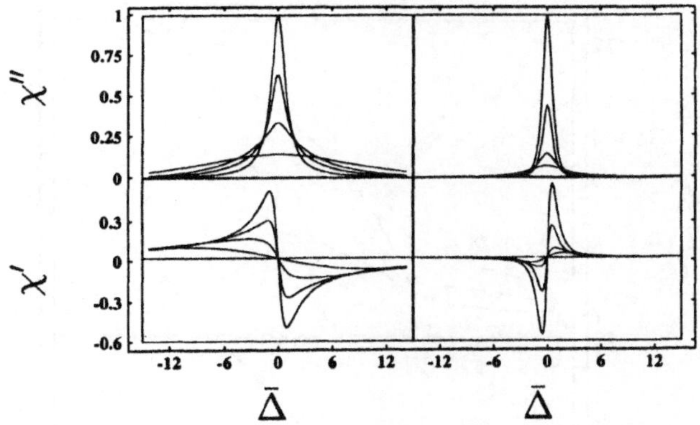

Fig. 1. Real and imaginary parts of the linear (left panel) and degenerate third-order (right panel) susceptibility for $A_{cr} = 9.5$. A narrowing of the lineshapes is apparent when the temperature is lowered as $\theta = 2, 1, 0.7, 0$ for the left, and $\theta = 1.1, 1, 0.5, 0$ for the right two panels. The vertical scale for $\chi^{(3)'}$ has been amplified by a factor 1.7 in order to resolve for the different plots. The vertical scale for $\chi^{(3)''}$ has been inverted.

the *degenerate third-order optical susceptibility* $\chi^{(3)}(\omega, -\omega, \omega)$. The result is

$$\bar{\chi}^{(3)} = -\frac{i}{2}\left\{ \left(\frac{\sqrt{\pi}\theta}{g_{3/2}(1)A_{cr}}\right)^2 \left(|\mathcal{W}|^2 - \mathcal{W}^{*2}\right) + (1-\theta^{3/2})^2 \left(\left|\frac{1}{\bar{\Delta}+i}\right|^2 - \left(\frac{1}{\bar{\Delta}+i}\right)^{*2} \right)_{\theta\le 1} \right\}. \tag{11}$$

Some typical examples are depicted in the right two panels of Fig. 1 corresponding to the same A_{cr} parameter as in the linear case. Narrowing of the lines similar to that of the linear susceptibility is apparent. In particular, the linewidth of the imaginary part of the susceptibility exhibits a break-point exactly at the critical temperature, $\theta = 1$, (see lower panel of Fig. 2) marking the beginning of the transition from the noncondensate to the condensate regime. As in the linear case, this effect is the result of the competition between the temperature-dependent contribution from the noncondensate atoms, and the contribution of the temperature-independent line of the condensate ones.

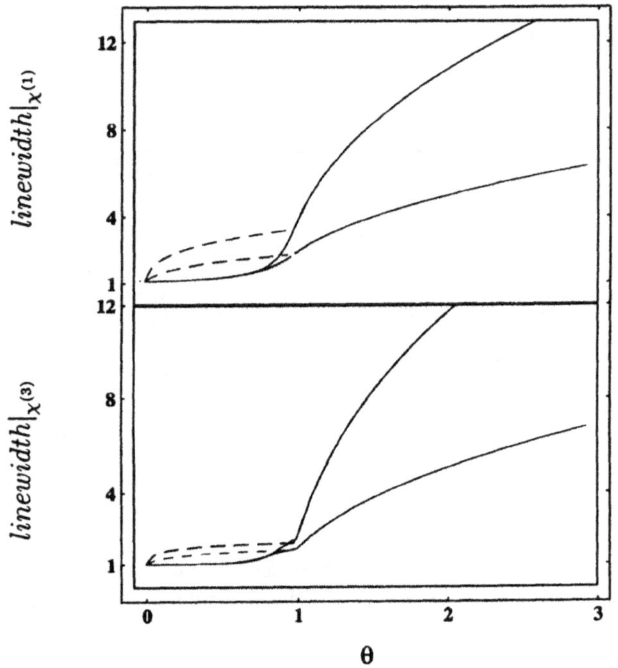

Fig. 2. Temperature dependence of the linewidth (FWHM) of the imaginary part of $\chi^{(1)}$ (upper panel), and of $\chi^{(3)}$ (lower panel). Inside each panel, the curves are plotted for $A_{cr} = 9.5$ (upper plots) and $A_{cr} = 4$ (lower plots). The dashed lines represent the contributions from the non-condensated atoms below the critical point. The linewidths are normalized to their minima, given by the atomic linewidth, i.e., 2Γ and 1.28Γ for $\chi^{(1)}$ and $\chi^{(3)}$, respectively.

3. DISCUSSION OF THE RESULTS

We conclude that the discontinuity in the density distribution, marking the critical point of the Bose-Einstein quantum statistical phase transition (see Eq. (6)), manifestes itself in the temperature dependence of the linewidth of the optical response of the gas. In particular, we studied the imaginary part of the susceptibility and found that the linewidth as a function of the temperature exhibits an inflection point in the linear and a break-point in the third-order case at the critical point. The larger the ratio of the inhomogeneous to the homogeneous linewidth at the critical temperature, the more pronounced these inflection/break-points. Imposing the condition of $A_{cr} \gg 1$ the line-narrowing effects can be applied to identify the critical point of the system and, in this way, diagnose the level of condensation in the gas. Note that Eq. (10) can be written

$$\frac{\lambda_l^6 \Gamma^3}{(c\lambda_C)^3}\left[\frac{1}{n\lambda_l^3}\right] = \frac{8\pi^{3/2}}{g_{3/2}(1)}\left(\frac{1}{A_{cr}^3}\right), \tag{12}$$

and hence, the left hand side of the above equation has to be small for $A_{cr} \gg 1$. On the other hand, in order to neglect collective excitations, we also have to assume that the gas is diluted so that $n\lambda_l^3 \ll 1$. These two conditions can be jointly written as

$$\frac{\lambda_l^6 \Gamma^3}{(c\lambda_C)^3} \ll n\lambda_l^3 \ll 1. \tag{13}$$

The above inequalities establish a consistency condition for the atomic transition with independence of the density, namely the line should be narrow enough to satisfy $\Gamma \ll \Gamma_c$ with $\Gamma_c := \hbar/(m\lambda_l^2)$. This condition sets the limits for the applicability of our results. It can be satisfied by Rydberg atoms, for instance for transitions to highly excited P levels of alkali atoms and hydrogen. In the optical domain, the experiments performed until now were in a regime where $A_{cr} \ll 1$. Nevertheless, new experiments are being planned to obtain BEC in a sample of strontium gas. In this case the method here proposed could be employed with the very narrow intercombination lines of strontium.

On the other hand, we also expect that several mechanisms, such as EIT [7], can be employed in the optical domain to reduce the natural linewidth. The details of such possibilities as well as the effects of trapping potential will be examined in future.

Acknowledgements

The authors thank H. Wallis and M. Naraschewski for their constructive comments. G.S.A. thanks Herbert Walther for hospitality at the Max-Planck-Institut für Quantenoptik. P.B. and J.M.-L. are grateful to the Max-Planck-Gesellschaft (Quantenoptik) for support. J.M.-L. is also supported in part by the Human Capital and Mobility Program under contract No. ERBCHBGCT930437.

REFERENCES

1. See, e.g., K. Huang, *Statistical Mechanics*, 2nd edn. (Wiley, New York, 1987).
2. D. R. Tilley and J. Tilley, *Superfluidity and Superconductivity*, 3rd edn. (Adam Hilger, Bristol, 1990).
3. M. H. Anderson, J. R. Ensher, M. R. Matthews, C. E. Wieman and E. A. Cornell, *Science* **269**, 198 (1995); C. C. Bradley, C. A. Sackett, J. J. Tollett and R. G. Hulet, *Phys. Rev. Lett.* **75**, 1687 (1995); K. B. Davis, M.-O. Mewes, M. R. Andrews, N. J. van Druten, D. S. Durfee, D. M. Kurn and W. Ketterle, *Phys. Rev. Lett.* **75**, 3969 (1995).
4. B. V. Svistunov and G. V. Shlyapnikov, *Sov. Phys. JETP* **70**, 460 (1990); H. D. Politzer, *Phys. Rev. A* **43**, 6444 (1991); J. Javanainen, *Phys. Rev. Lett* **72**, 2375 (1994); L. You, M. Lewenstein and J. Cooper, *Phys. Rev. A* **50**, R3565 (1994); J. Javanainen, Phys. *Rev. Lett.* **75**, 1927 (1995); O. Morice, Y. Castin and J. Dalibard, *Phys. Rev. A* **51**, 3896 (1995).
5. For formal expressions for linear and nonlinear susceptibilities, see, e.g., R.W. Boyd, *Nonlinear Optics* (Academic, New York 1992), Chap. 3.
6. M. Abramowitz and I. A. Stegun, *Handbook of mathematical functions* (Dover, New York, 1970), p. 297.
7. S. E. Harris, J. E. Field, and A. Imamoglu, *Phys. Rev. Lett.* **64**, 1107 (1990).

QUANTUM MEASUREMENT AND PRACTICE IN NUCLEAR PHYSICS

Z. Máté

Institute of Nuclear Research of the Hungarian Academy of Sciences, H-4001, Debrecen, POB. 51, Hungary
mate@moon.atomki.hu

Comparing real nuclear physics experiments with principles given by standard textbooks on quantum mechanics one can conclude that the 'measurement' concept is actually not used in the practical description of nuclear (and perhaps microphysical) phenomena.

Key words: quantum mechanics, measurement, practice, Oberaudorf.

Nuclear physics, similarly to other branches of microphysics, has also proved to be a field where ordinary quantum mechanics works excellently. Of course, the so called interpretational problems are also present in the nuclear physics applications of quantum mechanics. Presently there is no consensus on the new foundations of quantum theory. People use ordinary quantum mechanics both in research and education. The problems discussed in the present work are connected both with teaching of quantum mechanics and with the practice of experimental nuclear physics research. We try to clarify what is the real practical connection between the quantummechanical formalism and the experiments.

If we concentrate on a real nuclear physics problem, we are allowed to forget about the interpretational problems. As an example let us consider our experiment [1] on hyperdeformed states on ^{236}U. Here the target nucleus ^{235}U is bombarded with deuterons, which results in a ^{236}U nucleus in an excited state and an outgoing proton. Then ^{236}U undergoes fission process. In the experiment we measure the energy of protons by a magnetic spectrograph and also detect the fission products by parallel plate avalanche counter. It is an essential point in the experiment to select those events in which the proton appears in coincidence with the fission products. When we plan and perform this experiment we handle the system in a classical way in

M. Ferrero and A. van der Merwe (eds.),
New Developments on Fundamental Problems in Quantum Physics, 255–257.

many respects; the particles are considered point particles, we calculate the trajectories of deuterons, protons and fragment nuclei and it gives a perfect picture for the placement of the target and the detectors, set timing etc. At the same time we use quantummechanical formalism for the calculation of fission probability. By comparison of experimental and calculated cross sections we try to find out whether our states are really hyperdeformed states and what their properties are. In the course of searching for the solution of nuclear physical problems we are never disturbed by the open questions of foundations of quantum mechanics.

We meet the basic problems only when we try to speak about our work to the students. Considering the responsibilities of education one thinks that the use of the exact quantummechanical language is an indispensable requirement. Contradictory classical pictures should be eliminated using the concept of measurement and strictly keeping the rule that the observables do not have values before the act of the measurement. In this way the particles do not have trajectories except approximate ones. We should analyse the conditions of using classical approximation, and also we should determine where, when and how in the real process the quantummechanical measurement takes place. Finding the connection of our real measurement with the principal 'measurement' of quantum mechanics seems to be quite a difficult task.

Searching for the solutions of these problems in ordinary textbooks on quantum mechanics we can make an interesting observation. The concept of 'measurement' appears mainly in the general chapters of the textbooks, and it is almost absent when concrete examples are discussed. The specific physical systems and processes are mostly described using the concepts of trajectories, waves and no mention of 'measurement' occurs. Practically we can not get help for reaching our aim because in the concrete examples the textbooks use the same language that we want to avoid. These findings give the feeling that the measurement concept is not an essential part of quantum mechanics and, consequently, the practical succes of quantum mechanics does not verificate the adequacy of 'measurement'.

The most explicit and spectacular characterization of the practice of quantum mechanics is given by the founders themselves. Heisenberg and Bohr had an interesting discussion in the course of a skiing holiday near the village of Oberaudorf (Germany) in 1933. They were talking about a cloud-chamber photograph and a possible existence of a new particle now called positron. Let us quote a short part of this discussion according to Heisenberg's book [2]:

> ...After we had been looking for possible experimental mistakes for some time, I said to Niels: 'Isn't it odd that, throughout this discussion, no one should have mentioned

> quantum theory? We behave as if the electrically charged particles were an object like an electrically charged oil droplet, or like a pith ball in an old electroscope. We quite unthinkingly use the concepts of classical physics, as if we had never heard of the limitations of these concepts and of uncertainty relations. Isn't that bound to lead to errors? '
>
> 'No, certainly not,' Niels replied. 'After all, it is almost the essence of an experiment that the observations can be described with the concepts of classical physics. That is the whole paradox of quantum theory. On the one hand, we establish laws that differ from those of classical physics; on the other, we apply the concepts of classical physics quite unreservedly...'

It is well illustrated even by this short part that in the spirit of this discussion the use of classical quantities has much more importance and freedom than it would be allowed by the general principles connected with 'measurement'.

If we write 'M' for the considerations connected in the standard textbooks with the quantummechanical measurement concept, and 'O' for the practical language characterised by the quoted Oberaudorf discussion and used throughout in microphysics, we can summarise our observation as follows:

$$\text{ORDINARY Q.M.} = \text{FORMALISM} + \text{'O'} + \text{'M'},$$

$$\text{PRACTICAL Q.M.} = \text{FORMALISM} + \text{'O'},$$

where by "practical q.m." we mean that part of ordinary quantum mechanics that is enough to be 'fine for all practical purposes'.

The real connection between formalism and experiment is not 'M', as it is declared in the general chapters of the textbooks, but 'O'. In view of the principal debates around 'M' and the practical absence of it, it seems to be reasonable to teach for students only "practical q.m." until the basic problems of quantum mechanics will be clarified by the research in this field.

REFERENCES

1. A Krasznahorkay, M Hunyadi, M Csatlós, I Diószegi, J Gulyás and Z Máté, "Evidence for hyperdeformed nuclear shape in $^{236}U^*$," *Acta Phys. Pol. B* **27** 139 (1996).
2. W Heisenberg, *Physics and Beyond* (Allen & Unwin, London, 1971), p. 129.

QUANTUM PREDICTIONS WITHOUT NONLOCAL PROJECTIONS

D. J. Miller

School of Physics, The University of New South Wales
Sydney NSW 2052, Australia

A new formalism is proposed for calculating the properties of a quantum system between measurements. In the case of entangled quantum systems the probabilities can be tested by experiment and agree with orthodox quantum mechanics.

Key words: quantum theory, retrocausality, projection postulate.

1. INTRODUCTION

One of the motivations for the present suggestion of a new formalism for quantum mechanics is the difficulty with understanding the projection of the density density operator on measurement for entangled quantum systems which have separated in space. Consider the familiar case of a pair of particles which are prepared in a spin singlet state at time t_a and which separate along paths $i = u$ or v, without further interaction with each other. At time t_a the normalised basis vectors are $|i^\sigma, t_a; t_a\rangle$, where $\sigma = +$ or $-$ corresponds to spin up or down along what is taken to be the z-direction. The state vector at time t is

$$\begin{aligned}|\psi, t; t_a\rangle = \frac{1}{\sqrt{2}}(\hat{U}_{H_{u+}}(t; t_a)|u^+, t_a; t_a\rangle \otimes \hat{U}_{H_{v-}}(t; t_a)|v^-, t_a; t_a\rangle \\ + \hat{U}_{H_{u-}}(t; t_a)|u^-, t_a; t_a\rangle \otimes \hat{U}_{H_{v+}}(t; t_a)|v^+, t_a; t_a\rangle),\end{aligned} \tag{1}$$

where $\hat{U}_{H_{i\sigma}}(t; t_a)$ is the unitary time evolution operator and $H_{i\sigma}$ is the Hamiltonian, and $|i^\sigma\rangle$ is the state vector, on path i for a particle with spin σ.

Let us say that a local type-1 measurement is carried out on path u at time t_b with a result corresponding to projection operator $\hat{P}_{u^+}(t_b)$. In orthodox quantum mechanics the density operator $\hat{\rho}_a(t_b; t_a) = |\psi(t_b; t_a)\rangle\langle\psi(t_b; t_a)|$

M. Ferrero and A. van der Merwe (eds.),
New Developments on Fundamental Problems in Quantum Physics, 259–263.

is projected onto $|u^+, t_b; t_a\rangle\langle u^+, t_b; t_a| \otimes |v^-, t_b; t_a\rangle\langle v^-, t_b; t_a|$, which involves changes on both the paths u and v. In orthodox quantum mechanics, the density operator is taken purely as a means for making correct predictions about measurements and no further commentary is necessary [1]. An alternative view is that the correct predictions can be made from $\hat{\rho}_a$ because it corresponds in some way to a physical quantum system, which exists between the measurements at t_a and t_b, so that a change in one is reflected in a change in the other. The non-unitary nature of the projection is a problem in itself even for the local projection affecting path u. The non-local projection affecting path v is a separate, even more puzzling, problem if the entangled quantum system is thought of in physical terms in the way mentioned. One problem is that the projection initiated at the measurement of one particle affects the density operator describing both spatially separated particles irrespective of the potential in the region separating the two systems. In fact if the two systems are isolated from one another by regions of infinite potential, the projection must still occur in precisely the same way that it would if there were no intervening potential [2]. The other main problem, once again on the physical interpretation, is that the projection must occur instantaneously in both parts of the system for any observer. This raises difficulties with Lorentz invariance which have been considered recently by Cohen and Hiley [3,4].

2. FORMALISM

The formalism proposed now avoids these problems because the properties of the quantum system are taken to depend on the next measurement as well as the preparation state. For a quantum system prepared at time t_a with the density operator $\hat{\rho}_a$ and subsequently measured at time t_b to have eigenvalue b_1, the following operator is defined:

$$\hat{\rho}_{ab_1}(t; t_a, t_b) = \frac{\hat{\rho}_a(t; t_a)\hat{P}_{b_1}(t; t_b)}{\mathrm{Tr}(\hat{\rho}_a(t; t_a)\hat{P}_{b_1}(t; t_b))}, \quad t_a < t < t_b. \tag{2}$$

where $\hat{P}_{b_1}$ projects onto the subspace spanned by the eigenvector(s) corresponding to the eigenvalue b_1.

We next define a function $p^{tel}_{ab_1}(q,t) = \mathrm{Tr}(\hat{\rho}_{ab_1}(t; t_a, t_b)\hat{P}_q(t))$ which resembles the probability that a measurement of the observable Q would yield the eigenvalue value q. It has a teleological aspect because it depends on the next measurement outcome. We know from orthodox quantum mechanics that the probabilities that the eigenvalue(s) b_i will result from a measurement of B at t_b are $p^{oqm}_a(b_i, t_b) = \mathrm{Tr}(\hat{\rho}_a(t_b; t_a)\hat{P}_{b_i}(t_b; t_b))$. We now propose that the probability $p^{tel}_a(q,t)$ that the system has property q can be written in terms of the function $p^{tel}_{ab_i}(q,t)$:

$$p^{tel}_a(q,t) = \sum_{b_i} p^{tel}_{ab_i}(q,t) p^{oqm}_a(b_i,t) = \sum_{b_i} \frac{\mathrm{Tr}(\hat{\rho}_a(t;t_a)\hat{P}_q(t)\hat{P}_{b_i}(t;t_b))}{\sum_{b_i}\mathrm{Tr}(\hat{\rho}_a(t;t_a)\hat{P}_{b_i}(t;t_b))}. \tag{3}$$

If one is entirely ignorant of the outcome of the next measurement, or ignores it as in orthodox quantum mechanics, the $\{b_i\}$ form a complete set and $p^{tel}_a(q,t) = \mathrm{Tr}(\hat{\rho}_a(t;t_a)\hat{P}_q(t;t)) = p^{oqm}_a(q,t)$, showing that the present formalism is consistent with orthodox quantum mechanics. If one learns, say, that b_1 actually occurred, the above prediction is modified by restricting the range of the b_i to b_1, just as would be done in a classical calculation involving conditional probabilities. Thus an important conclusion is that the function in Eq. (3) reproduces the relationships of a classical probability space even when counterfactual properties are included.

This approach is similar to, but more formal and explicit than, the suggestions of de Beauregard [5] and Cramer [6] in the time-symmetric or transactional interpretations of quantum mechanics. It is also worth comparing the formalism with the double vector formalism of Aharanov, Bergmann and Lebowitz (ABL) [7] and subsequent elaborations [8]. In the present notation, the probability $p^{abl}_{ab_1}(q,t)$ that an intervening measurement of Q between preparation in $\hat{\rho}_a$ and final measurement in eigenstate b_1 corresponding to projection operator $\hat{P}_{b_1}$ is

$$p^{abl}_{ab_1}(q,t) = \frac{\mathrm{Tr}(\hat{P}_q(t)\hat{\rho}_a(t;t_a)\hat{P}_q(t)\hat{P}_{b_1}(t;t_b))}{\sum_q \mathrm{Tr}(\hat{P}_q(t)\hat{\rho}_a(t;t_a)\hat{P}_q(t)\hat{P}_{b_1}(t;t_b))}. \tag{4}$$

It has been suggested that the ABL expression can be used in the counterfactual sense when the measurement of Q is not actually carried out [9] but in that case the expression does not lead to the same expressions as orthodox quantum mechanics [10].

Before applying Eq. (3) to specific problems, it is appropriate to make two comments. The first is that the involvement of $\hat{P}_{b_i}$ evolved backwards in time in determining the properties of the quantum system is unconventional and counterintuitive but there is no formal problems with using the inverse of the time evolution operator in this way. Physically, it implies the "block universe" [11] view in which future events, although unknown to us, are otherwise as physically real as the past. The involvement of $\hat{P}_{b_i}$ also means that the contextuality of quantum mechanics appears explicitly in the formalism. Secondly, since the next measurement is involved in the expression in Eq. (3), it may appear that the expression applies only to counterfactual properties q by hypothesis. This is true for a single quantum system but for entangled systems, the measurement on one of the spatially separated systems determines the probabilities of outcomes of measurements of the other and these predictions can then be tested by measurement. Since the probabilities calculated in this way from Eq. (3) are the same as orthodox quantum mechanics, the predictions conform to experiment. Finally, the operator in Eq. (2) is not necessarily positive definite when counterfactuals are involved and therefore cannot be a probability measure in the conventional sense. Since by hypothesis probabilities for counterfactuals can never be tested by measurement, there is no contradiction with experiment. The involvement of unconventional "probabilities" in quantum mechanics has been reviewed by Mückenheim [12]. Therefore the does not seem to be any reason for rejecting them here without further investigation.

3. APPLICATIONS

For the example considered in the Introduction, knowledge that spin 1 is up in the z-direction means that $\hat{P}_{b_1}(t_b) = \hat{P}_{u^+}(t_b) \otimes \hat{I}$. This operator can be evolved back to t_a and the effect of this in Eq. (3) is that the properties of the particle on path v, on the basis of this partial knowledge of the final measurement state of the entangled system, are calculated from $\mathrm{Tr}_u(\hat{\rho}_{ab_1}(t; t_a, t_b)) = \hat{P}_{v^-}(t; t_a)$ where Tr_u is a partial trace over the u state. Substitution in Eq. (3), assuming field free regions, leads to $p_a^{tel}(q) = \sin^2 \alpha$ that spin 2 will have spin up in the direction 2α with respect to the z-axis in the xz-plane. Thus the same probabilities for measurement outcomes will be obtained from this formalism as from the instantaneous and non-unitary projection of the density operator across spacetime intervals discussed above. Here the formalism has involved only unitary evolution of the projection operator corresponding to the measurement back in time along the worldline of the particle that has been measured.

It is interesting to consider also the case of a single particle in a superposition of states. Consider a non-symmetric Mach-Zhender interferometer in which the particle is incident along path s and follows paths u or v with probabilities r^2 and t^2 respectively after the first beam splitter and follows paths c or d with additional probabilities t^2 and r^2 respectively after the second, to be detected at C or D respectively. The preparation density operator $\hat{\rho}_a = |a\rangle\langle a|$ and $|a\rangle \to |tv + iru\rangle \to |(t^2 - r^2)d + irtc\rangle$ with $r^2 + t^2 = 1$. If a location measurement is performed on path u, say, and it is null, the projection operator for the measurement outcome is $\hat{I} - \hat{P}_u = \hat{P}_v$. From Eq. (3) the probability $p_a^{tel}(u) = 0$ and $p_a^{tel}(v) = 1$ for all $t_a \le t \le t_b$, i.e. the counterfactual statement can be made that the particle follows path $a \to u$ which corresponds to the intuitive, realistic description. On the basis of the projection postulate over spacetime intervals, an attempt at a realistic description would involve paths u and v before the null measurement and path u afterwards.

If, instead, the measurement is detection at C, the projection operator for the measurement outcome is $\hat{P}_C = |c\rangle\langle c|$, where $|c\rangle = |-irv + tu\rangle$ and from Eq. (2) $p^{tel}_{aC}(u) = p^{tel}_{aC}(v) = \frac{1}{2}$. Similarly, if the detection is at D, $p^{tel}_{aD}(u) = \frac{1-t^2}{1-2t^2}$ and $p^{tel}_{aD}(v) = \frac{-t^2}{1-2t^2}$. Despite the presence of negative values, $p^{tel}_{ab}(u) + p^{tel}_{ab}(v) = 1$ in each case. Also the probability $p^{tel}_{a}(u) = p^{tel}_{aC}(u)p^{oqm}_{a}(C) + p^{tel}_{aD}(u)p^{oqm}_{a}(D)$, which applies when the final measurement is left open, is the orthodox value.

4. CONCLUSION

It has been shown that the expression in Eq. (3) which does not involve non-local projections of the density operator is is consistent with orthodox quantum mechanics. Thus problems of projection across spacelike intervals which arise in realistic interpretations of quantum mechanics are avoided because the density operator and projection operators involved in the proposed formalism are only evolved along the worldlines of the spatially separated parts of the quantum system. The unusual feature of the new formalism is the involvement of the projection operator of the next measurement evolved backwards in time and the calculation of probabilities in terms of a function which is not positive definite but which otherwise is the same as a conditional probability measure.

REFERENCES

1. A. Peres, *Ann. N. Y. Acad. Sci.* **755**, 445 (1995).
2. B. Mielnik, *Found. Phys.* **20**, 745 (1990).
3. O. Cohen and B.J. Hiley, *Phys. Rev. A* **52**, 76 (1995).
4. O. Cohen and B.J. Hiley, *Found. Phys.* **25**, 1669 (1995).
5. O. Costa de Beauregard, *Nuovo Cimento B* **42**, 41 (1977); **51**, 267 (1979).
6. J.G. Cramer, *Rev. Mod. Phys.* **58**, (1986) 647.
7. Y. Aharonov, P.G. Bergmann, and J.L. Lebowitz, *Phys. Rev.* **134**, B1410 (1964).
8. Y. Aharonov and L. Vaidman, *J. Phys. A* **24**, 2315 (1991).
9. L. Vaidman, *Phys. Rev. Lett.* **70**, 3369 (1993).
10. W.D. Sharp and N. Shanks, *Philos. Sci.* **60**, 488 (1993).
11. H. Price, *Time's Arrow and Archimedes' Point* (Oxford University Press, Oxford, 1996).
12. W. Mückenheim, *Phys. Rep.* **133**, 337 (1986).

THE EMERGENCE OF STATISTICAL LAWS IN QUANTUM MECHANICS

Peter Mittelstaedt

*Institute for Theoretical Physics, University of Cologne
Cologne, Germany*

Although the result of a quantum mechanical measurement is in general not only subjectively unknown but objectively undetermined, the statistics of measuring outcomes is strictly determined by the preparation of the system and the measured observable. This statistical behaviour can be fully explained by means of the quantum theory of measurement and by some recent mathematical results. Hence, a quantum mechanical probability law seems to emerge from objectively undetermined individual events as a "law without law" in the sense of Wheeler.

Key words: foundations of quantum mechanics, quantum probability, quantum theory of measurement.

1. INTRODUCTION

Since the beginning of quantum mechanics, it is a most discussed question whether this theory refers to individual quantum systems or to ensembles of identically prepared systems. The laws, which are primarily provided by the theory, are statistical laws, and these laws are confirmed experimentally to a very high degree of accuracy. On the other hand, our scientific language is concerned with individual systems. Indeed, we speak about the properties of a systems, its preparation, its time development, and the measurement of its objective observables. The difficulties to understand the physical behavior of an individual system on the basis of a theory the essential features of which are statistical ones, culminated in the fifties in the formulation of the most radical and rather bizarre consequence, the many worlds interpretation.

M. Ferrero and A. van der Merwe (eds.),
New Developments on Fundamental Problems in Quantum Physics, 265–274.

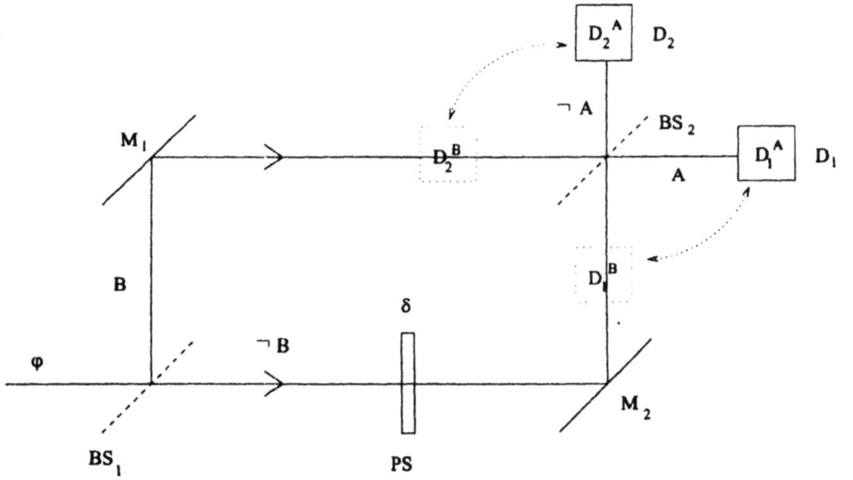

Fig. 1. Photon split-beam experiment with beam splitters BS_1 and BS_2, both half-reflecting mirrors, two fully reflecting mirrors M_1 and M_2, a phase shifter PS providing a phase shift δ and two detectors D_1 and D_2 in mutually exlusive positions (D_1^A, D_2^A) and (D_1^B, D_2^B).

2. OBJECTIVE INDETERMINACY VERSUS STATISTICAL DETERMINATION

Let us consider the split-beam experiment in Fig. 1, which has been realized both with photons and with neutrons. The state φ of the incoming photon is split by a half-transparent mirror, beam splitter BS_1, into two orthogonal components described by orthonormal states φ^B and $\varphi^{\neg B}$. The two parts of the split beam are reflected at two (fully reflecting) mirrors M_1 and M_2 and recombined with a phase shift δ at a second half-transparent mirror, beam splitter BS_2. In the experiment there are two mutually exclusive measuring arrangements: If the detectors D_1 and D_2 are in the positions (D_1^B, D_2^B) one observes which way (B or $\neg B$) the photon came. If the detectors are in the position (D_1^A, D_2^A) one observes the interference pattern, i.e., the intensities which depend on the phase δ.

In the latter case of position (D_1^A, D_2^A) and the special case $\delta = 0$ without phase shift the outcomes of the split-beam experiment is completely determined. Destructive interference destroys all radiation going to the second detector D_2^A, whereas the first detector D_1^A registers, in case of ideal sensitivity, every incoming photon. However, even in this situation of fully determined outcomes nothing can be said about the path of a single photon, neither which way the photon came nor that it traveled both routes.

In the general case of a nonvanishing phase shift $\delta \neq 0$ for an individual photon it is no longer determined whether it will be registered at detector D_1 or D_2. However, in spite of the indeterminacy in each individual case, for a large number of identically prepared photons the preparation φ and the observable A induce a probability for A (to register the photon in D_1^A) and for $\neg A$ (to register the photon in D_2^A) which reads [6]

$$p(\varphi, A) = \cos^2 \delta/2 \quad \text{and} \quad p(\varphi, \neg A) = \sin^2 \delta/2,$$

respectively. This means that the relative frequency of photons arriving at D_1^A is approximately given by $\cos^2 \delta/2$ and the relative frequency of systems arriving at D_2^A by $\sin^2 \delta/2$. Under these conditions one is tempted to assume that for each individual photon the state or the value after measurement can be attributed to the object even before the measurement, although these data are not yet known to the observer in this situation.

It is one of the most important results of quantum mechanics that this assumption is not correct. Neither an eigenstate of A nor the corresponding eigenvalue can be attributed to the photon system in the preparation state φ before the measurement of A. Indeed, if the observable A could be weakly objectified to the system S in state φ, i.e., if an A-value could be attributed to $S(\varphi)$, then the system S were in the mixed state [1]

$$W_S(\varphi, A) = p(\varphi, A)P[\varphi^A] + p(\varphi, \neg A)P[\varphi^{\neg A}]$$

and not in the pure state $P[\varphi]$. Here we have denoted the two eigenstates of A by φ^A and $\varphi^{\neg A}$, respectively. However, it can easily be shown that the states $P[\varphi]$ and $W_S(\varphi, A)$ are not generally equivalent. For this reason weak objectification of A in the state φ is not possible. This means that the result of an A-measurement performed at an individual photon is not only subjectively unknown for the observer but objectively undecided.

Summarizing the discussion of this section we arrive at the following two at the first glance incompatible results: On the one hand, for each individual photon the value of the observable A is objectively undetermined prior to the registration in one of the detectors. On the other hand, a sufficiently large number of photons fulfills the statistical law which states that the relative frequency of photons registered in detector D_1^A is approximately given by $\cos^2 \delta/2$. Hence we find that a large number of lawless quantum systems fulfills a strict probability law. Under these conditions one may wonder where this statistical law comes from. Wheeler posed the question whether we are confronted here with a "law without law," i.e., a lawlike behavior of a large number of elements which emerges without any causal reason [8]. Here we will take up Wheelers challenge and investigate the question whether the quantum probability law emerges by measuring a large ensemble of lawless single quantum systems.

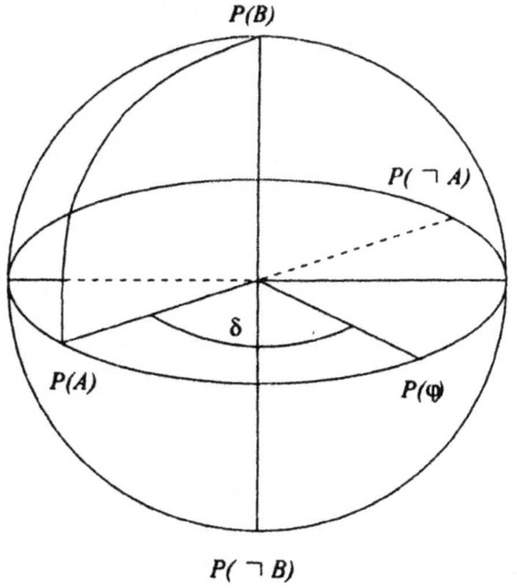

Fig. 2. Poincaré sphere of the photon split beam experiment shown in Fig. 1

3. THE QUANTUM MECHANICAL MEASUREMENT PROCESS

The split beam experiment of the proceeding section can be described in a two dimensional Hilbert space $H_2 = C^2$ which is conveniently illustrated by the Poincaré shere P. (Fig. 2) If we denote the orthogonal unit vectors of the sphere by $\boldsymbol{n}_1, \boldsymbol{n}_2, \boldsymbol{n}_3$ and make use of the Pauli matrices $\sigma_1, \sigma_2, \sigma_3$ then the projection operators in H_2 can be written as

$$P(\boldsymbol{n}) = 1/2(\mathbf{1} + n_i\sigma_i),$$

where $\boldsymbol{n} = (n_1, n_2, n_3)$ is a unit vector. Hence, projection operators correspond to points on the surface of the Poincaré sphere P. The properties B and A of the preceeding section can then be expressed by the projection operators

$$P(B) := P(\boldsymbol{n}_3), \quad P(\neg B) := P(-\boldsymbol{n}_3),$$

with eigenstates φ^B and $\varphi^{\neg B}$ and and eigenvalues $B_1 = 1, B_2 = 0$, and

$$P(A) := P(\boldsymbol{n}_1), \quad P(\neg A) := P(-\boldsymbol{n}_1),$$

with eigenstates φ^A and $\varphi^{\neg A}$ and eigenvalues $A_1 = 1, A_2 = 0$. The preparation φ of the system is then given by $\varphi = 1/\sqrt{2}(\varphi^B + e^{i\delta}\varphi^{\neg B})$ or by the projection operator

$$P[\varphi] = P(\boldsymbol{n}_\varphi) = 1/2(\mathbf{1} + \cos\delta\sigma_1 + \sin\delta\sigma_2), \tag{3.1}$$

with $\boldsymbol{n}_\varphi = (\cos\delta, \sin\delta, 0)$ and corresponds to a point on the equator of P. The angle between the vectors $\boldsymbol{n}_\varphi$ and $\boldsymbol{n}_1$ is then given by the phase shift δ.

For the probabilities of B and A we obtain

$$\begin{aligned} p(\varphi, B) &= \mathrm{tr}\,\{P[\varphi]P(B)\} = 1/2(1 + \boldsymbol{n}_\varphi \cdot \boldsymbol{n}_3) = 1/2, \\ p(\varphi, A) &= \mathrm{tr}\,\{P[\varphi]P(A)\} = 1/2(1 + \boldsymbol{n}_\varphi \cdot \boldsymbol{n}_1) = \cos^2\delta/2. \end{aligned} \tag{3.2}$$

In order to describe the measuring process of the observable $P(A)$ we consider the object system S with Hilbert space H_S and the measuring apparatus M with Hilbert space H_M. Let $\varphi \in H_S$ and $\Phi \in H_M$ be the preparations of S and M, respectively, i.e., the states prior to the measurement process. Here we consider a unitary and repeatable premeasurement of $P(A)$, which can be described by a unitary operator U_A acting on the tensor product $\varphi \otimes \Phi$ of the compound system $S + M$. The operator U_A is further determined by the calibration postulate. If the object system S is in one of the two eigenstates φ^A or $\varphi^{\neg A}$ of $P(A)$, then the unitary and repeatable premeasurement must reproduce this state. This means that for the special preparation $\varphi(\delta = 0) = \varphi^A$ and $\varphi(\delta = \pi) = \varphi^{\neg A}$ we have

$$\begin{aligned} \varphi^A \otimes \Phi &\to U_A(\varphi^A \otimes \Phi) = \varphi^A \otimes \Phi_A, \\ \varphi^{\neg A} \otimes \Phi &\to U_A(\varphi^{\neg A} \otimes \Phi) = \varphi^{\neg A} \otimes \Phi_{\neg A}, \end{aligned} \tag{3.3}$$

where Φ_A and $\Phi_{\neg A}$ are eigenstates of a pointer observable $Z = Z_A P[\Phi_A] + Z_{\neg A} P[\Phi_{\neg A}]$ whose eigenvalues Z_A and $Z_{\neg A}$ indicate the measuring results A and $\neg A$, respectively. From (3.2) it follows by the unitarity of U_A for an arbitrary preparation

$$\varphi = (\varphi^A, \varphi)\varphi^A + (\varphi^{\neg A}, \varphi)\varphi^{\neg A} = 1/2(1 + e^{i\delta})\varphi^A + 1/2(1 - e^{i\delta})\varphi^{\neg A}, \tag{3.4}$$

$$\varphi \otimes \Phi \to U_A(\varphi \otimes \Phi) = 1/2(1 + e^{i\delta})\varphi^A \otimes \Phi_A + 1/2(1 - e^{i\delta})\varphi^{\neg A} \otimes \Phi_{\neg A}.$$

The state of the object system after the premeasurement is then given by the reduced mixed state

$$W_S(\varphi, A) = \cos^2\delta/2\, P[\varphi^A] + \sin^2\delta/2\, P[\varphi^{\neg A}]. \tag{3.5}$$

The interpretation of this mixed state of the object system after the premeasurement is usually given by the probability reproducibility condition: The probability distribution $p(\varphi, A_i), A_i \in \{A, \neg A\}$ which is induced by the preparation φ and the measured observable $P(A)$, is reproduced in the statistics of the post-measurement values $(Z_A, Z_{\neg A})$ and states $(\Phi_A, \Phi_{\neg A})$ of the pointer. In case of repeatable measurements this means that $p(\varphi, A_i)$ is also reproduced in the statistics of the states $(\varphi^A, \varphi^{\neg A})$. On the basis of these arguments the main question of this paper can now be formulated as follows: Given an ensemble of (before the measurement) identically prepared systems S which are in the reduced state (3.5) after the premeasurement of A. Is it possible to justify that the (formal) probability $p(\varphi, A_i)$ is reproduced in the statistics of the measurement results A and $\neg A$, respectivly? A justification of this kind could presumably explain in which way a large number of objectively undetermined objects fulfill a strict statistical law.

4. THE PROBABILITY REPRODUCIBILITY CONDITION

Consider a large number of identically prepared systems S_i in states $\varphi^{(i)}$ which are not eigenstates of the observable $P(A)$. Let us further assume that the unitary operator U_A which is used for a measurement of the observable $P(A)$, fulfills the calibration postulate for repeatable measurements. Then we know that a measurement of the observable $P(A)$ in case of the particular preparation φ^A leads with certainty to the states Φ_A and φ^A showing the result A. On the basis of this probability free interpretation of quantum mechanics we want to show that for arbitrary preparations $\varphi \neq \varphi^A, \varphi \neq \varphi^{\neg A}$ the formal probability $p(\varphi, A_i)$, which is induced by φ and $P(A)$, is reproduced in the statistics of the measuring outcomes A_i. The probability reproducibility condition would then be a theorem of the probability free theory and no longer an additional postulate.

Let us consider N independent systems S_i with identical preparation $\varphi^{(i)}$ as a compound system $S^{(N)}$ in the tensor product state

$$(\varphi)^N = \varphi^{(1)} \otimes \varphi^{(2)} \otimes \ldots \otimes \varphi^{(N)}, \quad (\varphi)^N \in H_S^{(N)},$$

where $H_S^{(N)}$ is the tensor product of N Hilbert spaces $H(S_i)$. A premeasurement of A transforms the initial state $\varphi^{(i)}$ of each system S_i into the mixed state

$$W_{S'} = \sum p(\varphi, A_i) P[\varphi(A_i)],$$

with eigenstates $\varphi(A_i)$ of A corresponding to values A_i. If A is measured on each system S_i, then the measuring result is given by a sequence $\{A_{l(1)}, \ldots A_{l(N)}\}$ of system values $A_{l(i)}$ and states $\varphi(A_{l(i)})$, respectively, with an index sequence $l := \{l(1), l(2), \ldots l(N)\}$ such that $A_{l(i)} \in \{A_k\}$.

In the N-fold tensor product Hilbert space $H_S^{(N)}$ of the compound system $S^{(N)}$ the special states $(\varphi)_l^N = \varphi_{l(1)}^{(1)} \otimes \ldots \otimes \varphi_{l(N)}^{(N)}$ with $\varphi_{l(i)}^{(i)} := \varphi(A_{l(i)}) \in H(S_i)$ form an orthonormal basis. The relative frequency $f^N(k, l)$

of values A_k in the state $(\varphi)_l^N$ is then given by $f^N(k,l) = 1/N \sum \delta_{l(i),k}$. We can then define in $H_S^{(N)}$ an operator "relative frequency of systems with values A_k" by

$$f_k^N := \sum f^N(k,l) P[(\varphi)_l^N],$$

where the sum runs over all sequences l. The eigenvalue equation of this operator

$$f_k^N (\varphi)_l^N = f^N(k,l)(\varphi)_l^N$$

then shows that the relative frequency of the value A_k is an objective property of $S^{(N)}$ in the state $(\varphi)_l^N$ and given by $f^N(k,l)$. The eigenvalue equation can also be written in the equivalent form

$$\mathrm{tr}\left\{P\left[(\varphi)_l^N\right]\left(f_k^N - f^N(k,l)\right)^2\right\} = 0.$$

After a premeasurement of A a system S_i is in a mixed state $W_{S'}$. If N premeasurements of A are performed, then the state of the compound system $S^{(N)}$ is given by the N-fold tensor product state

$$(W_{S'})^N = W_{1'} \otimes W_{2'} \otimes \ldots \otimes W_{N'}$$

of these mixed states $W_{i'}$. One easily verifies that the expectation value of f_k^N in this product state is given by $p(\varphi, A_k)$. However, in general the state $(W_{S'})^N$ is not an eigenstate of the relative frequency operator f_k^N with eigenvalue $p(\varphi, A_k)$. This means that

$$T_k^N := \mathrm{tr}\left\{(W_{S'})^N \left(f_k^N - p(\varphi, A_k)\right)^2\right\} \neq 0,$$

and that the relative frequency of values A_k is not an objective property of the system $S^{(N)}$ in the state $(W_{S'})^N$.

In contrast to this somewhat unsatisfactory result one finds that for large values of N the post-measurement product state $(W_{S'})^N$ of the compound system $S^{(N)}$ becomes an eigenstate of the operator f_k^N and the value of the relative frequency of A_k approaches the probability $p(\varphi, A_k)$. Indeed, one finds after some tedious calculations [2], [4,5,6].

$$T_k^N = 1/N p(\varphi, A_k)\left(1 - p(\varphi, A_k)\right),$$

and thus one finally obtains the desired result

$$\lim_{N\to\infty} \mathrm{tr}\left\{(W_{S'})^N \left(f_k^N - p(\varphi, A_k)\right)^2\right\} = 0.$$

This means that in the limit of an infinite number N of systems the state $(W_{S'})^N$ is an eigenstate of the operator f_k^N of the relative frequency of values

A_k and that the compound system $S^{(N)}$ possesses the relative frequency $p(\varphi, A_k)$ of A_k as an objective property.

5. MATHEMATICAL CONSIDERATIONS

In classical probability theory it is well known that probabilities are not relative frequencies. According to the law of large numbers, for large N the relative frequency $f^N(k,l)$ for some index value k approaches the probability $p(k)$ for almost all sequences l, i.e., with a probability which is equal to one. This means that probability statements cannot be replaced in general by probability free statements, even if the number of systems is infinite. Hence it must be clarified for the above mentioned result that in quantum mechanics probabilities can in fact be completely eliminated, provided the number of systems or of measurement outcomes is infinite.

According to a recent investigation [3] for the problem of the present paper this can actually be shown. Let $S^{(\infty)}$ be a compound system which is composed of infinitely many copies of split beam photons S_i. The system $S^{(\infty)}$ can be described in the infinite tensor product space $H_S^{(\infty)}$ which is a nonseparable Hilbert space [7]. The eigenvalue equation of the observable $P(A)$, say, for the individual system S_i is written here as

$$P(A)^{(i)}\varphi_k^{(i)} = A_k^{(i)}\varphi_k^{(i)}, k \in \{1,2\}, A_1^{(i)} = 1, A_2^{(i)} = 0.$$

For an arbitrary state $\varphi \in H_S$ with $(\varphi, \varphi) = 1$ the tensor product state $(\varphi)^\infty$ is a state in $H_S^{(\infty)}$. If one performs $P(A)$-measurements on each system S_i one obtains a sequence $s_{\{l\}}$ of $P(A)$-eigenvalues $A_{l(i)}^{(i)} \in \{0,1\}, l = \{l_1, l_2, \ldots\}$ and $l_i \in \{1,2\}$. Let $\sum = \{s_{\{l\}}\}$ be the nondemumerable set of sequences of this kind. Any subset $\sum^{(\alpha)} \subseteq \sum$ describes a property. E.g., the set $\sum^{(p)}$ of sequences $s_{\{l\}}$ with the "probability **p** law of large number property" (with respect to $A^{(i)} = 1$) reads

$$\sum^{(p)} = \left\{ (s_{\{l\}}) : \lim_{N\to\infty} 1/N \sum_l^N A_{l(i)}^{(i)} = \mathbf{p} \right\}.$$

For any subset we introduce an indicator function $F^{(\alpha)}(s_{\{l\}})$ by $F^{(\alpha)}(s_{\{l\}}) = 1$ if $s_{\{l\}} \in \sum^{(\alpha)}$ and $F^{(\alpha)} = 0$ otherwise.

According to the spectral theorem by an indicator function $F^{(\alpha)}$ a projection operator $P^{(\alpha)}$ is uniquely defined in $H_S^{(\infty)}$. Hence, for the product state $(\varphi)_l^\infty$ we obtain

$$P^{(\alpha)}(\varphi)_l^\infty = F^{(\alpha)}(s_{\{l\}})(\varphi)_l^\infty.$$

In the nonseparable Hilbert space $H_S^{(\infty)}$ there are many product states $(\psi_k)^\infty = \psi_{k(1)}^{(1)} \otimes \psi_{k(2)}^{(2)\dots}$ which are not superpositions of the states $(\varphi)_l^\infty$. For these states one obtains

$$|P^{(\alpha)}(\psi)_k^\infty|^2 = \int F^{(\alpha)}(s_{\{l\}})d\mu, \tag{5.1}$$

where the measure μ depends on the state $(\psi)_k^\infty$ and on the measured observable [3].

Let$P^{(1/2)}$ be the projection operator of the "probability $p = 1/2$ law of large number property" of a sequence $s_{\{l\}}$:

$$F^{(1/2)}(s_{\{l\}}) = 1 \text{ if } \lim_{N\to\infty} 1/N \sum A_{l(i)}^{(i)} = 1/2 \text{ and } F^{(1/2)}(s_{\{l\}}) = 0 \text{ otherwise.}$$

This situation is realized, for exemple, if all systems S_i are prepared in states

$$\psi^{(n)} = 1/2(1 + e^{i\pi/2})\varphi_A^{(n)} + 1/2(1 - e^{i\pi/2})\varphi_{\neg A}^{(n)}$$

and $P(A)^{(n)}$ is measured. The post-premeasurement mixed states read in this case

$$W_S^{(n),} = 1/2P[\varphi_A^{(n)}] + 1/2P[\varphi_{\neg A}^{(n)}],$$

and the probability p to find the A-value $A^{(n)} = 1$ is then given by $p = | < \varphi_1^{(n)}|\psi^{(n)} > |^2 = 1/2$ for each n. The classical law of large numbers asserts in this case that the probability for the relative frequency 1/2 for the value $A = A_1 = 1$ in a sequence $s_{\{l\}}$ is equal to 1 , i.e.,

$$\int F^{(1/2)}(s_{\{l\}})d\mu(s_{\{l\}}) = 1.$$

This means that the relative frequency of $A = 1$ for almost all sequences amounts 1/2. Together with Eq. (5.1) we get the relation $|P^{(1/2)}(\psi)^\infty|^2 = 1$ and thus

$$P^{(1/2)}(\psi)^\infty = 1 \cdot (\psi)^\infty. \tag{5.2}$$

According to the realistic interpretation of quantum mechanics this eigenvalue Eq. (5.2) means that the compound system $S^{(\infty)}$ possesses the property given by $P^{(1/2)}$. Hence this property, or the relative frequency value 1/2 property, pertains to the system without any reference to probability.

It can be shown [3] that this way of reasoning can be applied to any "tail-property" with probability 1, e.g., to randomness. Within the context of the present problem this means, that in quantum mechanics the probabilistic way of speaking can be replaced by statements which *do not* refer to probability. This result justifies the interpretation of the more technical results of Sec. 4. Whereas in classical probability theory probabilistic statements like $p = 1/2$ can never be reduced to statements which, for an arbitrary large ensemble, hold with certainty but only to statements which are almost true, in quantum mechanics probability statements can be replaced by propositions which hold without reference to probability.

6. CONCLUDING REMARKS

In a photon split beam experiment for an individual system it is objectively undetermined whether the photon has the property A or the counter property $\neg A$. However, in spite of the objective indeterminacy of each individual system, for a sufficiently large ensemble of photons we observe a strict law. The relative frequency of systems with property A is given by $p_1(\delta) = \cos^2 \delta/2$ where δ is the phase in the preparation state.

This statistical law proves to be a "law without law". Indeed, for an ensemble $S^{(N)}$ of N identically prepared systems, in the post-premeasurement state the relative frequency of property A is in general not an objective property. However, in the limit $N \to \infty$ the relative frequency of A becomes an objective property of the compound system with the value $p(\varphi, A) = \cos^2 \delta/2$.

Formally this means that the probability for observing the relative A-frequency $\cos^2 \delta/2$ in a sequence of A-measurements is equal to 1. In terms of quantum mechanics this means that the "probability $p = \cos^2 \delta/2$ law of large numbers property" pertains to the system $S^{(\infty)}$ as an objective property. In this way the statistical laws of quantum mechanics emerge from the probability free theory.

REFERENCES

1. Busch, P., P. Lahti, and P. Mittelstaedt (1992), *Found. Phys.* **22**, pp. 949-962.
2. DeWitt, B.S. (1971), in *Foundations of Quantum Mechanics*, IL Corso, B. d'Espagnat, ed. (Academic, New York), pp. 167-218.
3. Gutmann, S. (1995), quant-ph/ 9506016.
4. Mittelstaedt, P. (1990), in *Symposium on the Foundations of Modern Physics 1990* (World Scientific, Singapore), pp. 261-279.
5. Mittelstaedt, P. (1993), *Int. J. of Theor. Phys.* **32**, 1763-1775.
6. Mittelstaedt, P. (1997), *The Interpretation of Quantum Mechanics and the Measurement Process* (Cambridge University Press, Cambridge).
7. von Neumann, J. (1938), *Compositio Mathematica* **6**, 1-77.
8. Wheeler, J.A. (1983), *Am. J. Phys.* **51**, 398-404.

CONSISTENT HISTORIES: A CRITIQUE

Roland Omnès

Laboratoire de Physique Thèorique et Hautes Energies
Laboratoire associé au CNRS
Université de Paris-Sud
91405 Orsay, France

The interpretation of quantum mechanics in terms of consistent histories is outlined. A critique is attempted, assessing achievements and pointing out open questions and unsolved technical problems.

Key words: interpretation, histories, decoherence, classical physics, epistemology.

1. INTRODUCTION

Consistent histories have been used for two different purposes: As initially conceived by Griffiths [1] and further implemented by the author [2, 3], they provide a convenient framework for reformulating the Copenhagen interpretation of quantum mechanics; as systematized by Gell-Mann and Hartle in a "generalized quantum mechanics" [4, 5], they are mainly turned towards the meaning of events in quantum cosmology. The two programmes have different ambitions and the present discussion will be restricted to the first category of questions.

In this approach, the "basic principles" (Hilbert space framework, Schrödinger dynamics and Bose/Fermi statistics) are shown to imply their own interpretation, i. e., the validity of classical physics (including its common sense logic) and the usual rules of "measurement theory". It relies heavily upon five ingredients that can be listed in the most convenient order for their synthesis: (i) quantum properties, (ii) consistent histories, (iii) their underlying logic, (iv) classical logic and determinism, (v) the decoherence effect.

These ingredients have been used to obtain a new derivation of the

M. Ferrero and A. van der Merwe (eds.),
New Developments on Fundamental Problems in Quantum Physics, 275–283.

Copenhagen interpretation, showing explicitly its logical consistency after minor changes [3]. No claim is made that this construction and its organisation are the only possible ones, since different mathematical formulations can often be found in the present domain. I believe however that any theory leaving the basic principles valid with no modification or addition will be more or less equivalent to the present one.

2. FROM PROPERTIES TO LOGIC

2.1. Theory

The "properties" entering the theory are mostly the so-called "elementary predicates" introduced by von Neumann [6]. Expressed in words, they state that the value of some observable (a self-adjoint operator) is in some real domain at some given time t. This is also expressed in mathematical terms by a projection operator $E(t)$. The definition of a property by such a "projector" is quite general since for instance the statement "A reaction $p + n \to d + \gamma$ has occurred before time t "is expressed by the projector upon the channel $d + \gamma$ in Hilbert space.

A "history" is a sequence of properties $E_1(t_1), \ldots, E_n(t_n)$ at times $0 \leq t_1 < \ldots < t_n$. Histories come in families, each one describing an alternative sequence of events. It will be convenient to introduce the product $C = E_1(t_1) \cdot E_2(t_2) \ldots E_n(t_n)$. Assuming the existence of a density operator at time zero, if only for the logical reasons entering in Gleason's theorem, the only possible probability for a history is $p = Tr(C^\dagger \rho C)$. Even so, this formula does not yield additive probabilities for alternative histories h and h', except when the "consistency conditions"

$$Re\{Tr(C^\dagger \rho C')\} = 0 \text{ for } h \neq h'. \tag{1}$$

are satisfied. The seminal discovery of these "consistent histories is due to Griffiths [1].

The reason why these consistent histories can be said to make sense is that ordinary (Aristotelian, Boolean) logic applies to them [2]. The propositions to be used belonging to a family of consistent histories (expressing either a full history, part of it or only a property), it is easy to give a meaning to the basic logical operations "and, or, not". As for logical implications between these propositions such as "a implies b", they are expressed by a conditional probability $p(b|a)$ equal to 1. One may even say that "a implies b" with a probability of error ε if $p(b|a) \geq 1 - \varepsilon$.

It was proposed as a unique and basic axiom for an interpretation that any description of a physical system should be made by propositions belonging to a consistent family of histories and all the logical consequences of this description should be substantiated by this kind of conditional probabilities [2].

2.2. Critique

This is perhaps the deepest part of the theory, asking us to change our usual habits of thinking. There are so many questions that I can only give a list for the reader to ponder over them.

Does one claim that the properties entering consistent histories are real events [7, 8]? In my opinion, the answer is "No". They can only pretend to be non-contradictory descriptions allowing sensible reasoning. The question of realism can only be discussed at a later stage in the construction.

The probabilities we have written in terms of a density operator are not time-reversal invariant. This means that there is a specific "logical" direction of time (that can be shown to coincide with the thermodynamical direction of time after introducing decoherence [3]). Griffiths proposed another expression for the probabilities, using explicit initial and final properties without a recourse to a density operator. This is time-reversal invariant, though I have the feeling that one would have to pursue histories infinitely back in time and infinitely far into the future to make it completely consistent. In any case, this is a deep question which is not completely mastered.

Several families of consistent histories can often be found to describe equivalently the same physical situation, though they are not compatible when put together (one of them stating a property about angular momentum, for instance, and the other one considering position, both for the same time). This is how Bohr's "principle" of complementarity is encountered in this approach. It can be shown however that no logical contradiction can result [3].

This does not mean that puzzling situations cannot be encountered and, as a matter of fact, the automatic calculations in the present approach make it much easier to exhibit such paradoxes (which "differ from common opinion"), though they are never apories ("a logical difficulty with no issue"). Up to now they can always be shown to be just as well present in the Copenhagen interpretation and expressible as non-paradoxical "gedanken measurements".

Nice examples of such paradoxes (though somewhat pretending to be apories) are given by Kent [9], but this matter will have to be treated at more length elsewhere.

There is finally an interesting question about logic. Most people easily accept that some propositions belonging to the description of a natural system could be true only up to some probability of error, as opposed to true propositions in pure mathematics. Many people are more reluctant to accept that the axioms of logic could also be applicable with a small probability of error when the consistency conditions are only approximate [10]. This is a rather new situation in logic and it would be interesting if logicians had their say on this.

3. CLASSICAL PHYSICS AND DETERMINISM

3.1. Theory

The relation between quantum physics and classical physics, in a joint conception of dynamics and logic, is certainly the most difficult step in an interpretation. Bohr had to avoid it, partly because of the techniques he had available. Significant progress have now been made [3, 11] because one can use rather recent powerful mathematics [12], and much more can still be expected. The theory proceeds roughly as follows:

Step 1: Considering a macroscopic object, one selects "collective coordinates" together with a multitude of microscopic coordinates describing the so-called environment. The total hamiltonian H is separated into three parts H(collective) + H(environment)+ H(coupling). The last term, which is responsible for friction, is conveniently left out in a first approach.

Step 2: The languages of classical and quantum physics are brought nearer by using Wigner-Weyl symbols $a(q,p)$ associated with quantum observables A. To the operator H(collective) corresponds a function $h(q,p)$, the "Hamilton function" giving a meaning to classical trajectories.

Step 3: A "classical property" states that the coordinates (q,p) are given up to some errors or, more generally, the point (q,p) is in some cell C in phase space. Improving slightly upon a theorem by Hörmander, one can show that there is a family of equivalent quantum projectors associated with this cell, if it is big enough and its boundary smooth enough. This allows one to express the classical language as a version of the quantum logical framework.

Step 4: Improving slightly upon a theorem by Egorov, one can prove the following result: Let C be a cell associated with some projector E, C' the transform of C by classical motion during a time t, E' a projector associated with C'. Then one has approximately $E' = UEU^{\dagger}, U = U(t)$ being the Schrödinger evolution operator. All the niceties are of course in the meaning of "equivalent" projectors, "approximate equality" between two of them, estimates of errors in various circumstances and so on.

Rather than a long discussion, one may say that the results are just as one would expect: One obtains in ordinary circumstances a justification of common sense logic, which turns out to be a special case of a quantum-history logic. Determinism holds true up to quantum fluctuations (Notice that determinism is nothing else than the logical equivalence of two classical properties holding at different times). For chaotic systems however, the errors may become out of control.

3.2. Critique

Step 1, introducing collective coordinates, is equivalent to giving a meaning for the notion of "object" in quantum mechanics. Its derivation from first principles, quantitative assignments of "collectiveness" and similar questions remain practically untouched. The reintroduction of friction (because of decoherence) is still at a stage of hand waving.

The results in the next steps are so natural and so close to well-known important mathematical theorems that there is no doubt about them. The proofs, however, are in a mathematician's style, with gross overestimates of errors. More numerical examples would bring clarity.

The great difficulty is, once again, to change our modes of thought in a drastic way, ceasing to oppose probabilism and determinism and seeing the first one as a special circumstance of the first one. Still more difficult is the change of status for common sense logic, to see it as a secondary consequence and not necessarily a basic premise.

This part of the theory has received few criticisms if any. Perhaps this is because it is obviously true, or perhaps because it stands upon unfamiliar though unavoidable mathematics.

4. DECOHERENCE

4.1. Theory

Decoherence is essentially the dynamical disappearance of macroscopic interferences because of microscopic destructive interferences in the "environment". It seems that Heisenberg had a rather clear idea of this effect [13], which has been described in detail by various people working with different models [14]. Since the subject is dealt with by Zeh elsewhere in this issue [15], one may pass directly to its critique

4.2. Critique

Though decoherence is widely believed to be a universal effect in the presence of even a slight dissipation, there is no general proof of it. One may attribute this to our poor handling of phases in a many-body system and the difficulty is mostly mathematical. The main problem is of course to obtain definitive experimental evidence for the effect, which is probably not too far away in the future.

An essential question is to decide whether the effect is by itself a sufficient explanation for the absence of macroscopic interferences or whether it holds only "for all practical purposes" [16]. A promising approach in support of the first point of view is to take into account the errors in classical statements that were mentioned in the previous section [3]. Take for instance a system S with N degrees of freedom with a superposition of wave functions that is hidden by decoherence when one looks at collective observables. Another system S' with N' degrees of freedom could measure in principle an observable showing the persistence of interferences. The probability p of the answer "yes" for such a measurement is very small and it must be compared with the probability p' for the same result occurring wrongly through a quantum fluctuation in S'. Simple examples give $N' \sim \exp N^{2/3}$. Take $N \sim 10^{24}$ and think of it, of the size of the universe, the time a signal must take to cross a monstrous apparatus or the end of M' into a black hole if one makes it small enough. This line of thought, where the notion of complexity of a system seems to be essential, would be worth a deeper analysis.

It should also be mentioned that decoherence is the main responsible for the validity of consistency conditions. Time-reversed consistency conditions are not satisfied and this is why thermodynamics and logic end up with the same direction of time.

5. MEASUREMENT THEORY

5.1. Theory

Let us consider only perfect measurements (Pauli's Type I) for brevity. An observable of a quantum system Q is measured by an apparatus M, which decoheres (and must accordingly be dissipative). The "datum" (e.g., a counter clicks) and the "result" (e. g., a spin component of Q is $+1/2$) can be shown to be logically equivalent. The very notion of properties shows that the result is necessarily an eigenvalue of A. The probability of data is given by Born's rule. When two measurements are made in succession, one can dispose of the history baggage and show that the probability of the second measurement is given by the rule of wave function reduction, though there is no physical effect in action except for decoherence. All the familiar results of measurement theory are therefore recovered as so many theorems, which shows the logical consistency of Bohr's interpretation and gives a completely different approach to it.

5.2. Critique

Some essential points in the critique will be left for the last Section. Let us mention only a few interesting aspects that have not yet been published. They deal with a large number N of identical measurements. It should be stressed that the probabilities we deal with have been only logical tools up to that point. One says: If some system is so prepared (past history) and with such data, then one can assert some properties with logical probability 1. One may show that in the limit $N \to \infty$, the logical probability is 1 for the following statements: The results of successive individual measurements are uncorrelated, i.e., random. The frequency of each datum is equal to its logical probability for an individual measurement. In other words, the theory seems to demand the randomness of quantum events though this was nowhere included in the basic principles! The proof I had before this conference was clumsy but the beautiful paper by Mittlestaedt [17] seems to provide all the tools to make it perfect.

6. EPISTEMOLOGY

Is the theory realist? Does it tell us what actually occurs in the quantum world? The answer is "no" for two reasons. First, because many quantum properties one would like to assert are questioned by complementarity. True enough, as Bohr said, the experimental devices strongly reduce complementarity by picking out some subclasses of consistent histories, but not a unique family. Histories are very convenient, everybody uses them implicitly when describing an experiment (though they should better be consistent), they can be used reliably to justify a calculation of systematic errors in an experiment. They do not tell us however what actually occurs, what is real if this has a meaning.

Histories belong to what d'Espagnat call "weak realism", namely they can describe and utilize every experiment one can make or any natural event that can occur. They do claim to reach the marrow of Nature.

This question is strongly linked with the notion of truth in quantum mechanics. Dowker and Kent [10] have convincingly shown that a definition of truth I gave earlier [3] was too restrictive and should be revised. Rather than mending it quickly (which is not too difficult), I believe it should be reconstructed from the bottom with a look at the overall status of logic in physics (which may take time).

The second reason why the theory is not "strongly realist" brings us into deep waters. One has recovered all along all the aspects of ordinary, experimental reality except one: At the end of a measurement, only one actual datum occurs. All those that were also possible are condemned. One does not explain the uniqueness of reality whereas reality is defined by philosophers through its uniqueness: Wittgenstein, for instance, gives a substance to language games by pointing out a finger at something, saying "that", that which is thereby defined because and only because it is unique.

Leaving out hidden variables or spontaneous reduction, keeping only quantum mechanics as it exists presently, there are only two answers known to this question of "actualization". One is Everett's many-worlds metaphysical assumption. Another is the one I prefer, namely that quantum mechanics is such a deep theory that it has reached the limit of the Cartesian programme, which assumes that there can be a complete mapping of a mathematical model on reality. Every discussion of realism should begin by taking a stand upon this too often implicit programme.

Perhaps a more interesting epistemological consequence of the theory comes from the status of secondary construct it gives to classical physics and its common sense logic. This is where the distance with Bohr's epistemology is largest. One may reconsider for instance such questions as: When can one rely on causality, on locality, on separability, on a visual representation of the world and its description by ordinary language? When are these familiar philosophical primaries valid? The answer is "when one deals with macroscopic objects (except maybe SQUID's or analogous devices and for full chaos, at least for some of these statements). As the objects become smaller, the probability of error when applying the familiar primaries increases. It is of the order of 1, giving completely foolish primaries, for an electron in an atom. This means that one should better not start from this kind of primaries if the aim is to build a philosophy of knowledge encompassing quantum mechanics. More bluntly said, quite a few philosophical approaches to the foundations of physics are unsuited to their aim from the start. There are fortunately much more general, more reliable principles, those of physics itself. They easily encompass the ordinary primaries to which they assert a precise but limited domain of validity. An interesting development of the present trend of interpretation is therefore to propose a change of premiss in the philosophy of knowledge [18], a rather amusing consequence if one remembers that everything started from purely technical investigations.

REFERENCES

1. R. Griffiths, *J. Stat. Phys.* **36**, 219 (1984).
2. R. Omnès, J. *Stat. Phys.* **53**, 893 (1988), *Ann. Phys. (N. Y)* **201**, 354 (1990).

3. R. Omnès, *Rev. Mod. Phys.* **64**, 339 (1992), *The Interpretation of Quantum Mechanics* (Princeton University Press, Princeton, 1994).
4. M. Gell-Mann and J. B. Hartle, in *Complexity, Entropy and the Physics of Information*, W. Zurek, ed. (Addison-Wesley, Redwood City, 1991).
5. C. J. Isham, *J. Math. Phys.* **35**, 2157 (1994).
6. J. von Neumann, *Mathematische Grundlagen der Quantummechanik* (Springer, Berlin, 1932).
7. B. d'Espagnat, *J. Stat. Phys.* **56**, 747 (1989).
8. R. Griffiths, *Consistent histories and quantum reasoning*, preprint, Carnegie Mellon University.
9. A. Kent, these proceedings.
10. F. Dowker and A. Kent, *J. Stat. Phys.* **82**, 1575 (1996).
11. R. Omnès, *Quantum-classical correspondence using projection operators*, to appear in *J. Math. Phys.*
12. L. Hörmander, *The Analysis of Linear Partial Differential Operators* (Springer, Berlin, 1985).
13. W. Heisenberg, *Physics and Philosophy* (Allen, London, 1958), pp. 153-154.
14. K. Hepp and E. H. Lieb, *Helv. Phys. Acta* **46**, 573 (1974), W. Zurek, *Phys. Rev. D* **26**, 1862 (1982), E. Joos and H. D. Zeh, *Z. Phys. B* **59**, 223 (1985).
15. H. D. Zeh, these proceedings.
16. J. S. Bell, *Helv. Phys. Acta* **48**, 93 (1975).
17. P. Mittlestaedt, these proceedings.
18. R. Omnès, *Philosophie de la science contemporaine* (Gallimard, Paris, 1994).

IMMATERIAL INTERPRETATION OF QUANTUM THEORY IN THE CONTEXT OF QUANTUM COSMOLOGY

Z. Onyszkiewicz

Faculty of Physics, A. Mickiewicz University
61-614 Poznań, Poland

An immaterial (non-material) interpretation of quantum theory has been suggested, which explains the genesis of the principles of quantum theory and provides a simple solution to the problem of quantum measurement in the context of quantum cosmology.

Key words: principles of quantum cosmology, quantum measurement problem.

1. MOTIVATION

In the scenario proposed by quantum cosmology [1,2], the universe simply comes out of nowhere, in full accordance with the laws of quantum theory, and creates along the way all the matter and energy needed to build the universe we now see. It thus incorporates the creation of all physical things, including space and time. The quantum cosmological models attempt to explain everything entirely within the context of the principles of quantum theory. Quantum cosmology can perhaps explain the origin and organization of the universe, but not the laws of physics themselves. Quantum laws have to exist (in some sense) so that a quantum transition could have generated the universe in the first place. They have to be "there" to start with so that the universe could have come to being. So, the question arises: whether the quantum theory principles can be derived from even more profound principles? We believe that the answer to this question may be positive. In this paper a model of universe, close to the ontology of Berkeley [3] which explains the genesis of the quantum cosmology and provides a simple solution to the problem of quantum measurement, is proposed.

M. Ferrero and A. van der Merwe (eds.),
New Developments on Fundamental Problems in Quantum Physics, 285–289.

2. BERKELEY'S MODEL OF THE UNIVERSE

We start from the general model of the universe suggested in 1710 by the Irish philosopher George Berkeley [3]. After Berkeley, let's take the following assumptions:

(a) There is a set of human consciousness. As human consciousness is a non-physical concept it is hard to give its precise definition although everyone of us has their own clear intuitive understanding of what it means to be conscious. Consciousness is at least a realization of its own existence. It is useful to introduce a symbolic description of consciousness (C) in the form: $C_n[I_1, I_2, \ldots]$, where n = 1, 2,... number consciousnesses of particular conscious beings, and the symbols in the square parentheses represent the knowledge (ideas $I_1, I_2, \ldots$) belonging to the n-th consciousness.

(b) There is a certain consciousness $C[\ ]$ showing the properties analogous to those of human consciousness which transmits the information between individual $C_n[\ ]$ according to the scheme

$$C_n[\ ] \leftrightarrow C[\ ] \leftrightarrow C_{n'}[\ ], \tag{1}$$

where $n \neq n' = 1, 2, \ldots$, and arrows represents the information transmission.

(c) All human consciousness from the set $\{C_n[\ ]\}$ believe that they are in the same universe as they receive from $C[\ ]$ mutually correlated information.

(d) The physical laws apply only to the rules of the information transmission between $C_n[\ ]$ and $C[\ ]$ but they do not apply to the nature of $C_n[\ ]$ and $C[\ ]$. Thus, the nature of consciousnesses and the way of the information transmission is not cognizable by the method of physics.

(e) The universe is not build of the interacting matter but of sequences of informations (ideas) transmitted by $C[\ ]$ to $C_n[\ ]$.

Thus the universe is immaterial (non-material). Apart from the observations the universe does not contain anything else. According to Berkeley the universe is only a dynamic, correlated and ordered sequence of ideas of the consciousness $C[\ ]$ communicated to the set of individual consciousness $C_n[\ ]$. The image of the universe in Berkeley's ontology is simple but extremely difficult to accept because of its non-material character.

The elementary manifestation of will of the consciousness is the act of choice. In order to perform this act the consciousness $C[\ ]$ must have access to a set of mutually exclusive possibilities $\{p\}$, where p represents one of them. Of these possibilities, the consciousness $C[\ ]$ actuates one $p' \in \{p\}$ in a jumpwise process, which can be symbolically written as

$$C[\{p\} \rightarrow p'], \tag{2}$$

where the arrow represents the process of actuation of the possibility p'.

There comes a question about the mathematical structure of the set $\{p\}$. For obvious reasons it should satisfy the following conditions: i) Both p and the set $\{p\}$ represent states of the consciousness $C[\quad]$, therefore the set $\{p\}$ must be a certain analytical function f of many variables $p_1 \in \{p\}, p_2 \in \{p\}, \ldots$. Hence $\{p\} = f := f(p_1, p_2, \ldots)$. ii) The set of possibilities $\{p\}$ can be either finite or infinite, however, it must consist of at least two mutually exclusive possibilities, otherwise there will be no possibility of making a choice. iii) We are not able to follow the act of choice by $C[\quad]$. The only thing we can do is to predict a probability of actuating a given possibility p. The probability should be a functional of the arguments f and p, so

$$prob_p = prob_p[f, p]. \tag{3}$$

Our purpose now is to find a functional (3) which will satisfy the conditions (i) – (iii). In order to find such a functional we introduce a symbol of the possibility $\bar{p}$ conjugated to p, defined as:

$$(zp)^* = z^*\bar{p} \quad \text{and} \quad \bar{f} = f^*(\bar{p}_1, \bar{p}_2, \ldots), \tag{4}$$

where the star stands for the same conjugation as for the complex numbers, and z is an arbitrary complex number. We have to do it as otherwise it will be impossible to describe the act of choice of one of the possibilities. We also have to define an interior product of the possibility symbols as a functional (function of numerical values) whose argument is an ordered pair of possibility symbols of the following main properties:

$$\bar{p}p' = \delta_{p,p'}, \quad \bar{p}(z_1 p' + z_2 p'') = ((z_1^* \bar{p}' + z_2^* \bar{p}'')p)^* = (z_1^* \bar{p}' p + z_2^* \bar{p}'' p)^*, \tag{5}$$

where z_1, z_2 are an arbitrary complex numbers and $\delta_{p,p'}$ is the Kronecker delta.

Since f and $\bar{f}$ represents the state of the consciousness $C[\quad]$ prior to the actuation, according to the properties (i) and (5) we have

$$\bar{f} f = 1. \tag{6}$$

Then, expanding functions f and $\bar{f}$ in equation (6) and taking into account (5) we obtain

$$f = \sum_p \bar{p} f p, \quad \text{and} \quad \bar{f} = \sum_p \bar{f} p \bar{p}, \tag{7}$$

$$\sum_p (\bar{f} p)(\bar{p} f) = 1, \quad \text{and} \quad \sum_p p \bar{p} = 1. \tag{8}$$

Considering Eq. (8), we find that the product $(\bar{f}p)(\bar{p}f)$ meets the conditions required for probability and thus can be identified as functional (3) of probability:

$$prob_p = (\bar{f}p)(\bar{p}f). \tag{9}$$

Substituting (7) and (2) to expression (1) we find

$$\begin{aligned} C_n[p'] &\leftrightarrow C\left[\sum_p \bar{p}fp \to p'\right] \leftrightarrow C_{n'}[p'], \\ C_n[\bar{p}'] &\leftrightarrow C\left[\sum_p \bar{f}p\bar{p} \to \bar{p}'\right] \leftrightarrow C_{n'}[\bar{p}']. \end{aligned} \tag{10}$$

Those formulas are a concise expression of the Berkeley's model of the universe and mean that the state of consciousness prior to the act of choice is a linear superposition of all admissible possibilities.

3. THE GROUNDS OF THE QUANTUM THEORY PRINCIPLES

Let's assume that the process of actualization (10) is a physical process of measuring the value of an observable a. As a result of the measurement of this observable we obtain its specific value expressed (in the simplest case) by a single real number $a' = (a')^*$. Assuming the notation $p = |a\rangle$, $\bar{p} = \langle a|$, $f = |\phi\rangle$ and $\bar{f} = \langle\phi|$, we can rewrite Eqs. (10) in the form:

$$\begin{aligned} C_n[|a'\rangle] &\leftrightarrow C[|\phi\rangle \to |a'\rangle] \leftrightarrow C_{n'}[|a'\rangle], \\ C_n[\langle a'|] &\leftrightarrow C[\langle\phi| \to \langle a'|] \leftrightarrow C_{n'}[\langle a'|], \end{aligned} \tag{11}$$

where

$$|\phi\rangle \to |a'\rangle = \sum_a \langle a|\phi\rangle|a\rangle \to |a'\rangle, \quad \langle\phi| \to \langle a'| = \sum_a \langle\phi|a\rangle\langle a| \to \langle a'|. \tag{12}$$

The state of the universe before the measurement $|\phi\rangle$ is a linear complex superposition of states $|a\rangle$ which determine all possible values of the observable a. At the moment of measurement a certain state $|a'\rangle$ corresponding to the value a' of the observable a is actualized in a jumpwise process (non-spacetime in nature). From expression (9) it follows that the probability of such an actualization is given by the equation $prob_{a'} = \langle\phi|a'\rangle\langle a'|\phi\rangle$ and the expected value $\bar{a}$ of the observable a is $\bar{a} = \sum_a a\, prob_a = \langle\phi|\hat{a}|\phi\rangle$ where the operator $\hat{a} = \hat{a}^+$ satisfies the eigenequation $\hat{a}|a\rangle = a|a\rangle$ for all possible eigenstates of $|a\rangle$ and $\langle a|a'\rangle = \delta_{a,a'}$, $\sum_a |a\rangle\langle a| = \hat{1}$. Moreover, to ensure the possibility of choice, there must be at least one pair of observables a, b whose linear operators $\hat{a}$ and $\hat{b}$ do not commute $[\hat{a}, \hat{b}]_- \neq 0$. We note that a dynamical equation in the quantum cosmology of a closed universe has the time independent form [2] (quantum cosmology suggested that time is a semiclassical concept).

4. CONCLUSIONS

The above expressions make the basic skeleton of the mathematical apparatus of the quantum theory (and quantum cosmology). However, in the quantum theory they are postulated but here they are a natural consequence of the Berkeley's model of reality. Therefore, this model seems to justify the fundamentals of the quantum theory. Furthermore, this model explains ontological sense of the state vector of the universe and its collapse taking place in $C[\quad]$ at the moment of measurement, which is a non-spacetime process in nature. At the moment when $C_n[\quad]$ asks what is actually happening i.e. $C_n[\quad]$ performs a quantum measurement with the help of appropriate macroscopic measuring devices, $C[\quad]$, giving the answer, causes a collapse of linear superposition of states. Each following answer given by $C[\quad]$ is correlated with the previous one according to the quantum theory laws. This model also explains the necessity of the reduction of the linear superposition of states at the contact between the micro- and macrocosmos which is implied by the necessity of expressing the results of quantum measurement (the answers of $C[\quad]$) in the language of classical physics as only such a language is understood by a macroscopic man. (The origin and organization of the macroscopic (quasi-classical) domains in the universe can be explained in the framework of quantum cosmology for example by the method proposed in papers [4]). In this manner we explain the quantum measurement process for the universe as a whole without the necessity of introducing any changes in the mathematical formalism of the quantum theory.

REFERENCES

1. A. Vilenkin, *Phys. Lett. B* **117**, 25 (1982); *Phys. Rev.* D **33**, 3560 (1982).
2. J.B. Hartle and S.W. Hawking, *Phys. Rev. D* **28**, 2960 (1983). S.W. Hawking and R. Penrose, *The Nature of Space and Time* (Princeton University Press, Princeton, N.J., 1996).
3. The Works of George Berkeley, A.C. Frasera, ed. (Clarendon Press, Oxford, 1901).
4. Z. Onyszkiewicz, *Phys. Lett. A* **163**, 450 (1992); *Acta Cosmologica*, **18**, 7 (1992).

ATOM INTERFEROMETRY FOR QUANTUM GRAVITY?

Ian C. Percival and Walter T. Strunz

Department of Physics, Queen Mary and Westfield College
University of London
Mile End Road, London E1 4NS, United Kingdom

Matter interferometry puts quantum gravity within reach of laboratory experiment. We demonstrate how quantum state diffusion processes on Planck scales can have measurable effects on atomic scales. Our approach is by analogy with Einstein's results on Brownian motion, where measurements on a macroscopic scale can be used to determine quantities on the atomic scale. We show implications for quantum foundations.

Key words: fluctuating space-time, atom interferometry, quantum foundations.

1. INTRODUCTION

Later we shall describe a *fluctuation hypothesis.* Most of the article, including this section, depends on the truth of that hypothesis. We think that the evidence for the hypothesis is strong, but you must judge for yourself.

We are near to a new revolution in physics, comparable in magnitude to the revolution that established without question the reality of atoms and the revolution that established quantum dynamics. We believe that the crucial advance will be experimental, but that the experimenters will need a lot of theoretical help. We would like some of the theorists who read this article to provide that help.

The experiments will open up quantum gravity and quantum foundations to laboratory investigation *together*, turning them into a single research field, with a continual interaction between theory and experiment. Both these phenomena are on Planck scales.

We are using *quantum gravity* in its broadest sense, to include superstrings, the Ashtekar program, non-commutative space-time geometries,

M. Ferrero and A. van der Merwe (eds.),
New Developments on Fundamental Problems in Quantum Physics, 291–300.

etc. Theories of quantum gravity have been inspired by Dirac's principles of beauty and elegance. But beauty is in the eye of the beholder, it is subjective, and in science it is a guide to performance, not a guarantee. Experiment is the final test.

Quantum foundations means the dynamics of the elusive boundary between the quantum and the classical domains. Modern developments in quantum measurement theory were stimulated by Einstein's ideas on physical realism, as modified by Bell [1]. Bell's crucial contribution to the field was to show that quantum non-locality could be tested experimentally. But many of the recent theories, such as spontaneous localization [2] or primary state diffusion [3], have never been put to any effective experimental test.

It was principally Einstein's theory of Brownian motion that first established without question the reality of atoms. Brownian motion is a diffusion process, and because it is a diffusion process, measurements on a macroscopic scale can be used to determine quantities on the atomic scale. This indirect method is based on much easier experiments than the more direct methods like X-ray diffraction, scattering experiments with beams, or observation using ultra-microscopes. Brown published his research in 1828, long before such methods were possible.

As it was with atoms, so it could be with quantum gravity, which describes the physics of the Planck scales, the Planck
time $T_{P\ell} = (\hbar G c^{-5})^{1/2} \approx 5 \times 10^{-44}$s,
length $cT_{P\ell} \approx 10^{-35}$m, and
mass $\approx 10^{-8}$kg.

When compared with an atom, the time and length are extremely small, and the mass extremely large. This is what makes experiments on quantum gravity so difficult.

Direct methods of observation or experiment are beyond our present means. We would need to accelerate particles to near 10^{16}TeV, or demonstrate the wave properties of a mass near to that of a flea, and none of this is going to be possible in the next decade or two.

So, by analogy with Brownian motion, we should look for a diffusion process which enables us to determine quantities on the Planck scale by experiments on the atomic scale.

2. ITÔ CALCULUS AND QUANTUM STATE DIFFUSION

Itô has developed a very elegant theory and differential calculus for stochastic processes like Brownian motion that satisfy Langevin equations, described for physicists in the books of Gardiner [4]. For example when a Brownian particle with displacement s diffuses in one dimension with no drift, the mean of the distance traversed is zero and the mean square is proportional to the time. So for small displacements ds,

$$\mathrm{M}\mathrm{d}s(t) = 0, \qquad \mathrm{M}[\mathrm{d}s(t)]^2 = a^2\mathrm{d}t, \tag{1}$$

where a is a constant and M represents a mean over the ensemble of all possible processes. For more complicated stochastic processes, it is convenient to introduce a standard normalized *stochastic* differential dw, with zero mean and mean square *equal* to the time, so that

$$\mathrm{M}\mathrm{d}w = 0, \qquad \mathrm{M}(\mathrm{d}w)^2 = \mathrm{d}t. \tag{2}$$

For Brownian motion with drift velocity v and diffusion constant a^2, the Langevin-Itô differential equation is written as

$$\mathrm{d}s = v\mathrm{d}t + a\mathrm{d}w = v(\mathrm{d}t + \tau_0^{1/2}\mathrm{d}w), \tag{3}$$

where the coefficient of dt gives the drift and the coefficient of dw the diffusive part of the motion. The time constant τ_0 represents the approximate boundary between time intervals for which the diffusion dominates and time intervals for which the drift dominates. For a time interval $T \ll \tau_0$, the diffusion dominates and the drift may be negligible, but for a time interval $T \gg \tau_0$ the drift dominates, and the diffusion may be difficult to see.

Quantum state diffusion (QSD) [5] represents the stochastic evolution of individual quantum systems of an ensemble, when the ensemble as a whole is usually represented by a density operator ρ. In quantum state diffusion theory, unlike the usual density operator theory, care must be taken to distinguish the quantum expectation $\langle \ldots \rangle$ for an individual system from the mean M over the ensemble.

The general form of the quantum state diffusion equation is

$$|\mathrm{d}\psi\rangle = (\hat{\mathrm{dri}}\ \mathrm{d}t + \hat{\mathrm{dif}}\ \mathrm{d}\xi)|\psi\rangle, \tag{4}$$

where $|\psi\rangle$ represents one state of the ensemble, and $d\xi$ is the complex differential with properties

$$\mathrm{M}\mathrm{d}\xi = 0, \qquad \mathrm{M}(\mathrm{d}\xi)^2 = 0, \qquad \mathrm{M}|\mathrm{d}\xi|^2 = \mathrm{d}t. \tag{5}$$

The drift operator $\hat{\mathrm{dri}}$ and the diffusion operator $\hat{\mathrm{dif}}$ in (4) are usually nonlinear. The Schrödinger evolution is in the drift term, and for atomic systems dominates for times that are long compared with the Planck time. For macroscopic systems the diffusion term dominates, and results in a localization of the quantum wave in phase space. In short times it becomes a wave packet that is negligible outside a region of phase space that is comparable to a cell determined by Planck's constant of action $2\pi\hbar$.

There are now several alternative quantum theories based on a diffusion of quantum states. These modify the Schrödinger equation, so in principle they are distinguishable from ordinary quantum theory by experiment. Important advances were made by Bohm and Bub [6], by Pearle [7], by Gisin [8], and by Diósi [9]. The first complete theory of Ghirardi, Rimini and Weber in 1986 was based on quantum jumps [10], see also Bell [1]. It was reformulated as a quantum state diffusion theory by Diósi, Gisin and Pearle. A general quantum state diffusion theory is presented in [5].

3. THE FLUCTUATION HYPOTHESIS

This is the key hypothesis on which the most of the rest of this article depends. In the form in which we shall state it, space-time is treated classically. The hypothesis has two parts. One refers to magnitudes, the other to geometry.

FH1. The metric of space-time has Planck-scale fluctuations which contribute to the action integrals in the exponents of all quantum path summations. These fluctuations may be considered as zero-point gravitational fluctuations.

FH2. The geometry of the fluctuations can be determined by experiment.

It is useful to reconsider the analogy with Brownian motion, for which the Langevin equations do not apply much below the atomic scale, but it is simpler theoretically to use the equations as if they applied to arbitrarily small scales. Details of the dynamics of the collisions between atoms and Brownian particles are ignored. Einstein had to ignore them, as in his time no-one understood them sufficiently. We shall use Langevin-Itô equations for space-time fluctuations in the same way *as if* they applied at and below the Planck scale. 'Details' like space-time foam are ignored. We have to ignore them, as in our time no-one understands them sufficiently.

From FH1, a time-like interval in space-time is represented by

$$ds = d\bar{s} + \tau_0^{1/2} dw(x), \tag{6}$$

where τ_0 is approximately the Planck time $T_{P\ell}$. The barred metric is the Minkowski metric. For finite intervals that are much shorter than τ_0 the space-time fluctuations are dominated by diffusion, whereas for intervals that are much longer they are dominated by drift, and the space-time will appear to be smooth.

Given a geometry for the fluctuations $dw(x)$, a particle of rest mass

m has an action over a path in space-time, given by

$$S = mc^2 \int \mathrm{d}s = mc^2 \Big[\int \mathrm{d}\bar{s} + \tau_0^{1/2} \int \mathrm{d}w(x) \Big], \tag{7}$$

and a corresponding quantum dynamics given by Feynman path sums.

We do not know the geometry of the fluctuations $\mathrm{d}w(x)$. There are various 'reasonable' assumptions, for example

GF1. Independent fluctuations in space-time cells of volume given by the Planck length and time.

GF2. Gravitational waves: fluctuations that propagate with the velocity of light, so that they are correlated along light cones, with ordinary space-time geometry (gravitons).

GF3. Gravitational waves in non-commutative space-time, following Connes [11].

It would be interesting to use superstring theory or the quantization of gravity following the Ashtekar program to determine directly whether the space-time fluctuations, if they exist, have one of these geometries, or some entirely different geometry.

If the space-time fluctuations are *complex*, so that $\mathrm{d}w$ in equation (6) is replaced by $\mathrm{d}\xi$, and some nonlinear terms are introduced into the equations (7), then the state vectors obey the equations of primary state diffusion [3]. This is a development of quantum state diffusion theory in which the diffusion results from space-time fluctuations, and relates the foundations of quantum mechanics to gravity.

Feynman in his 1962 lectures [12] was probably the first to suggest a possible connection between the foundations of quantum theory and quantum gravity. The theme was taken up in 1966 by Károlyházy [13], whose theory required imprecision in space-time structure. Penrose in 1986 based his ideas on the entropy relations of Hawking, and the need for gravitational entropy [14]. Diósi in 1987 proposed the first gravitational quantum state diffusion theory [15]. Ellis and his collaborators have been looking since 1984 for connections between gravity and violations of quantum theory, and have proposed experimental tests, including matter interferometry [16].

4. ATOM INTERFEROMETERS

In an atom interferometer an atomic wave packet is split into a linear combination of two or more wave packets which follow different paths, which are then brought together again to make an interference pattern. Such interferometers are excellent at measuring the difference in interval (or proper time) between the two time-like paths. First consider the precision of such an interferometer for systematic changes in phase, such as those produced by the gravitational effect of the Earth. The change in phase along a path is given by

$$\mathrm{d\ phase} = \mathrm{d}S/\hbar = \frac{mc^2}{\hbar}\mathrm{d}s = \Omega \mathrm{d}s, \tag{8}$$

where $\Omega = mc^2/\hbar$ is the quantum angular frequency associated with the rest mass m. For a hydrogen atom, the quantum angular frequency is

$$\Omega_{\mathrm{H}} \approx 1.4 \times 10^{24}\mathrm{s}^{-1}. \tag{9}$$

The phase change of an atom of atomic number A over time interval T is

$$\text{phase change} = \Omega_{\mathrm{H}} AT. \tag{10}$$

For the 1992 experiment of Kasevich and Chu [17] the atom was Na, with $A = 23$ and the time interval $T = 0.2$s, giving a phase change of about 7×10^{24} radians. Since a phase change of about 0.1 radians can be detected, this represents a precision of about 1 part in 10^{26}. However, if the phase change in one arm is the same as that in the other, there is no change in the interference pattern, so matter interferometers are excellent detectors only of small *relative* changes in time-like intervals.

There is another way of looking at a matter interferometer, which relates it to Einstein's use of clocks in developing both the theory of special and of general relativity. An atom can be considered as a quantum clock which beats with a frequency $2\pi A\Omega_{\mathrm{H}}$. The best way to compare proper time intervals over two different paths is to use the *same* clock in both paths, a coherent linear combination of the form

$$\frac{1}{\sqrt{2}}|\text{clock here, vacuum there}\rangle + \frac{1}{\sqrt{2}}|\text{clock there, vacuum here}\rangle. \tag{11}$$

To find the effect of space-time fluctuations on the phase of one arm of the interferometer, replace $\int \mathrm{d}\tilde{s}$ by $\tau_0^{1/2} \int \mathrm{d}w$, giving for fluctuations

$$\begin{aligned} \text{phase change} &= A\Omega_{\mathrm{H}}\tau_0^{1/2} \int \mathrm{d}w \\ &\approx A\Omega_{\mathrm{H}}(\tau_0 T)^{1/2}. \end{aligned} \tag{12}$$

The time T_{F} for which the phase change in one arm due to the fluctuation is of order 1 is

$$T_{\mathrm{F}} = \frac{1}{A^2\Omega_{\mathrm{H}}^2\tau_0}. \tag{13}$$

Note that Ω_{H} enters as a square, and it is *this* which makes the indirect method of detecting space-time fluctuations better than the direct methods.

For the experiment of Kasevich and Chu, with $\tau_0 = T_{P\ell}$

$$T_F \approx 4\text{ps}, \tag{14}$$

which is very much less than the time interval $T = 0.2s$ for this experiment. So for this simplified model of the experiment, if the fluctuations in one arm were independent of the other, no interference should have been observed.

But they *did* observe interference. The above description is too crude to conclude that such fluctuations do not exist, but it does demonstrate that the small numerical value of the Planck time does not alone prevent experimental access to Planck-scale phenomena. The small value is no impassable barrier to the precision of modern matter interferometers. This is negative evidence against some types of space-time fluctuation. It would be much better to see positive evidence of space-time fluctuations through suppression of interference, which could be tried by increasing the atomic number A, the atomic drift time and the separation of the wave packets.

5. QUANTUM REALITY

Galileo helped to found modern physics by insisting that experiment and observation, aided by instruments like the telescope, were superior to ancient wisdom. Before him most natural philosophers had divided the cosmos into an imperfect Earth and a perfect heaven. Those like Newton who followed him discovered that the same dynamical laws applied to both, and so united heaven and Earth into a single universe. Maxwell and Boltzmann applied this classical dynamics of macroscopic systems to atoms and molecules, so before 1900 it appeared that all physics might be explained by a natural extension of these classical laws.

But the spectroscope sent out a different message from the telescope. The quantized dynamics of atoms is essentially different from the classical dynamics of planets. Seeing this difference, Bohr and his colleagues in Copenhagen and Göttingen divided the universe again, this time into the classical and quantum domains. By following Galileo's principles they were led in the opposite direction from Newton.

In Bohr's time, and for most physics since then, the precise location of the boundary between the classical and quantum domains was of no importance. Atoms, molecules, baryons and leptons are quantum systems, whereas stars, planets, spectroscopes and accelerators are classical. The instruments are around 10^{30} times more massive than the systems they are used to study, so the location of the boundary is irrelevant in practice and ordinary quantum mechanics serves us well. It is our most successful theory, in which classical and quantum systems obey different deterministic dynamics, but there is an essential indeterminacy where they interact.

Bohr's Copenhagen interpretation was a glorious and successful compromise. In 1935, those like Einstein and Schrödinger who were dissatisfied with this division of the universe found themselves working at the margins of physics, because their enquiries made no contact with experiment or observation.

Throughout most of the 19th century there was doubt in the scientific community about the reality of atoms. According to the chemist Ostwald in 1895, atomic theory was just a convenient model that could be used to solve chemical and physical problems, but a belief in the reality of atoms was misguided. By the end of the century most physicists had been convinced of their reality by the kinetic theories of Maxwell and Boltzmann, but there was still a core of mainly older scientists who disagreed. Einstein, Perrin and others finally resolved the controversy with theory and experiments on Brownian motion. Atoms are real.

Throughout most of the 20th century there has been doubt about the reality of quantum waves. According to the quantum theory of many textbooks, these waves are just a convenient mathematical model that can be used to answer questions about experiments and observations, but a belief in the reality of quantum waves is misguided. This opinion is part of the very language of those textbooks that talk about wave functions, but not about waves.

There is today no experimental evidence like Brownian motion for the reality of quantum waves, although many modern alternative quantum theories rely on diffusion of quantum states, which is a kind of Brownian motion in Hilbert space. But there is nothing to stop us from looking for this evidence, provided that we know where to look and that we have the experimental tools. As we shall see, there is other evidence.

6. THE DIMENSIONAL REVOLUTION

Einstein and Schrödinger were both realists.

Einstein made this clear in his 1935 paper with Podolsky and Rosen on elements of physical reality. That paper included also a belief in the locality of physical influences. Bell showed that simultaneous reality and locality could be tested experimentally, and subsequent experimental tests by Aspect and others forced every realist, including Bell, to accept non-locality.

Schrödinger was happy with his picture of an electron in a hydrogen atom as a wave in 3-dimensional physical space, but he could not accept that the two electrons of a helium atom were waves in a real 6-dimensional configuration space. Because he was a realist, no real wave could exist in a space that was not also real, and Schrödinger could not accept the reality of the higher-dimensional configuration spaces.

So both Einstein and Schrödinger were prevented from developing a realistic quantum theory by their *other* unshakeable beliefs.

Now that we have tests of the violation of Bell's inequality and we have quantum cryptography, there is a strong practical argument for the reality of waves in higher-dimensional spaces. Otherwise how could we use them to provide distant correlations or to keep secrets? The indeterminacy principle of Heisenberg showed that we can sometimes do less with a quantum system than with the corresponding classical system. Non-locality experiments and quantum cryptography show that we can sometimes do *more*.

This is evidence for the reality of quantum waves, but the issue is not finally settled, for if quantum waves are real and classical systems are real, we need a common dynamics for both, and the nature of this dynamics cannot be finally resolved until we are able to experiment at the quantum-classical boundary. The 'measurement hypothesis' of the usual theory is just not enough. Schrödinger's equation in smooth space-time has to be modified if it is to represent localized particles that obey classical dynamics, and primary state diffusion [3] which can be derived from the fluctuation hypothesis does this. Again we come back to the fluctuation hypothesis and experiments to test it.

Previous revolutions in physics involved changes in *magnitude*. For example, in this century the lifetime of our universe has (in our view) expanded to more than 10^{10} years, and its size to as many light-years. Its basic components have shrunk from the size of atoms to the size of quarks and possibly to Planck lengths and times.

Now we face a change in *dimension*. This requires a recognition that our space-time and everything in it is no more than a projection from the huge-dimensional space that carries real quantum waves. The universe appears to be 4-dimensional to our senses only because they cannot cope, and do not need to cope, with the higher dimensions of reality.

ACKNOWLEDGEMENTS

We are very grateful to Nicolas Gisin, Todd Brun, Rüdiger Schack, John Charap, Tulio Regge, Chris Hull, Wafiq Sabra, Christopher Foot, Erling Riis, Roger Penrose, Bob Jones and Chris Isham for helpful discussions and to the Alexander von Humboldt Foundation for financial support.

REFERENCES

1. J. S. Bell, *Speakable and Unspeakable in Quantum Mechanics* (Cambridge University Press, Cambridge, 1987).

2. P. Pearle, *Phys. Rev. A* **48**, 913-923 (1993).
3. I. C. Percival, *Proc. Roy. Soc. A* **447**, 189-209 (1994); **451**, 503-513 (1995); I.C. Percival and W.T. Strunz., submitted to *Proc. Roy. Soc. A.*
4. C. W. Gardiner, *Handbook of stochastic methods* (Springer, New York 1994); *Quantum noise* (Springer, New York, 1991).
5. N. Gisin and I. C. Percival, *J. Phys. A* **25**, 5677-5691 (1992); **26**, 2233-2260 (1993). I. C. Percival, *J. Phys. A* **27**, 1002-1020 (1994).
6. D. Bohm and J. Bub, *Rev. Mod. Phys.* **38**, 453-469 (1966).
7. P. Pearle, *Phys. Rev. D* **13**, 857-868 (1976).
8. N. Gisin, *Phys. Rev. Lett.* **52**, 1657-1660 (1984).
9. L. Diósi, *J. Phys. A* **21**, 2885-2898 (1988), *Phys. Lett. A* **129**, 419-442 (1988).
10. G.-C. Ghirardi, A. Rimini and T. Weber, *Phys. Rev. D* **34**, 470-491 (1986).
11. A. Connes, *J. Math. Phys.* **36**, 6194-6231 (1995).
12. R. P. Feynman, F. B. Moringo and W. G. Wagner *1962-3 Lectures on Gravitation* (California Institute of Technology, 1973).
13. F. Károlyházy, *Nuovo Cimento A* **42**, 390-402 (1966).
14. R. Penrose, *Quantum Concepts in Space and Time* (Clarendon, Oxford, 1986) .
15. L. Diósi, *Phys. Lett. A* **120**, 377-381 (1987). See also G.-C. Ghirardi, R.Grassi, and A. Rimini, *Phys. Rev. A* **42**, 1057-1064 (1990).
16. J. Ellis, J. S. Hagelin, D. V. Nanopoulos and M. Srednicki, *Nuc. Phys B* **241**, 381-405 (1984). J. Ellis, N. E. Mavromatos and D. V. Nanopoulos, *Phys. Lett. B* **293**, 37-48 (1992). Adler et al. (CPLEAR collaboration), Ellis, Lopez, Mavromatos, and Nanopoulos, *Phys. Lett. B* **364**, 239-245 (1995).
17. M. Kasevich and S. Chu, *Appl. Phys.* **B54**, 321-332 (1992).

QUANTUM NONLOCALITY AND INSEPARABILITY

Asher Peres

Department of Physics
Technion—Israel Institute of Technology
32 000 Haifa, Israel

A quantum system consisting of two subsystems is *separable* if its density matrix can be written as $\rho = \sum w_K \, \rho'_K \otimes \rho''_K$, where ρ'_K and ρ''_K are density matrices for the two subsytems, and the positive weights w_K satisfy $\sum w_K = 1$. A necessary condition for separability is derived and is shown to be more sensitive than Bell's inequality for detecting quantum inseparability. Moreover, *collective* tests of Bell's inequality (namely, tests that involve several composite systems simultaneously) may sometimes lead to a violation of Bell's inequality, even if the latter is satisfied when each composite system is tested separately.

Key words: nonlocality, quantum inseparability, collective test.

1. INTRODUCTION

From the early days of quantum mechanics, the question has often been raised whether an underlying "subquantum" theory, that would be deterministic or even stochastic, was viable. Such a theory would presumably involve additional "hidden" variables, and the statistical predictions of quantum theory would be reproduced by performing suitable averages over these hidden variables.

A fundamental theorem was proved by Bell [1], who showed that if the constraint of *locality* was imposed on the hidden variables (namely, if the hidden variables of two distant quantum systems were themselves be separable into two distinct subsets), then there was an upper bound to the correlations of results of measurements that could be performed on the two distant systems. That upper bound, mathematically expressed by Bell's inequality [1], is violated by some states in quantum mechanics, for example

M. Ferrero and A. van der Merwe (eds.),
New Developments on Fundamental Problems in Quantum Physics, 301–310.

the singlet state of two spin-$\frac{1}{2}$ particles.

A variant of Bell's inequality, more general and more useful for experimental tests, was later derived by Clauser, Horne, Shimony, and Holt (CHSH) [2]. It can be written

$$|\langle AB\rangle + \langle AB'\rangle + \langle A'B\rangle - \langle A'B'\rangle| \leq 2. \tag{1}$$

On the left-hand side, A and A' are two operators that can be measured by an observer, conventionally called Alice. These operators do not commute (so that Alice has to choose whether to measure A or A') and each one is normalized to unit norm (the norm of an operator is defined as the largest absolute value of any of its eigenvalues). Likewise, B and B' are two normalized noncommuting operators, any one of which can be measured by another, distant observer (Bob). Note that each one of the *expectation* values in Eq. (1) can be calculated by means of quantum theory, if the quantum state is known, and is also experimentally observable, by repeating the measurements sufficiently many times, starting each time with identically prepared pairs of quantum systems. The validity of the CHSH inequality, for *all* combinations of measurements independently performed on both systems, is a necessary condition for the possible existence of a local hidden variable (LHV) model for the results of these measurements. It is not in general a sufficient condition, as will be shown below.

Note that, in order to test Bell's inequality, the two distant observers independently *measure* subsytems of a composite quantum system, and then *report* their results to a common site where that information is analyzed [3]. A related, but essentially different, issue is whether a composite quantum system can be *prepared* in a prescribed state by two distant observers who receive *instructions* from a common source. For this to be possible, the density matrix ρ has to be separable into a sum of direct products,

$$\rho = \sum_K w_K\, \rho'_K \otimes \rho''_K, \tag{2}$$

where the positive weights w_K satisfy $\sum w_K = 1$, and where ρ'_K and ρ''_K are density matrices for the two subsystems. A separable system always satisfies Bell's inequality, but the converse is not necessarily true [4–7]. I shall derive below a simple algebraic test, which is a necessary condition for the existence of the decomposition (2). I shall then give some examples showing that this criterion is more restrictive than Bell's inequality, or than the α-entropy inequality [8].

2. SEPARABILITY OF DENSITY MATRICES

The derivation of the separability condition is easiest when the density matrix elements are written explicitly, with all their indices [3]. For example, Eq. (2) becomes

$$\rho_{m\mu,n\nu} = \sum_K w_K \, (\rho'_K)_{mn} \, (\rho''_K)_{\mu\nu}. \tag{3}$$

Latin indices refer to the first subsystem, Greek indices to the second one (the subsystems may have different dimensions). Note that this equation can always be satisfied if we replace the quantum density matrices by classical Liouville functions (and the discrete indices are replaced by canonical variables, **p** and **q**). The reason is that the only constraint that a Liouville function has to satisfy is being non-negative. On the other hand, we want quantum density matrices to have non-negative *eigenvalues*, rather than non-negative elements, and the latter condition is more difficult to satisfy.

Let us now define a new matrix,

$$\sigma_{m\mu,n\nu} \equiv \rho_{n\mu,m\nu}. \tag{4}$$

The Latin indices of ρ have been transposed, but not the Greek ones. This is not a unitary transformation but, nevertheless, the σ matrix is Hermitian. When Eq. (2) is valid, we have

$$\sigma = \sum_A w_A \, (\rho'_A)^T \otimes \rho''_A. \tag{5}$$

Since the transposed matrices $(\rho'_A)^T \equiv (\rho'_A)^*$ are non-negative matrices with unit trace, they can also be legitimate density matrices. It follows that *none of the eigenvalues of σ is negative*. This is a necessary condition for Eq. (2) to hold [9].

Note that the eigenvalues of σ are invariant under separate unitary transformations, U' and U'', of the bases used by the two observers. In such a case, ρ transforms as

$$\rho \to (U' \otimes U'')\, \rho \,(U' \otimes U'')^\dagger, \tag{6}$$

and we then have

$$\sigma \to (U'^T \otimes U'')\, \sigma \,(U'^T \otimes U'')^\dagger, \tag{7}$$

which also is a unitary transformation, leaving the eigenvalues of σ invariant.

As an example, consider a pair of spin-$\frac{1}{2}$ particles in an impure singlet state, consisting of a singlet fraction x and a random fraction $(1 - x)$ [10].

Note that the "random fraction" $(1-x)$ also includes singlets, mixed in equal proportions with the three triplet components. We have

$$\rho_{m\mu,n\nu} = x\, S_{m\mu,n\nu} + (1-x)\,\delta_{mn}\,\delta_{\mu\nu}\,/4, \tag{8}$$

where the density matrix for a pure singlet is given by

$$S_{01,01} = S_{10,10} = -S_{01,10} = -S_{10,01} = \frac{1}{2}, \tag{9}$$

and all the other components of S vanish. (The indices 0 and 1 refer to any two orthogonal states, such as "up" and "down.") A straightforward calculation shows that σ has three eigenvalues equal to $(1+x)/4$, and the fourth eigenvalue is $(1-3x)/4$. This lowest eigenvalue is positive if $x < \frac{1}{3}$, and the separability criterion is then fulfilled. This result may be compared with other criteria: Bell's inequality holds for $x < 1/\sqrt{2}$, and the α-entropic inequality [8] for $x < 1/\sqrt{3}$. These are therefore much weaker tests for detecting inseparability than the condition that was derived here.

In this particular case, it happens that this necessary condition is also a sufficient one. It is indeed known that if $x < \frac{1}{3}$ it is possible to write ρ as a mixture of unentangled product states [11]. This suggests that the necessary condition derived above (σ has no negative eigenvalue) might also be sufficient for any ρ. A proof of this conjecture was indeed recently obtained [12] for composite systems having dimensions 2×2 and 2×3. However, for higher dimensions, the present necessary condition was shown *not* to be a sufficient one.

As a second example, consider a mixed state consisting of a fraction x of the pure state $a|01\rangle + b|10\rangle$ (with $|a|^2+|b|^2=1$), and fractions $(1-x)/2$ of the pure states $|00\rangle$ and $|11\rangle$. We have

$$\rho_{00,00} = \rho_{11,11} = (1-x)/2, \tag{10}$$

$$\rho_{01,01} = x|a|^2, \tag{11}$$

$$\rho_{10,10} = x|b|^2, \tag{12}$$

$$\rho_{01,10} = \rho^*_{10,01} = xab^*, \tag{13}$$

and the other elements of ρ vanish. It is easily seen that the σ matrix has a negative determinant, and thus a negative eigenvalue, when

$$x > (1+2|ab|)^{-1}. \tag{14}$$

This is a lower limit than the one for a violation of Bell's inequality, which requires [7]

$$x > [1+2|ab|(\sqrt{2}-1)]^{-1}. \tag{15}$$

An even more striking example is the mixture of a singlet and a maximally polarized pair:

$$\rho_{m\mu,n\nu} = x\, S_{m\mu,n\nu} + (1-x)\,\delta_{m0}\,\delta_{n0}\,\delta_{\mu 0}\,\delta_{\nu 0}. \tag{16}$$

For any positive x, however small, this state is inseparable, because σ has a negative eigenvalue $(-x/2)$. On the other hand, the Horodecki criterion [13] gives a very generous domain to the validity of Bell's inequality: $x \leq 0.8$.

3. COLLECTIVE TESTS FOR NONLOCALITY

The weakness of Bell's inequality as a test for inseparability is due to the fact that the only use made of the density matrix ρ is for computing the probabilities of the various outcomes of tests that may be performed on the subsystems of a *single* composite system. On the other hand, an experimental verification of that inequality necessitates the use of *many* composite systems, all prepared in the same way. However, if many such systems are actually available, we may also test them collectively, for example two by two, or three by three, etc., rather than one by one. If we do that, we must use, instead of ρ (the density matrix of a single system), a *new* density matrix, which is $\rho\otimes\rho$, or $\rho\otimes\rho\otimes\rho$, in a higher dimensional space. It will now be shown that there are some density matrices ρ that satisfy Bell's inequality, but for which $\rho\otimes\rho$, or $\rho\otimes\rho\otimes\rho$, etc., violate that inequality [14].

The example that will be discussed is that of the Werner states [4] defined by Eq. (8). Let us consider n Werner pairs. Each one of the two observers has n particles (one from each pair). They proceed as follows. First, they subject their n-particle systems to suitably chosen local unitary transformations, U, for Alice, and V, for Bob. Then, they test whether each one of the particles labelled 2, 3, ..., n, has spin up (for simplicity, it is assumed that all the particles are distinguishable, and can be labelled unambiguously). Note that any other test that they can perform is unitarily equivalent to the one for spins up, as this involves only a redefinition of the matrices U and V. If any one of the $2(n-1)$ particles tested by Alice and Bob shows spin down, the experiment is considered to have failed, and the two observers must start again with n new Werner pairs.

A similar elimination of "bad" samples is also inherent to any experimental procedure where a failure of one of the detectors to fire is handled by discarding the results registered by all the other detectors: only when *all* the detectors fire are their results included in the statistics. This obviously requires an exchange of *classical* information between the observers. (There is a controversy on whether a violation of Bell's inequality with postselected data [15] is a valid test for nonlocality [16]. I shall not discuss this issue here; I only examine whether or not Bell's inequality is violated by the postselected data.)

The calculations shown below will refer to the case $n = 3$, for definiteness. The generalization to any other value of n is straightforward. Spinor indices, for a single spin-$\frac{1}{2}$ particle, will take the values 0 (for the "up" component of spin) and 1 (for the "down" component). The 16 components of the density matrix of a Werner pair, consisting of a singlet fraction x and a random fraction $(1-x)$, are, in the standard direct product basis:

$$\rho_{mn,st} = x\, S_{mn,st} + (1-x)\, \delta_{ms}\, \delta_{nt} /4, \tag{17}$$

where I am now using only Latin indices, contrary to what I did in Eq. (8); this is because Greek indices will be needed for another purpose, as will be seen soon. Thus, now, the indices m and s refer to Alice's particle, and n and t to Bob's particle.

When there are three Werner pairs, their combined density matrix is a direct product $\rho \otimes \rho' \otimes \rho''$, or explicitly, $\rho_{mn,st}\, \rho_{m'n',s't'}\, \rho_{m''n'',s''t''}$. The result of the unitary transformations U and V is

$$\rho \otimes \rho' \otimes \rho'' \to (U \otimes V)(\rho \otimes \rho' \otimes \rho'')(U^\dagger \otimes V^\dagger). \tag{18}$$

Explicitly, with all its indices, the U matrix satisfies the unitarity relation

$$\sum_{mm'm''} U_{\mu\mu'\mu'',mm'm''}\, U^*_{\lambda\lambda'\lambda'',mm'm''} = \delta_{\mu\lambda}\, \delta_{\mu'\lambda'}\, \delta_{\mu''\lambda''}. \tag{19}$$

In order to avoid any possible ambiguity, Greek indices (whose values are also 0 and 1) are now used to label spinor components *after* the unitary transformations. Note that the indices without primes refer to the two particles of the first Werner pair (the only ones that are not tested for spin up) and the primed indices refer to all the other particles (that are tested for spin up). The $V_{\nu\nu'\nu'',nn'n''}$ matrix elements of Bob's unitary transformation satisfy a relationship similar to (19). The generalization to a larger number of Werner pairs is obvious.

After the execution of the unitary transformation (18), Alice and Bob have to test that all the particles, except those labelled by the first (unprimed) indices, have their spin up. They discard any set of n Werner pairs where that test fails, even once. The density matrix for the remaining "successful" cases is thus obtained by retaining, on the right hand side of Eq. (18), only the terms whose primed components are zeros, and then renormalizing the resulting matrix to unit trace. This means that only two of the 2^n rows of the U matrix, namely those with indices 000... and 100..., are relevant (and likewise for the V matrix). The elimination of all the other rows greatly simplifies the problem of optimizing these matrices. We shall thus write, for brevity,

$$U_{\mu 00,mm'm''} \to U_{\mu,mm'm''}, \tag{20}$$

where $\mu = 0, 1$. Then, on the left hand side of Eq. (19), we effectively have two unknown row vectors, U_0 and U_1, each one with 2^n components (labelled by Latin indices $mm'm''$). These vectors have unit norm and are mutually orthogonal. Likewise, Bob has two vectors, V_0 and V_1. The problem is to optimize these four vectors so as to make the expectation value of the Bell operator [17],

$$C := AB + AB' + A'B - A'B', \tag{21}$$

as large as possible.

The optimization proceeds as follows. The new density matrix, for the pairs of spin-$\frac{1}{2}$ particles that were *not* tested by Alice and Bob for spin up (that is, for the first pair in each set of n pairs), is

$$\begin{aligned}(\rho_{\mathrm{new}})_{\mu\nu,\sigma\tau} &= \\ N\, U_{\mu,mm'm''}\, V_{\nu,nn'n''}\, &\rho_{mn,st}\, \rho_{m'n',s't'}\, \rho_{m''n'',s''t''}\, U^*_{\sigma,ss's''}\, V^*_{\tau,tt't''},\end{aligned} \tag{22}$$

where N is a normalization constant, needed to obtain unit trace (N^{-1} is the probability that all the "spin up" tests were successful). We then have [13], for fixed ρ_{new} and all possible choices of C,

$$\max\left[\mathrm{Tr}\left(C\rho_{\mathrm{new}}\right)\right] = 2\sqrt{M}, \tag{23}$$

where M is the sum of the two largest eigenvalues of the real symmetric matrix $T^\dagger T$, defined by

$$T_{pq} := \mathrm{Tr}\left[(\sigma_p \otimes \sigma_q)\, \rho_{\mathrm{new}}\right]. \tag{24}$$

(In the last equation, σ_p and σ_q are the Pauli spin matrices.) Our problem is to find the vectors U_μ and V_ν that maximize M.

At this point, some simplifying assumptions are helpful. Since all matrix elements $\rho_{mn,st}$ are real, we can restrict the search to vectors U_μ and V_ν that have only real components. Furthermore, the situations seen by Alice and Bob are completely symmetric, except for the opposite signs in the standard expression for the singlet state:

$$\psi = \left[\binom{1}{0}\binom{0}{1} - \binom{0}{1}\binom{1}{0}\right] / \sqrt{2}. \tag{25}$$

These signs can be made to become the same by redefining the basis, for example by representing the "down" state of Bob's particle by the symbol $\binom{0}{-1}$, *without* changing the basis used for Alice's particle. This unilateral change of basis is equivalent a substitution

$$V_{\nu,nn'n''} \to (-1)^{\nu+n+n'+n''}\, V_{\nu,nn'n''}, \tag{26}$$

on Bob's side. The minus signs in Eq. (9) also disappear, and there is complete symmetry for the two observers. It is then plausible that, with the new basis, the optimal U_ν and V_ν are the same. Therefore, when we return to the original basis and notations, they satisfy

$$V_{\nu,nn'n''} = (-1)^{\nu+n+n'+n''}\, U_{\nu,nn'n''}. \tag{27}$$

We shall henceforth restrict our search to pairs of vectors that satisfy this relation.

After all the above simplifications, the problem that has to be solved is the following: find two mutually orthogonal unit vectors, U_0 and U_1, each one with 2^n real components, that maximize the value of $M(U)$ defined by Eqs. (23) and (24). This is a standard optimization problem which can be solved numerically. Since the function $M(U)$ is bounded, it has at least one maximum. It may, however, have more than one: there may be several distinct local maxima with different values. A numerical search leads to one of these maxima, but not necessarily to the largest one. The outcome may depend on the initial point of the search. It is therefore imperative to start from numerous randomly chosen points in order to ascertain, with reasonable confidence, that the largest maximum has indeed been found.

4. NUMERICAL RESULTS

In all the cases that were examined, $M(U)$ turned out to have a local maximum for the following simple choice:

$$U_{0,00\ldots} = U_{1,11\ldots} = 1, \tag{28}$$

and all the other components of U_0 and U_1 vanish. Recall that the "vectors" U_0 and U_1 actually are two rows, $U_{000\ldots}$ and $U_{100\ldots}$, of a unitary matrix of order 2^n (the other rows are irrelevant because of the elimination of all the experiments in which a particle failed the spin-up test). In the case $n = 2$, one of the unitary matrices having the property (28) is a simple permutation matrix that can be implemented by a "controlled-NOT" quantum gate [18]. The corresponding Boolean operation is known as XOR (exclusive OR). For larger values of n, matrices that satisfy Eq. (28) will also be called XOR-transformations.

It was found, by numerical calculations, that XOR-transformations always are the optimal ones for $n = 2$. They are also optimal for $n = 3$ when the singlet fraction x is less than 0.57, and for $n = 4$ when $x < 0.52$. For larger values of x, more complicated forms of U_0 and U_1 give better results. The existence of two different sets of maxima may be seen in Fig. 1: there are discontinuities in the slopes of the graphs for $n = 3$ and 4, that

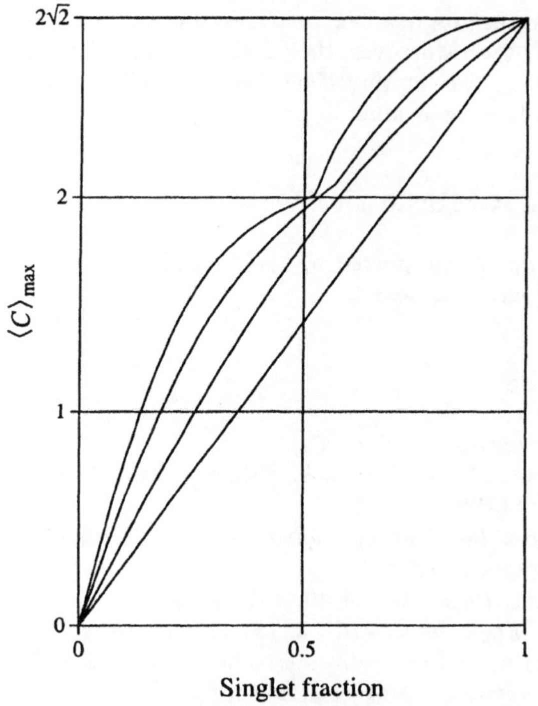

Fig. 1. Maximal expectation value of the Bell operator, versus the singlet fraction in the Werner state, for collective tests performed on several Werner pairs (from bottom to top of the figure, 1, 2, 3, and 4 pairs, respectively). The CHSH inequality is violated when $\langle C \rangle > 2$.

occur at the values of x where the largest value of $\langle C \rangle$ jumps from one local maximum to another one.

For $n = 5$, a complete determination of U_0 and U_1 requires the optimization of 64 parameters subject to 3 constraints, more than my workstation could handle. I therefore considered only XOR-transformations, which are likely to be optimal for $x \lesssim 0.5$. In particular, for $x = 0.5$ (the value that was used in Werner's original work [4]), the result is $\langle C \rangle = 2.0087$, and the CHSH inequality is violated. This violation occurs in spite of the existence of an explicit LHV model that gives correct results if the Werner pairs are tested one by one.

These results prompt a new question: can we get stronger *inseparability* criteria by considering $\rho \otimes \rho$, or higher tensor products? It is easily seen that no further progress can be achieved in this way. If ρ is separable as in Eq. (2), so is $\rho \otimes \rho$. Moreover, the partly transposed matrix corresponding to $\rho \otimes \rho$ simply is $\sigma \otimes \sigma$, so that if no eigenvalue of σ is negative, then $\sigma \otimes \sigma$ too has no negative eigenvalue.

ACKNOWLEDGMENT

This work was supported by the Gerard Swope Fund, and the Fund for Encouragement of Research.

REFERENCES

1. J. S. Bell, *Physics* **1**, 195 (1964).
2. J. F. Clauser, M. A. Horne, A. Shimony, and R. A. Holt, *Phys. Rev. Lett.* **23**, 880 (1969).
3. A. Peres, *Quantum Theory: Concepts and Methods* (Kluwer, Dordrecht, 1993), Chapters 5 and 6.
4. R. F. Werner, *Phys. Rev. A* **40**, 4277 (1989).
5. S. Popescu, *Phys. Rev. Lett.* **72**, 797 (1994); **74**, 2619 (1995).
6. N. D. Mermin, in *Perspectives on Quantum Reality*, R. K. Clifton, ed. (Kluwer Academic, Dordrecht, 1996), pp. 57–71.
7. N. Gisin, *Phys. Lett. A* **210**, 151 (1996).
8. R. Horodecki, P. Horodecki, and M. Horodecki, *Phys. Lett. A* **210**, 377 (1996).
9. A. Peres, *Phys. Rev. Lett.* **77**, 1413 (1996).
10. J. Blank and P. Exner, *Acta Univ. Carolinae, Math. Phys.* **18**, 3 (1977).
11. C. H. Bennett, G. Brassard, S. Popescu, B. Schumacher, J. Smolin, and W. K. Wootters, *Phys. Rev. Lett.* **76**, 722 (1996).
12. M. Horodecki, P. Horodecki, and R. Horodecki, "Separability of mixed quantum states: necessary and sufficient conditions" (e-print archive: quant-ph/9605038).
13. R. Horodecki, P. Horodecki, and M. Horodecki, *Phys. Lett. A* **200**, 340 (1995).
14. A. Peres, *Phys. Rev. A* **54**, 2685 (1996).
15. A. Peres, in *Quantum Potentiality, Entanglement, and Passion-at-a-Distance: Essays for Abner Shimony*, R. S. Cohen, M. Horne, and J. Stachel, eds. (Kluwer Academic, Dordrecht, 1996).
16. E. Santos, *Phys. Rev. Lett.* **66**, 1388 (1991).
17. S. L. Braunstein, A. Mann, and M. Revzen, *Phys. Rev. Lett.* **68**, 3259 (1992).
18. A. Barenco, D. Deutsch, A. Ekert, and R. Jozsa, *Phys. Rev. Lett.* **74**, 4083 (1995).

DECOHERENCE LIMITS TO QUANTUM FACTORING

M. B. Plenio and P. L. Knight

Optics Section, Blackett Laboratory, Imperial College
London SW7 2BZ, United Kingdom

Soon after the idea of quantum computation became an active part of current research through the seminal work of Shor on factorization [1], decoherence was recognised as a major problem that cannot be ignored, at least when one is interested in practical applications including especially the factorization of large numbers [2,3]. In this paper we present fundamental limits imposed by spontaneous emission on the size of the numbers that can be factorized on a quantum computer. We also include the use of error correction methods in our estimates and show that this does not substantially change our pessimistic predictions of the maximum size of factorizable numbers on a quantum computer.

Key words: Decoherence, quantum computation, error correction.

1. ESTIMATES WITHOUT ERROR CORRECTION

In the following we assume that our quantum computer is realized by a string of ions localised in a linear ion trap as proposed by Cirac and Zoller [4]. Other possible realizations lead to similar predictions as they rely on the interaction between atoms and light with essentially equivalent dynamics. In this implementation the controlled-NOT (CNOT) gate (one of the most important gates in quantum computation) can be realized using an interaction in which the excitation of an ion leads to the absorption of a phonon of the centre-of-mass mode. In the RWA, the Hamiltonian for this interaction reads as

$$H = \frac{\eta}{\sqrt{5L}} \frac{\Omega_{12}}{2} \left[|e\rangle\langle g| a + |g\rangle\langle e| a^{\dagger} \right] \quad + O(\eta^2), \tag{1}$$

where $\eta \ll 1$ is the Lamb-Dicke parameter (essentially the ratio between the confinement distance of the trapped ion and the optical wavelength),

M. Ferrero and A. van der Merwe (eds.),
New Developments on Fundamental Problems in Quantum Physics, 311–316.

L the number of qubits that is required to represent the number that is to be factorized (the factor 5 is due to the necessary overhead in qubits in the practical implementation of Shor's algorithm [5]) and Ω_{12} is the Rabi frequency on the qubit transition. To implement a CNOT gate, one effectively has to perform a 4π rotation according to the Hamiltonian Eq. (1). As the factorization of an L bit number requires the equivalent of ϵL^3 CNOT gates [5] the total factorization will require the time

$$T = 4\pi \frac{\sqrt{5L}}{\eta\Omega_{12}} \epsilon L^3 \ . \tag{2}$$

The result of the calculation will, however, contain useful information only if the coherence is maintained and if no spontaneous emission has taken place during the calculation. If $2\Gamma_{22}$ equals the Einstein coefficient on the qubit transition then the spontaneous lifetime of the whole quantum computer containing $5L$ ions will be

$$\tau_{dec} \cong \frac{1}{5L\Gamma_{22}} \tag{3}$$

as on average half of the ions are excited. τ_{dec} has to be longer than the total computation time T so that we have to satisfy the inequality

$$\frac{1}{5L\,\Gamma_{22}} \gg \frac{4\,\pi\epsilon}{\Omega_{12}} \sqrt{\frac{5L^7}{\eta^2}} \ . \tag{4}$$

Now we use the important fact that the Rabi frequency Ω_{12} and the Gamma coefficient Γ_{22} on a transition are *not* independent. One has

$$\frac{\Omega_{12}^2}{\Gamma_{22}} = \frac{6\pi c^3 \epsilon_0}{\hbar \omega_{12}^3} E^2 \ . \tag{5}$$

where E is the electric field of the laser and ω_{12} the transition frequency. Inserting Eq. (5) into Eq. (4) we obtain

$$\frac{1}{\Gamma_{22}} \gg \frac{2000\pi^2\epsilon^2}{\eta^2} \frac{\Gamma_{22}}{\Omega_{12}^2} L^9 \ , \tag{6}$$

and furthermore, using Eqs. (5) and (6), in Eq. (2)

$$T \gg 400\pi^2 \left(\frac{\epsilon}{\eta}\right)^2 \frac{\Gamma_{22}}{\Omega_{12}^2} L^8 \ . \tag{7}$$

Assuming $\eta = 0.1, \Omega_{12}^2/\Gamma_{22} = 10^{14} s^{1/2}$ and $\epsilon = 500$ we obtain for a four bit number $T \gg 64s$ and $\Gamma_{22}^{-1} \gg 13 \cdot 10^4 s$. These are large values which show that factorization will not be possible even for small numbers unless we were able to increase the strength of the lasers significantly. However, we will show that one cannot increase the laser-power indefinitely as for strong lasers the two-level approximation breaks down. In the following we include this effect quantitatively. We assume that the laser that drives the qubit transition also couples the lower level 1 of the qubit to a third extraneous level 3. As the laser is far detuned from this transition the effective spontaneous emission rate on the $3 \leftrightarrow 1$ transition will be

$$R = 2\Gamma_{33}\frac{\eta^2\Omega_{13}^2}{5L}\frac{1}{8\Delta_{13}^2}\,, \tag{8}$$

where $2\Gamma_{33}$ equals the Einstein coefficient, Ω_{13} the Rabi frequency and $\Delta_{13} = \omega_{13} - \omega_{12}$ the detuning on that $3 \leftrightarrow 1$ transition. There should be no spontaneous emission from the auxiliary level during the whole calculation time T, i.e., $RT \ll 1$. Using Eq. (5) we find

$$L \ll \left\{\frac{\Delta_{13}^2}{40\pi^2\epsilon^2\,\Gamma_{33}^2}\left(\frac{\lambda_{12}}{\lambda_{13}}\right)^3\right\}^{1/7}\,. \tag{9}$$

From Table 1 we observe that the limits on the bit size of the numbers that can be factorized on a quantum computer is very small. However, choosing two-level systems to store qubits does not seem to be the best possible method as during the whole calculation a substantial amount of population will be in the upper level which then leads to spontaneous decay. A system that seems to be more promising is a configuration in which the qubit is stored in two Zeeman sub levels 1 and $1'$ of a stable ground state. Population transfer will be performed using Raman pulses far detuned from an upper level 2 so that the population in that level 2 will be small at all times. An analysis similar in spirit to the one given above for the two-level case then yields the following limitation for the bit size of the numbers that can be factorized:

$$L \ll \left\{\frac{\Delta_{13}^2}{256\pi^2\epsilon^2\Gamma_{33}^2}\left(\frac{\lambda_{12}}{\lambda_{13}}\right)^3\right\}^{1/6}\,. \tag{10}$$

Using this expression we find the values shown in the second to last column of Table 2. It should be noted that the estimates for L are practically unaffected by our different way of storing the qubits.

Table 1. Equation (9) is evaluated for four different ions. Obviously the maximum size of the number that can be factorized on a quantum computer is small compared to the fact that classical computers can deal with $L = 400$ bit numbers.

Ion	level	atomic level	$\lambda_{12}\,[nm]$	$\lambda_{13}\,[nm]$	$\Gamma_{33}\,[s^{-1}]$	$L \ll$
Ca^+	1	$4s\,^2S_{1/2}$	729	397	$6.5 \cdot 10^7$	13
	2	$3d\,^2D_{5/2}$				
	3	$4s\,^2P_{1/2}$				
Hg^+	1	$5d^{10}6s^2\,^2S_{1/2}$	282	194	$2.2 \cdot 10^8$	9
	2	$5d^9 6s^2\,^2D_{5/2}$				
	3	$5d^{10}6p^2\,^2P_{1/2}$				
Ba^+	1	$6s\,^2S_{1/2}$	1761	455	$5.8 \cdot 10^7$	20
	2	$5d\,^2D_{5/2}$				
	3	$6s\,^2P_{3/2}$				
Yb^+	1	$4f^{14}6s^2\,^2S_{1/2}$	467	370	$5.6 \cdot 10^7$	9
	2	$4f^{13}6s^2\,^2F_{7/2}$				
	3	$4f^{14}6p^2\,^2P_{1/2}$				

2. ESTIMATES INCLUDING ERROR CORRECTION

The analysis in Sec. 1 has clearly shown that spontaneous emission already imposes very strong limits on the size of the number that can be factorized on a quantum computer. Obviously noise can not be ignored in a quantum computer and therefore ways have to be found to correct for the errors due to that noise. In fact such methods exist and it has been shown that it is possible to encode qubits in such a way that a certain number of general errors (including spontaneous emission) can be corrected [6]. However, many qubits are required to implement these codes and performing quantum gates on encoded qubits is much more involved than on unencoded qubits [7,8]. We introduce two parameters: q which describes the factor by which the required number of qubits rises and the factor c by which the number of required operations grows. It is difficult to estimate these parameters properly as it is not perfectly known how to implement quantum gates on the most efficient quantum codes. Nevertheless, we can assume that $c = q = 10$ is an optimistic choice for a single error correcting code. One can then follow a similar analysis as in Sec. 1. One additional ingredient has to be put into the analysis and that is the fact that if there is a spontaneous emission from an extraneous level then the population will in most cases be lost irretrievably.

Table 2. Equation (10) is evaluated for the same ions and levels as in Table 1. Again only small numbers can be factorized on a quantum computer. In the last column of this table the corresponding values for Eq. (11) are calculated.

Ion	$\lambda_{12}\,[nm]$	$\lambda_{13}\,[nm]$	$\Gamma_{33}\,[s^{-1}]$	$L \ll$	$L_{EC} \ll$
Ca^+	729	397	$6.5 \cdot 10^7$	14	21
Hg^+	282	194	$2.2 \cdot 10^8$	9	12
Ba^+	1761	455	$5.8 \cdot 10^7$	25	43
Yb^+	467	370	$5.6 \cdot 10^8$	10	13

In this case even the implementation of an error correction code is of no help as we cannot get the population back into the system. Therefore, assuming a single error correcting code we obtain the following estimate for the maximal number of bits L_{EC} that the input may have.

$$L_{EC} \ll \left\{ \frac{1}{64\pi^2 c^2 q \epsilon^{3/2}} \left(\frac{\Delta_3}{\Gamma_{33}}\right)^2 \left(\frac{\lambda_{12}}{\lambda_{13}}\right)^3 \right\}^{2/9} . \tag{11}$$

In the second column of Table 2 we assume $c = q = 10$ and observe that the estimates do not improve substantially. Even with error correction the numbers that can be factorized are of more or less trivial size.

3. CONCLUSION

In this paper we have investigated upper limits imposed by spontaneous decay upon the size of numbers that can be factorized on a quantum computer. We found that these upper limits exclude any number of appreciable size from being factorizable on a realistic quantum computer even if quantum error correction can be implemented. We should stress that these limits are of fundamental nature. In any realizable implementation many other sources of error would lead to further limitations making factorization on a quantum computer even more unrealistic.

This work was supported by the European Community Network "Nonclassical Light", by the UK Engineering and Physical Sciences Research Council and by a Feodor-Lynen grant of the Alexander von Humboldt foundation.

REFERENCES

1. P. W. Shor, in *Proceedings of the 35th Annual Symposium on the Foundations of Computer Science, Los Alamitos* (IEEE Computer Society

Press, New York, 1994), p. 124.
2. M.B. Plenio and P.L. Knight, *Phys. Rev. A* **53**, 2986 (1996).
3. M.B. Plenio and P.L. Knight, submitted to *Proc. Roy. Soc. A*.
4. J. I. Cirac and P. Zoller, *Phys. Rev. Lett.* **74**, 4091 (1995).
5. V. Vedral, A. Barenco, and A. Ekert, *Phys. Rev. A* **54**, 147 (1996).
6. A.M. Steane, *Phys. Rev. Lett.* **77**, 793 (1996).
7. P.W. Shor, lanl e-print quant-ph/9605011.
8. D.P. Vincenzo and P.W. Shor, lanl e-print quant-ph/9605031.

QUANTUM ENGINEERING WITH TRAPPED IONS

J.F. Poyatos,[a] S.A. Gardiner,[b] R. Walser,[b]
J.I. Cirac,[a] P. Zoller,[b] and R. Blatt,[c]

*a. Departamento de Fisica, Universidad de Castilla-La Mancha
13071 Ciudad Real, Spain
b. Institut für Theoretische Physik, Universität Innsbruck
Technikerstrasse 25, A-6020 Innsbruck, Austria
c. Institut für Experimental Physik, Universität Innsbruck
Technikerstrasse 25, A-6020 Innsbruck, Austria*

We describe how to manipulate, and characterize the motional state of a trapped ion. We also give a method to measure an arbitrary motional observable in a single shot.

Key words: trapped ions, non-classical states, decoherence, tomography.

1. INTRODUCTION

The detailed study of quantum systems, in particular the synthetization, controlled evolution, and measurement of quantum states, has been in the scope of physicists for the last decades. While in quantum optics the achievement of this aim seemed to be restrected to cavity QED systems, it turned out recently that trapped ions can reproduce such studies, having the possibility of going even further [1].

In this work we will focus on the trapped ion system. We will describe several fundamental studies that can be performed with trapped ions. The first one consists of performing single shot measurements of an arbitrary motional observable of the trapped ion. The second one consists of engineer the coupling between the motion of the ion and a reservoir so that the evolution is governed by a desired master equation. Finally, we will show how to completely characterize the state of motion of a single trapped ion, that can be repeatedly prepared on the same state.

*M. Ferrero and A. van der Merwe (eds.),
New Developments on Fundamental Problems in Quantum Physics,* 317–323.

2. THE MODEL

We consider a single trapped ion confined in a harmonic potential. We take into account the internal and external degrees of freedom, related respectively with the electronic and the center of mass motion of the ion. The internal and external structures are coupled in the presence of laser light, since in every absorption or emission event a recoil occurs. Typically, the internal structure will be characterized by two different transitions: a short-lived one, inducing dissipative phenomena, and two long-lived ones used to manipulate the ion in a coherent way. We will assume that at a given time the laser is tuned to only one transition which allows to reduce the description of the internal structure from a three- to a two-level atom, i.e., ground, $|g\rangle$, and excited, $|e\rangle$, states. When the laser excites the long-lived transition the evolution of the ion is described by the usual Schrödinger equation, where the interacting Hamiltonian in a rotating frame at the laser frecuency ω_L reads:

$$H_{\rm i} = \frac{\Omega}{2} \sin\left[\eta(a+a^\dagger)+\phi\right](\sigma_+ + \sigma_-). \tag{1}$$

It describes the interaction of the ion with a laser beam in standing wave configuration. Here, Ω is the Rabi frequency, ϕ is related to the relative position of the center of the trap in the laser standing wave, $\eta = \sqrt{\hbar k_L^2/(2M\nu)}$ (with $k_L = \omega_L/c$, the laser wavevector) is the Lamb–Dicke parameter corresponding to this specific transition, $a(a^\dagger)$ is the annihilation (creation) operator related to the harmonic oscillator, and the σ's are the usual spin operators associated to the internal transition.

On the contrary, when the short-lived transition is excited the system cannot be described by a pure state but by a density operator. The evolution is given by the following master equation

$$\dot{\rho} = -i[H,\rho] + \mathcal{L}\rho, \tag{2}$$

where H is the system Hamiltonian, i.e., $H = H_{\rm free} + H_{\rm i}$, being $H_{\rm free}$ the external and internal free Hamiltonians, and

$$L\rho = \Gamma\sigma_-\tilde{\rho}\sigma_+ - \frac{1}{2}\Gamma(\sigma_+\sigma_-\rho + \rho\sigma_+\sigma_-). \tag{3}$$

Here, Γ is the spontaneous emission rate from level $|e\rangle$, and

$$\tilde{\rho} = \int_{-1}^{1} du N(u) e^{-i\eta u(a+a^\dagger)} \rho e^{i\eta u(a+a^\dagger)}, \tag{4}$$

where $N(u)$ is the dipole emission pattern corresponding to the transitions from $|e\rangle$ to $|g\rangle$.

3. QUANTUM ENGINEERING

Here, we describe some fundamental studies that can be performed with trapped ions.

3.1. Measurement of Arbitrary Observables

Suppose that we wish to perform a single shot measurement of an observable $\mathcal{A}$ on an ion, which is represented by a Hermitian operator $\hat{A}$. In practice, only very few observables can be measured. Here we illustrate how this general measurement procedure can be carried out [2]. We assume the ion to be in the unknown state $|\phi\rangle$. The measurement consists of checking sequentially whether the ion is in the state $|\psi_k\rangle$, $(k = 0, 1, \ldots)$, (the basis associated to the chosen observable) in a filtering way. That is, check if it is in $|\psi_0\rangle$. If the result is positive, we have measured a_0; if not, the state is projected onto $|\phi_1\rangle \equiv k_1(1 - |\psi_0\rangle\langle\psi_0|)|\phi\rangle$, where k_1 is a normalization constant, and we continue the procedure. We check if the ion is in the state $|\psi_1\rangle$. If the result is positive, we have measured a_1; if not, the state is projected onto $|\phi_2\rangle \equiv k_2(1 - |\psi_1\rangle\langle\psi_1|)|\phi_1\rangle$ and continue in the same vein.

Let us now show more in detail how the k–th filtering measurement of the resulting state $|\phi_k\rangle$ takes place. We denote by $\hat{U}_k$ the unitary time evolution operator which generates the eigenstate $|\psi_k\rangle = \hat{U}_k|0\rangle$. We first transform the state of the ion by the inverse transformation to $|\chi_k\rangle = \hat{U}_k^\dagger|\phi_k\rangle$, and measure whether the ion is in the ground motional state or not, after which $\hat{U}_k$ is applied. If the ion was found to be in the ground motional state, the state after this procedure will be $|\psi_k\rangle$, which implies that we have measured a_k. The probability for this to happen is

$$|\langle 0|\chi_k\rangle|^2 = |\langle\psi_k|\phi_k\rangle|^2 \tag{5}$$

which, by induction and using the fact that in the k previous measurements have been negative, one can easily check that is exactly P_k. In the case that we do not measure the ion in $|0\rangle$, its state is projected onto

$$|\phi_{k+1}\rangle = k_{k+1}\left(1 - |\psi_k\rangle\langle\psi_k|\right)|\phi_k\rangle. \tag{6}$$

In practise this procedure must be carried out in a finite number of steps, so one has to restrict himself to a finite dimensional Hilbert space.

According to the above discussion, in order to carry out an arbitrary measurement, two requirements have to be fulfilled: first, one has to be able to generate arbitrary states starting from the vacuum; second, one has to be able to perform filtering measurements of the ground state of the trapped ion. The first requirement corresponds to quantum state engineering, and it can be attained with a combination of diagonal and vertical pulses using the long–lived transition [3]. Let us show now how it is possible to perform filtering measurements of the state of the ion. First we shift all of the population in states $|g, n+1\rangle$ to the states $|e, n\rangle$ by adiabatic passage (where $n \geq 0$), so that all amplitudes with the exception of that for $|g, 0\rangle$ are transferred unchanged (being $|n\rangle$ a Fock state). Next we use quantum jump techniques to determine whether $|g, 0\rangle$ is populated. Finally, if we do not detect population in state $|g, 0\rangle$, we use adiabatic passage to restore the amplitudes in states $|e, n\rangle$ back to the states $|g, n+1\rangle$. Adiabatic passage is required so that the population can be transferred independent of the Rabi frequencies. (see, for example, Cirac *et al.* [4] and references therein).

3.2. Reservoir Engineering

Another intriguing possibility with a trapped ion is to engineer the coupling between its motion and a reservoir, such as it fulfills a prescribed master equation in the form:

$$\dot{\rho} = \gamma(2f\rho f^\dagger - f^\dagger f\rho - \rho f^\dagger f), \tag{7}$$

where ρ is the reduced density operator for the ion motion in the interaction picture after tracing over the reservoir, and f is a function of the annihilation and creation operators a and $a^\dagger$. Master equations of the form (7) have been studied theoretically by numerous authors in order to analyze the process of decoherence.

Here, we illustrate how this coupling can be engineered to yield a master equation of the form (7) [5]. We consider that the short–lived transition of the trapped ion is excited with laser light. The laser is in standing wave configuration, and the center of the ion coincides with either one node or one antinode of the standing wave. The laser is tunned to the k–the motional sideband. In particular: (i) For excitation at the node ($\phi = 0$), $\delta = (2k+1)\nu$ ($k = 0, \pm 1, \ldots$) (ii) For excitation at the antinode ($\phi = \pi/2$), $\delta = 2k\nu$ ($k = 0, \pm 1, \ldots$). According to the above discussion, in the Lamb–Dicke limit, i.e. $\eta \ll 1$, and under the low intensity and strong confinement regimes, i.e $\Gamma \ll \nu$, the dynamics of the ion is described by master equation (2), where now $H_i = \frac{\Omega'}{2}(\sigma_+ f + f^\dagger \sigma_-)$. Here both Ω' and the form of the operator f depend on the frequency of the laser. For example, for $\delta = -\nu$, we have $f = a$, and $\Omega' = \Omega\eta$, whereas for $\delta = -2\nu$, $f = a^2$, and $\Omega' = -\Omega\eta^2/2$. In the low intensity limit and provided that the internal transition is not saturated, one can eliminate the internal excited state $|e\rangle$ using standard procedures of quantum optics. Physically, since $\Omega' \ll \Gamma$ the ion practically spends no time in the excited level and therefore we can eliminate it. The resulting master equation is precisely (7).We emphasize that the master equation will be valid for times such that these corrections are not important, that is for times $t \ll (\gamma\eta^2\bar{n})^{-1}$ where $\bar{n}$ is the typical phonon

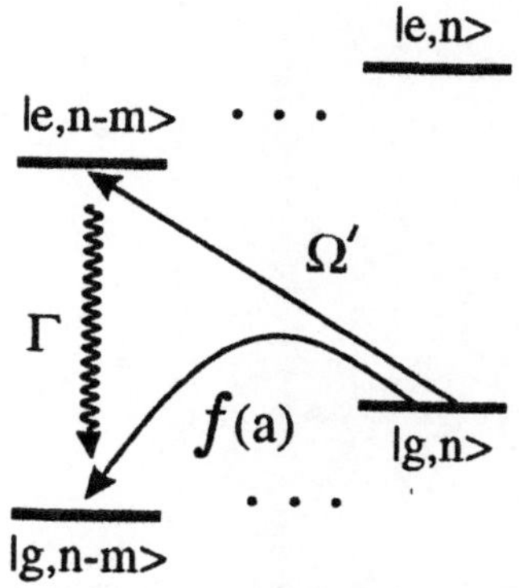

Fig. 1. Adiabatic elimination leading different f couplings. In this case there is only one laser tuned to the m-lower sideband.

number of the state of the ion. Nevertheless in the Lamb–Dicke limit this time can be much longer than the time required to reach the steady state using the approximated master equation.

According to our analysis, by varying the laser frequency we obtain the master equation (7) with different coupling operators f. For example, the laser is tuned to the so-called "lower motional sideband," $\delta = -\nu$, and the ion is located at the node of the standing wave field which leads to a coupling operator $f = a$. This can be easily understood by noting that in each absorption and spontaneous emission cycle one phonon is annihilated on a time scale given by the optical pumping time. Similarly, if the laser is tuned to the "second lower sideband" $\delta = -2\nu$ at the antinode of the laser standing wave which gives the two–phonon coupling $f = a^2$. Using more elaborated laser configurations one can readily show that more complicated f operators can be engineered.

3.3. Quantum Tomography

In previous sections we have shown how to prepare and destroy at will a quantum state of motion of the trapped ion. Here, we show a method for the experimental reconstruction of the quantum state of motion for a single trapped ion, by reapeated measurements on the ion assuming that the state of the motion can be reproduced [6,7]. Specifically, we present two methods to measure the $Q(\alpha)$ function, and the quadrature distribution function $P(x, \theta)$ (see Fig. 2). Recent theoretical advances have shown how to recover the full information about the state of a system by knowing these

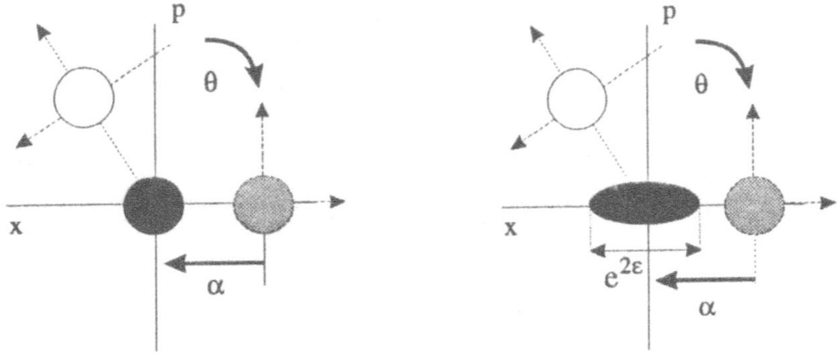

Fig. 2. Phase-space representations of the operations required to measure the Q and $P(x,\theta)$ functions.

probabilities [8]. In order to show how the Q function can be measured, we reexpress it in a more suitable way

$$Q(\alpha) = \frac{1}{\pi}\langle\alpha|\rho|\alpha\rangle = \frac{1}{\pi}\langle 0|\tilde{\rho}|0\rangle, \tag{8}$$

where $\tilde{\rho} = U(|\alpha|,\theta)\rho U^{\dagger}(|\alpha|,\theta)$ is related to the original ρ by a unitary transformation $U^{\dagger}(|\alpha|,\theta) = \mathcal{R}^{\dagger}(\theta)\mathcal{D}(|\alpha|)$. Here, $\mathcal{D}(|\alpha|)$ is the displacement operator and $\mathcal{R}(\theta)$ is the phase shifting operator that acting on the vacuum, create a coherent state, i.e., $||\alpha|e^{i\theta}\rangle = U^{\dagger}(|\alpha|,\theta)|0\rangle$.

According to the previous unitary evolution all we have to do is wait a given time while the ion evolves freely in the trap, then suddenly displace the trap a given distance a finally measure the probability of the ion to be in the lowest motional state $|0\rangle$ as in previous subsections.

Alternatively, for the measurement of the $P(x,\theta)$ function we recall the property of the squeezed states to behave in the limit of high squeezing as line states and reexpress the quadrature distribution function in the form

$$P(x,\theta) = \langle x,\theta|\rho|x,\theta\rangle = \lim_{|\epsilon|\to\infty} |N_{\epsilon}|^2\langle 0|\tilde{\rho}|0\rangle, \tag{9}$$

here $|N_{\epsilon}|^2$ is a normalization constant, $\tilde{\rho} = U(|\epsilon|,|\alpha|,\theta)\rho U^{\dagger}(|\epsilon|,|\alpha|,\theta)$ is related to the original ρ by a unitary transformation $U^{\dagger}(|\epsilon|,|\alpha|,\theta) = \mathcal{R}^{\dagger}(\theta)\mathcal{D}(|\alpha|)\mathcal{J}(|\epsilon|)$, and $\mathcal{J}(|\epsilon|)$ is the squeezing operator. In the same way as before, we associate a given physical process with the previous unitary evolution. While the first two are the same for the determination of the Q, the last one is associated with the creation of squeezed states, a process that is attainable in a trap by opening or closing suddenly the harmonic potential. Finally, all we have to do is to measure the population in the ground state as before.

REFERENCES

1. C. Monroe, D.M. Meekhof, B.E. King, and D.J. Wineland, *Science* **272**, 1131 (1996)
2. S. A. Gardiner, J.I. Cirac, and P. Zoller, unpublished.
3. C. K. Law and J. H. Eberly, *Phys. Rev. Lett.* **76**, 1055 (1996).
4. J. I. Cirac, A. S. Parkins, R. Blatt, and P. Zoller, in *Advances in Atomic and Molecular Physics*, in press.
5. J. F. Poyatos, J.I. Cirac, and P. Zoller, *Phys. Rev. Lett.*, in press, 1996.
6. S. Wallentowitz and W. Vogel, *Phys. Rev. Lett.* **75**, 2932 (1995); J. F. Poyatos, R. Walser, J.I. Cirac, R. Blatt, and P. Zoller, *Phys. Rev. A* **53**, R1966 (1996).
7. D. Leibfried, D.M. Meekhof, B.E. King, C. Monroe, W.M. Itano, and D.J. Wineland, *Phys. Rev. Lett.*, in press, 1996.
8. K. Vogel and H. Risken, *Phys. Rev. A* **40**, R2847 (1989).

MANY-VALUED INTERPRETATION OF THE LOGIC OF QUANTUM MECHANICS

Jarosław Pykacz

Institute of Mathematics
University of Gdańsk
Wita Stwosza 57, 80-952 Gdańsk, Poland
pykacz@ksinet.univ.gda.pl

Quantum logic, usually interpreted as non-distributive but two-valued logic, is reinterpreted as many-valued logic. This allows to explain the logical background of the wave-particle duality exhibited in the double-slit experiment in a better way than it can be done with the use of classical two-valued logic.

Key words: quantum logic, many-valued logic, double-slit experiment.

1. STANDARD INTERPRETATION OF THE LOGIC OF QUANTUM MECHANICS

Elementary ("yes-no") propositions about quantum objects do not obey all laws of classical logic, i.e., "quantum logic", whatever it means, is non-classical. This idea is due to Birkhoff and von Neumann [1] who argued that elementary propositions in quantum mechanics correspond to (projections onto) closed subspaces of a Hilbert space associated with a quantum system. Families of all such subspaces are orthomodular (non-distributive) orthocomplemented lattices. Birkhoff and von Neumann interpreted orthocomplementation as representing negation of a proposition and lattice operations of meet and join as representing, respectively, conjunction and disjunction of propositions, although they were aware that this interpretation becomes doubtful when one deals with noncompatible propositions. The only two eigenvalues (0 and 1) of a projection operator were interpreted in [1] as corresponding to the falsity or truth of a respective proposition, therefore, according to the standard interpretation, quantum logic is a two-valued logic and it differs from the classical logic only because distributivity law is replaced by a weaker orthomodular law (see [2] for all necessary definitions).

M. Ferrero and A. van der Merwe (eds.),
New Developments on Fundamental Problems in Quantum Physics, 325–328.

Probably the overwhelming success of a Hilbert space formulation of quantum mechanics caused that Hilbert–space–based standard interpretation of quantum logic as two–valued but nondistributive logic became a canon and attempts to interpret quantum logic as many–valued logic never gained popularity comparable to the standard Birkhoff–von Neumann interpretation.

2. VIA FUZZY SETS TO MANY-VALUED LOGIC

Mączyński [3] proved that any set of functions L from a set M into $[0,1]$ satisfying the conditions:

(i) the zero function belongs to L,
(ii) $f \in L$ implies $1 - f \in L$,
(iii) For any sequence $f_1, f_2, \ldots$ of members of L satisfying $f_i + f_j \leq 1$ for $i \neq j$ we have $f_1 + f_2 + \ldots \in L$,

is a quantum logic and that any quantum logic with a full set of states S can be isomorphically represented as a set of functions from S into $[0,1]$ satisfying (i)–(iii). It was noticed in [4] that these functions can be treated as membership functions of fuzzy subsets of the set M or S and in [5] that conditions (i)–(iii) can be expressed in terms of Giles [6] operations on fuzzy sets.

Giles noticed in [6] that fuzzy sets are as closely related to many–valued Lukasiewicz logic as traditional sets to the classical two–valued logic. The standard fuzzy complement is generated by the Lukasiewicz negation:

$$\tau(\neg\tilde{a}) = 1 - \tau(\tilde{a}) \tag{1}$$

($\tau(\tilde{a})$ denotes truth value of a proposition $\tilde{a}$ and it can be equal to any number within the interval $[0,1]$), and Giles union and intersection are generated, respectively, by Frink [7] arithmetic disjunction:

$$\tau(\tilde{a} \oplus \tilde{b}) = \min[\tau(\tilde{a}) + \tau(\tilde{b}), 1] \tag{2}$$

and conjunction:

$$\tau(\tilde{a} \odot \tilde{b}) = \max[\tau(\tilde{a}) + \tau(\tilde{b}) - 1, 0]. \tag{3}$$

It was shown in [8] that with any quantum logic L with a full set of states S we can associate a family $\mathcal{L}$ of infinite-valued propositional functions with arguments ranging over the set S such that:

(i) The always-false propositional function belongs to $\mathcal{L}$.

(ii) If a propositonal function belongs to $\mathcal{L}$, its negation also belongs to $\mathcal{L}$.

(iii) Arithmetic disjunction of any countable family of propositional functions which exclude each other belongs to $\mathcal{L}$.

(iv) The always-false propositional function is the only propositional function in $\mathcal{L}$ which excludes itself.

3. MANY-VALUED INTERPRETATION OF A LOGIC OF DOUBLE-SLIT EXPERIMENT

From the very definition of arithmetic operations (2), (3) follows [8]

Theorem. For any two many-valued propositions $\tilde{a}$, $\tilde{b}$ either
(1a) $\tau(\tilde{a} \odot \tilde{b}) = 0$, and $\tau(\tilde{a} \oplus \tilde{b}) < 1$, or
(1b) $\tau(\tilde{a} \odot \tilde{b}) = 0$, and $\tau(\tilde{a} \oplus \tilde{b}) = 1$, or
(2a) $0 < \tau(\tilde{a} \odot \tilde{b}) < 1$, and $\tau(\tilde{a} \oplus \tilde{b}) = 1$, or
(2b) $\tau(\tilde{a} \odot \tilde{b}) = \tau(\tilde{a} \oplus \tilde{b}) = 1$.

This theorem allows to reveal the logical background of the famous wave-particle duality exhibited in the double-slit experiment in a better way than it can be done with the use of the classical two-valued logic.

Let $\tilde{a}$ (resp. $\tilde{b}$) denotes a proposition "a quantum object passes through the slit A (resp. B)". If we interpret arithmetic conjunction $\tilde{a} \odot \tilde{b}$ as representing compound proposition "a quantum object passes through the slit A *and* through the slit B" and arithmetic disjunction $\tilde{a} \oplus \tilde{b}$ as representing compound proposition "a quantum object passes through the slit A *or* through the slit B", then truth-values of these compound propositions can be used to decide whether a quantum object reveals its wave-like or particle like properties:

- If $\tau(\tilde{a} \odot \tilde{b}) = 0$ and $\tau(\tilde{a} \oplus \tilde{b}) = 1$ we can never state experimentally that an object passes simultaneously through the slit A *and* through the slit B although we are sure that it passes *either* through A *or* through B, therefore an object reveals purely particle-like behaviour.
- If $\tau(\tilde{a} \odot \tilde{b}) = 1$ and $\tau(\tilde{a} \oplus \tilde{b}) = 1$ we are sure that an object passes simultaneously through the slit A *and* through the slit B, so it behaves like a pure wave.

These two cases, corresponding to possibilities (1b) and (2b) can be equally well described with the aid of the classical two-valued logic. However, in many–valued logic there are also possibilities (1a) and (2a) which can be interpreted as follows:

(1a) We are not sure whether an object passes through the slit A *or* B although we are sure that it does not pass simultaneously through A *and*

through B. This seems to correspond to a non-perfect experiment in which an object behaving like a particle can pass *either* through the slit A *or* through the slit B, although it is not sure whether it passes at all.
(2a) We are sure that an object passes through the slit A *or* through the slit B (or through both slits) which allows both wave-like and particle-like behaviour. However, contrary to the case (2b) we cannot be sure that it passes through both slits, therefore its behaviour is not 100% wave-like. On the other hand its behaviour is also not 100% particle-like since we are also not sure that it does not pass through both slits simultaneously. Therefore, an object reveals partial wave-like and partial particle-like behaviour. This kind of behaviour is not allowed by the orthodox Copenhagen paradigm based on classical two-valued logic but it was experimentally confirmed by Mittelstaedt, Prieur and Schieder [9]. This experimental result strongly supports the proposed many-valued interpretation of a quantum logic.

I am very grateful to my friend Dr. M. Czachor for his help in preparing the TEX file of this paper. University of Gdańsk grant BW/5100-5-0065-6 is gratefully acknowledged.

REFERENCES

1. G. Birkhoff and J. von Neumann, *Ann. of Math.* **37** (1936) 823-843.
2. E. Beltrametti and G. Cassinelli, *The Logic of Quantum Mechanics* (Addison-Wesley, Reading, 1981).
3. M. J. Mączyński, *Studia Mathematica* **47** (1973) 253-259.
4. J. Pykacz, "Quantum logics as families of fuzzy subsets of the set of physical states," in *Preprints of the Second IFSA Congress, Tokyo, July 20-25, 1987*, Vol. 2, pp. 437-440.
5. J. Pykacz, *Int. J. Theor. Phys.* **33** (1994) 1419-1432.
6. R. Giles, *Int. J. Man-Machine Studies*, **8** (1976) 313-327.
7. O. Frink, "New algebras of logic," *Amer. Math. Monthly* **45** (1938) 210-219.
8. J. Pykacz, "Algebras and logics of physical systems," to be published in: M. Dalla Chiara, R. Giuntini, and F. Laudisa, eds., *Philosophy of Science in Florence 1995* (Kluwer Academic, Dordrecht, 1997).
9. P. Mittelstaedt, A. Prieur, and R. Schieder, *Found. Phys.* **17** (1987) 891-903.

NO POINT PARTICLES, DEFINITELY NO WAVES

C. Rangacharyulu

Department of Physics and Engineering Physics
University of Saskatchewan
Saskatoon, SK., Canada, S7N 5E2
chary@sask.usask.ca

It is argued that in microscopic physics the discussions of wave-particle duality are not meaningful. This reasoning is based on the observation that waves are a conglomeration of coherent disturbances in a many body system and as such they do not represent individual entities. Newtonian kinematics have no predictive power and they do not offer a physical description of participants in an interaction, except to say that they obey the conservation laws. The point-particle concept is an unnecessary complication in physics.

Key words: wave-particle duality, quantum physics

It is perhaps no exaggeration to say that in physics literature and in classroom discussions, there is no topic as popular and as fascinating as the question of wave-particle duality. The prescription for classroom teaching is relatively simple. It is to point out that interference and diffraction are wavephenomena and argue that the photoelectric effect and Compton scattering can not be explained by wave theories. This statement is not quite correct but most of the teachers do not know this. Then we conclude that photons exhibit particle-wave duality. Immediately, we bring in the matter-waves *à la de Broglie* and talk about electron and neutron diffraction. In the class rooms, teachers can be satisfied that they have a *fait accompli.* However, the student reaction is initially bewilderment, followed by a state of resignation and perhaps then they get indoctrinated into physics. It is clearly not a satisfactory situation. After several years of struggle with this problem, I conclude that this entire discussion is not meaningful. Below, I elaborate my reasoning.

In the discussions of kinematics of reactions or decay processes, one

M. Ferrero and A. van der Merwe (eds.),
New Developments on Fundamental Problems in Quantum Physics, 329–333.

concerns with conservation principles. There is no reference to the sizes of participants. If one wants to determine the sizes of microscopic objects, one has to resort to measurements such as elastic scattering cross sections as momentum transfers are varied and determine matrix elements $< f||O(q)||i >$. Make some assumptions about $O(q)$ and deduce sizes. The most famous example is the discovery of atomic nucleus. Currently, the smallest particle viz. electron is of size $< 10^{-15}$ cm, and yet it may be a humongous object on Planck scale ($\approx 10^{-33}$cm). We do not know.

From the classical standpoint, it should be clear that objects do not exist in a waveform, nor they follow wavy paths. Newton's laws of motion do not allow that. Classical waves are manifestations of complex interactions. In classrooms, one usually gives an example of water waves created by a pebble striking a pond. The discussion does not tell the whole story. A pebble following a rectilinear path imparts momentum to water molecules. Those molecules acquire momentum radially and they in turn impart to the molecules along their paths. If we analyze each individual encounter, we will find that they obey energy-momentum conservation laws either in Newtonian or relativistic description. Thus, the wave formation is a result of many complex interactions and wave equations simply reproduce the temporal/spatial averages of these processes and they do not concern with individual entities or encounters. The fate of the pebble is completely ignored. This aspect of wave description was recognized by Einstein [1].

The basic premise of particle-like behavior in analytical mechanics is that they obey the principle of least action. In quantum mechanical formulations, one loses one-to-one correspondence between the equations of motion and the action principle [2]. Unlike classical mechanics which emphasizes rectilinear propagation for objects in the absence of external forces, quantum mechanics allows for *zitterbewegung* around the classically allowed path. The fact an object satisfies quantum mechanical wave equation or that it exhibits diffraction or interference patterns does not necessarily make it a wave of the classical mechanical language. Furthermore, as a particle is not obliged to follow a rectilinear path, one is led to the conclusion that the quantum mechanics does not describe individual entities in isolation. Put another way, there is no such thing as quantum analog of a classical free particle. Perhaps, it causes no problem because quantum mechanical objects are probed only by interacting with them.

We are in a situation where we find that particle-like description is unnecessary and wave-like description deals only with averages of ensembles. The characteristics of individual entities in isolation are inaccessible to quantum mechanical treatments.

We would like to illustrate the above reasonings with the help of a well celbrated example, viz. Compton effect [3]. A photon of energy E is incident on a target of mass mc^2 (say, an electron at rest). After elastic scattering, a photon of energy E' emerges at an angle θ with respect to the incident photon direction. It is a standard exercise in elementary modern physics courses to determine the kinematics of the final state in its entirety, i.e., kinetic energy T and emerging angle (ϕ) of the recoil electron and the scattered photon energy E':

$$\frac{1}{E'} = \frac{1}{E} + \frac{1}{mc^2}(1 - \cos\theta),$$
$$T = E - E',$$
$$\tan\phi = \frac{\cot\theta/2}{1 + E/mc^2}.$$

It is important to recognize that the ingredients that go into this calculation are simple energy and momentum conservation laws. The initial parameters could be determined in the infinite past, when the target and projectile are very far apart from the interaction region and the final quantities are again determined when the recoil electron and scattered photon are very far away from each other. All one requires here is that the conservation laws are obeyed. The success of this theory is a vindication of the conservation principles. It has no knowledge about the nature of partners in the process. It does not know the directions nor the energies of objects in the final state till we specify one of the observables. If we observe the scattered photon at angle θ, then we know all the kinematical variables involved in the process. It is essential that we provide one parameter of the final state and the theory then fixes the other kinematical observables, in good agreement with the experiment. In fact, we should recognize that these equations hold good for photon elastic scattering off complex targets, which are clearly identified as material bodies of finite physical extensions. In the recent years, Compton scattering has been studied with various nuclear targets from hydrogen to lead. The above equations, with appropriate parameter for mc^2 will suffice to determine the kinematical conditions. In fact, as mentioned above, no point-like particle characteristics are implied in our conservation laws, except perhaps that the region of interaction is much smaller than the distances over which the observations are made and the distance of closest approach of the target and projectile is larger than the sum of their sizes.

The first correct account of the intensity distribution of scattered photons was provided by Klein and Nishina [4]. It is of interest to note that this theoretical account treats the problem as the scattering of *monochromatic waves* by a stationary electron. It becomes clear from the formula for cross section,

$$\frac{d\sigma}{d\Omega} = \frac{1}{2}r_e^2\left(\frac{\omega'}{\omega}\right)^2\left(\frac{\omega'}{\omega} + \frac{\omega}{\omega'} - \sin^2\theta\right),$$

where we employ the commonly used notation. The r_e represents the classical electron radius and the ω, ω' are the frequency of the incident and scattered radiation and θ is the scattering angle.

It should be emphasized that this equation is correct even today. This formula gives the probability that a scattered photon may be observed along a specific direction or the fraction of scattered photons that would follow a specific path. It thus speaks of average properties. It should be noted that this treatment accounts for data over a wide range of energies and scattering centers, be they electrons or nucleons or complex nuclei.

In the above description of Compton scattering, we do not make any specific use of the particle-like behavior of photons, except for the fact energy and momentum conservations are guaranteed at all points of observation that constitute physically accessible space-time points. These conservation principles along with the wave-like descriptions of Klein and Nishina account for all the observables of this scattering process. We might argue that the later developments of quantum electrodynamics, via the second quantization reenforce the particle concept for cross sections. However, these treatments suffered from the divergency problems, which in turn was cured by the renormalization procedures. Sticking to our current example of Compton scattering, these mathematical gymnastics give us back the Klein-Nishina formula. In these treatments, the spatial exensions are subject to Heisenberg's uncertainity relations, which render our knowledge of the system obscure for spatial extensions $\Delta x \leq h/\Delta p$, with Δp as the momentum uncertainty, not to enable us to discern it. We are back to the situation where we can not insist upon the point-like particles. We should also allow for virtual states not accessible to direct physical observations. Furthermore, as mentioned above, one has to bring in charge and mass renormalization to make the physical particles "dressed," with the point particles as mathematical idealizations. Needless to say, a "dressed" particle is not a free particle, as it knows about its surroundings and is influenced by them.

In the last decade or so, the string theories seem to try to come to grips with this problem. One no longer considers particles as point objects. Accordingly, one might say that the point-particle concept of the Greeks is now abandoned [2]. However, the physical phenomena are consistent with the idea of localization of energies and momenta over narrow regions of space. It is a separate question whether these dimensions are of the scale of Planck 's length (10^{-33}cm) or not. Despite the fact that the string theory has its own serious short-comings, the clear message is that the particles are not points after all.

To conclude, waves do not represent individual entities. Individual entities that we call particles are not point objects. The mathematical idealizations of point-particles give conceptual problems of infinities, such as infinite charge and mass densities. Renormalization procedures not only sweep the problems under the rug and the concept of a free particle is also lost. Asymptotically, one might get away with such treatments and language. A purist, which I think all basic physicists are, should avoid insisting on this terminology as they are conceptually unsound. In fact, the standard model assumes that the entire physical universe can be described by two prescriptions: a) fundamental particles as constituents and b) fundamental interactions as the driving mechanisms, each completely independent of the other. One should question this very assumption. But this discussion is beyond the scope of the present work.

REFERENCES

1. A. Einstein(1905), *Collected Papers of Albert Einstein, Vol.2*, Anna Beck and Peter Havas, eds. (Princeton University Press, Princeton, 1989).
2. M. Kaku, *Introduction to Superstrings* (Springer, New York, 1988).
3. See, for example, H. Ohanian, *Modern Physics* (Prentice-Hall, New York, 1987). C. Rangacharyulu, *Elements of Modern Physics*, in preparation.
4. O. Klein and Y. Nishina, *Z. Phys.* **52** (1929) 853; English translation in *O. Klein Memorial Lectures* (World Scientific, Singapore, 1994), Vol. 2, p. 113.

INTERFERENCE AND VIOLATION OF BELL'S INEQUALITIES USING SEPARATE SOURCES

J. G. Rarity and P. R. Tapster

DRA Malvern
St Andrews Rd, Malvern
Worcs., United Kingdom WR14 3PS

Interference effects between separate particles occur when they are made indistinguishable. Here we demonstrate the simplest experiment that proves this statement by creating a source of one-photon states and mixing with an independent weak coherent source at a beamsplitter. We discuss experimental configurations which would allow a violation of Bell's inequality to be demonstrated.

Key words: quantum interference, non-locality.

1. INTRODUCTION

Non-classical two photon interference effects were first predicted for a two photon field falling on a screen [1] and then in the interference of two 'one-photon' states input into opposite ports of a beamsplitter [2,3]. In the simple particle picture a single photon input into a beamsplitter will appear randomly in one or other of the beamsplitter outputs. Extending this to one photon in each port we naively expect no correlation in the output taken by each photon of the pair and thus a 50% probability that one photon will appear in each output port. In the quantum mechanical picture the beamsplitter is represented as in Fig. 1 with input modes $|\rangle_a, |\rangle_b$ and output modes $|\rangle_c, |\rangle_d$. The transmission t and reflection ir amplitudes include a phase shift $i = \exp^{i\pi/2}$ to satisfy energy conservation (r and t are real). Input photons are indistinguishable from measurements made on the output modes c, d. An input state containing two single photon states is thus transformed by the beamsplitter to give

$$|1>_a |1>_b \rightarrow \left[t^2|1\rangle_c|1\rangle_d - r^2|1\rangle_c|1\rangle_d + irt|2\rangle_c|0\rangle_d + irt|0\rangle_c|2\rangle_d\right] . \quad (1)$$

M. Ferrero and A. van der Merwe (eds.),
New Developments on Fundamental Problems in Quantum Physics, 335–345.

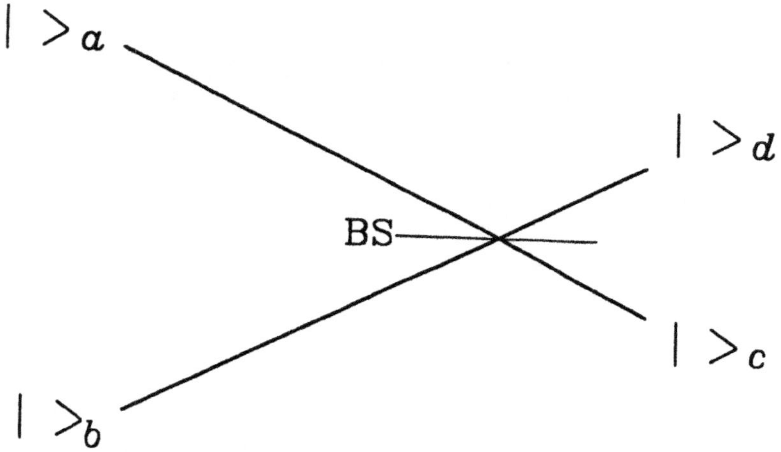

Fig. 1. The input $a, b-$ and output $c, d-$modes of a beamsplitter (BS). Coincidences are measured between photon counting detectors placed in the c and d modes.

The first two terms show a destructive interference effect and the probability P_{cd} of seeing photons simultaneously in both beamsplitters outputs becomes

$$P_{cd} = t^4 + r^4 - 2r^2t^2. \tag{2}$$

In fact when $r^2 = t^2 = 1/2$ (a 50/50 beamsplitter) $P_{cd} = 0$ and the photons always appear as *pairs* in random outputs of the beamsplitter.

2. EXPERIMENTS WITH PHOTON PAIRS CREATED IN PARAMETRIC DOWNCONVERSION

2.1. Degenerate Photon Pairs at Beamsplitters

We can create a source of simultaneous photon pairs using a suitably cut parametric down-conversion crystal pumped by a short wavelength laser. These pair beams can be made identical in wavelength simply by selecting them at equal (and opposite) angles to the pump beam and filtering through identical interference filters. Directing the beams into opposite input ports of a beamsplitter allows the above effect to be experimentally demonstrated [4,5] (Fig. 2) as a reduction of the pair detection rate at the beamsplitter outputs. By scanning the path length difference from the crystal to the beamsplitter inputs one can show that the pairs are coincident to within their coherence length which can be much less than 100fs (Fig. 3). This

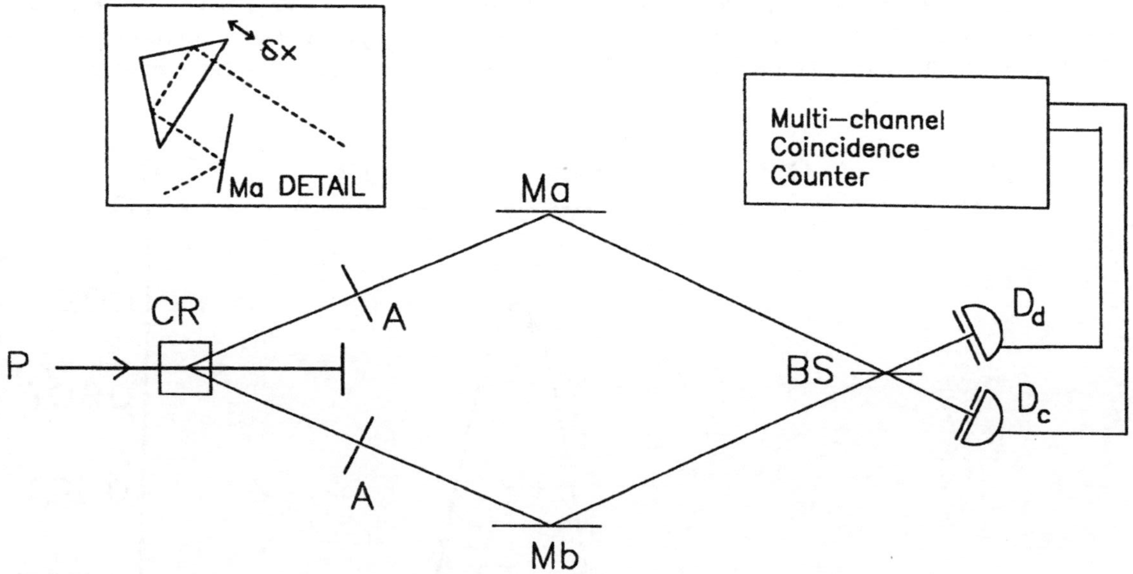

Fig 2. Experiment showing suppression of coincidences when correlated photon pairs are mixed at a beamsplitter BS. Pump beam P illuminates a downconversion crystal CR. Apertures A select degenerate photon pairs which are directed via mirror arrangements Ma and Mb onto opposite sides of beamsplitter BS. Coincident detections are measured in D_c and D_d as a function of pathlength difference δx varied by moving the prism in Ma.

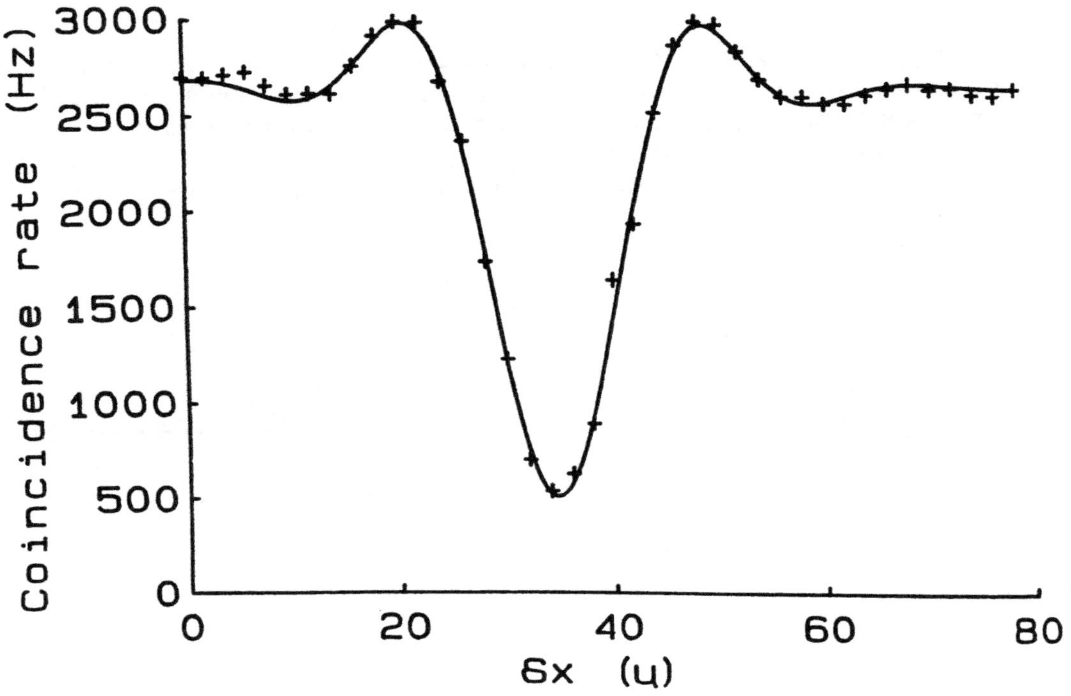

Fig. 3. Coincidence rate as a function path length difference δx (in microns) showing the reduction of coincidences when the pairs overlap. The width of the coincidence reduction reflects the coherence time of the light which is less than 40fs here. The shape of the dip resembles a sinc function reflecting the "top hat" shape of the spectrum of the downconverted light.

time is several orders of magnitude less than the fastest resolving time of typical photon counting detectors.

2.2. Two-Colour Pairs and Non-Local Interference

This experiment was extended by choosing two sets of pair beams as shown in Fig. 4 [6]. Phase matching requires that the pairs are created either in the $ac-$ and $bd-$ mode OR in the $ad-$ and $bc-$ mode. We superpose these two possibilities when we recombine the $c-$ and $d-$ wavelengths at separate beamsplitters (or separate parts of the same beamsplitter). When the two possibilities are indistinguishable we see the interference effects associated with the entanglement. The wavefunction before the beamsplitters is in the entangled state

$$|\Psi\rangle = 2^{-1/2}\left[\exp^{i\phi_d}|1\rangle_{ac}|1\rangle_{bd} + \exp^{i\phi_c}|1\rangle_{ad}|1\rangle_{bc}\right]. \quad (3)$$

The $bd-$ and $ad-$modes combine at one beamsplitter and the $bc-$ and $ac-$modes at the other. The beamsplitters (reflection and transmission amplitudes $1/\sqrt{2}, i/\sqrt{2}$) transform the wavefunction to

$$\begin{aligned} |\Psi\rangle_{cd} = 2^{-3/2}[& i(\exp^{i\phi_d} + \exp^{i\phi_c})|1\rangle_{c0}|1\rangle_{d0} \\ & + i(\exp^{i\phi_d} + \exp^{i\phi_c})|1\rangle_{c1}|1\rangle_{d1} \\ & + (\exp^{i\phi_d} - \exp^{i\phi_c})|1\rangle_{c0}|1\rangle_{d1} \\ & - (\exp^{i\phi_d} - \exp^{i\phi_c})|1\rangle_{c1}|1\rangle_{d0}]. \end{aligned} \quad (4)$$

For clarity we ignore the vacuum terms in the above. Clearly the coincidence probabilities now show interference as a function of phase difference $\phi_c - \phi_d$:

$$\begin{aligned} P_{c0d0} = P_{c1d1} &= \frac{1}{4}(1 + cos(\phi_c - \phi_d)), \\ P_{c0d1} = P_{c1d0} &= \frac{1}{4}(1 - cos(\phi_c - \phi_d)), \end{aligned} \quad (5)$$

As the $c-$ and d modes are separated and the interference visibility remains unity for any fixed ϕ_c and variable ϕ_d (and vice versa) this system shows non-local effects. In the experiment [6] an interference visibility of 82% was measured and a violation of Bell's inequality [10] by 10 standard deviations was measured.

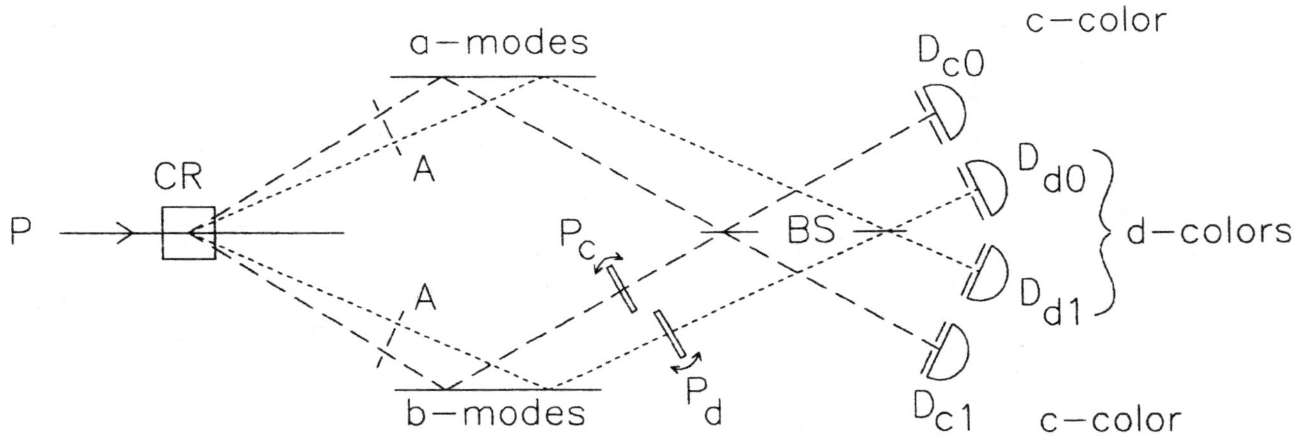

Fig.4. Schematic of the momentum state superposition apparatus. A 413nm pump beam (P) illuminates a KD*P non-linear crystal (CR). Double apertures (A) are used to select *ac*− , *ad*− and *bc*−, *bd*−modes. The *c*− and *d*− wavelengths (angles) are combined at separate beamsplitters. After mixing in the beamsplitter we cannot distinguish between $ac - bd$ and $ad - bc$ photon pairs and we see interference effects. Phase plates P_c and P_d placed in the *b*−modes alter the rate of coincident detections between detectors $D_{c0,c1}$ and $D_{d0,d1}$.

3. EXPERIMENTS WITH SEPARATE SOURCES

3.1. Interference of a Gated Single Photon with a Weak Coherent Pulse

In our latest experiments we use uncorrelated pairs of photons which in principle could have no common history. We can gate out single photons from a parametric down-conversion source [7], but we cannot guarantee overlap of the photons until their inverse bandwidth (therefore time uncertainty) is longer than the detector resolution time. With present technology this bandwidth is about 1GHz which can be achieved using Fabry-Perot etalons but leads to low single photon rates. Here we choose to time gate using a short pump pulse. The minimum bandwidth associated with a pulse length of a few hundred femtoseconds is then a few nanometres which is reached by narrow band interference filters. The experiment is shown in Fig. 5 [8,9]. A frequency-doubled mode-locked laser (407.5nm wavelength) pumps a thin parametric downconversion crystal cut for non-degenerate operation. Signal and idler photons satisfying energy conservation are emitted spontaneously in a broad band cone behind the crystal and apertures are placed to select 815nm wavelength beams from opposite ends of a cone diameter. Detection of an idler (or gate) photon in one beam with time resolution better than the pump pulse separation time essentially localises a single signal ($b-$mode) photon within a pulse length which in the experiment is around 200fs. The signal photon so selected is a good approximation to a one-photon state and thus must have random phase when we measure it with respect to the original near infra-red beam. The two can be thought of as separate sources. We expect to see no first order interference fringes when we mix this single photon source with coherent pulses from the undoubled mode-locked laser. However when we reduce the intensity of the coherent pulses to the point where the energy per pulse is much less than one quantum we again approximate a one photon state diluted by a large amount of vacuum. We find that the dominant terms in the state behind a beamsplitter are just those highlighted in Eqs. 1 and 2. When the single photon and coherent pulses overlap there is a second order interference effect and a suppression of coincidences after the beamsplitter. A typical result showing the gated coincidence rate (triple coincidence) as a function of path length difference altered by an optical trombone in the weak pulse path is shown in Fig. 6. The interference dip of 63% is beyond the maximum (50%) expected from the second order interference of two random-relative-phase classical pulses.

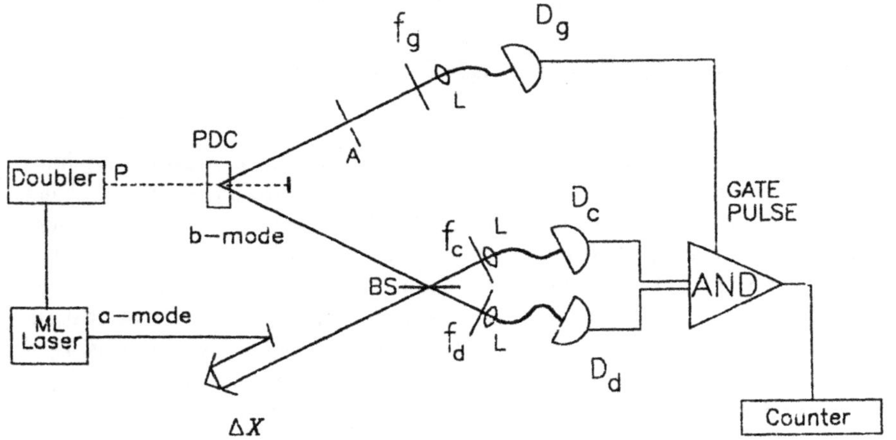

Fig. 5. The experimental apparatus used to demonstrate interference between separate sources.. Key: (ML) Mode locked Ti-Sapphire laser, (PDC) parametric downconversion crystal, (A) aperture, (f_g, f_c, f_d) interference filters with 815nm centre wavelength, (L) lenses coupling light to optical fibres, (BS) beamsplitter, ($D_{g,c,d}$) photon counting avalanche photodiodes. Triple coincidences are measured by a gated AND gate. Solid lines indicate light of 815nm wavelength, dashed lines indicate light of 407.5nm wavelength and thick curved solid lines represent optical fibres.

3.2. Viloating Bell's Inequality By Entangling Initially Separate Sources

As the pumping laser in these experiments is made up of short pulses it necessarily has a broad spectrum (≃3nm). Hence even when we select a single wavelength in the gating channel there is an uncertainty in the wavelength and hence direction of our gated single photons. We can thus extend our apparatus in a similar way to earlier two-colour interference experiments where downconverted photons were mixed at the beamsplitter. The proposed experiment is shown in Fig. 7 where two possible $b-$modes are selected by filters with slightly differing wavelengths. These different parts of the gated single photon are then mixed at separated beamsplitters with identical selected wavelengths from the weak coherent pulses. The input state to the beamsplitters is now approximated by the product state

$$|\Psi\rangle = \frac{1}{2}\left[(\exp^{i\phi_d}|1\rangle_{ac} + \exp^{i\phi_c}|1\rangle_{ad})(|1\rangle_{bc} + |1\rangle_{bd})\right]. \qquad (6)$$

Using again the beamsplitter reflection and transmission coefficients $1/\sqrt{2}, i/\sqrt{2}$

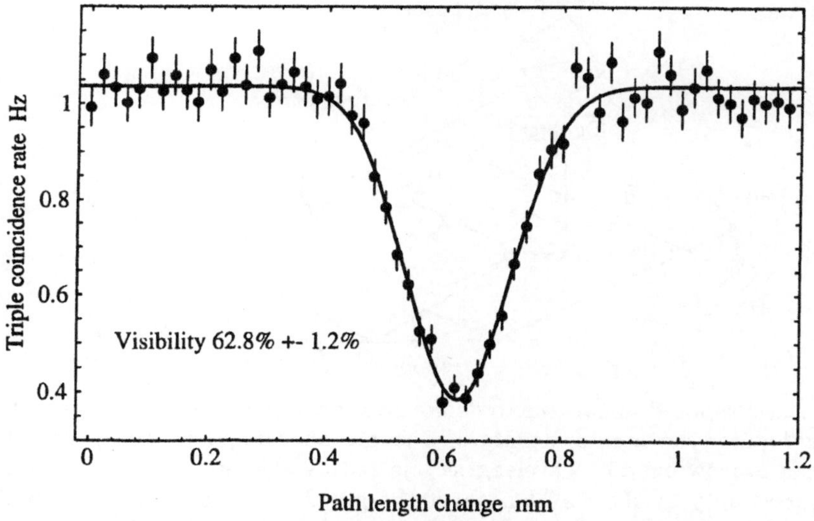

Fig. 6. Triple coincidence rate measured as a function of delay ΔX in the weak laser path. Solid line shows a least squares fit based on a Gaussian dip with visibility 64% and 1/e width 133μm.

the wavefunction is transformed to

$$\begin{aligned} |\Psi\rangle_{cd} = \frac{1}{4}[&i(\exp^{i\phi_d} + \exp^{i\phi_c})|1\rangle_{c0}|1\rangle_{d0} \\ &+ i(\exp^{i\phi_d} + \exp^{i\phi_c})|1\rangle_{c1}|1\rangle_{d1} \\ &+ (\exp^{i\phi_d} - \exp^{i\phi_c})|1\rangle_{c0}|1\rangle_{d1} \\ &- (\exp^{i\phi_d} - \exp^{i\phi_c})|1\rangle_{c1}|1\rangle_{d0}] \\ +i\exp^{i\phi_d}(|2\rangle_{d0} + |2\rangle_{d1}) &+ i\exp^{i\phi_c}(|2\rangle_{c0} + |2\rangle_{c1})]. \end{aligned} \tag{7}$$

Again empty modes (vacuum terms) are omitted for clarity. When we measure only coincidences between the gate, $c-$ and $d-$ detectors the final terms in $|2\rangle_{d0}$ etc. are ignored. As a result we are selecting a state approximating to Eq. 3 from the initial product state. Operating at zero pathlength difference the gated probability of seeing a coincidence (per pulse) in the $c0$ and $d1$ (or the $c1$ and $d0$) detectors is again

$$P_{cdg} = C/8\,[1 - \cos(\phi_c - \phi_d)]\,, \tag{8}$$

where C is the probability of a coincidence per pulse in the absence of beamsplitters ($C \ll 1$). For coincidences between $c0, d0$ or $c1, d1$ detectors the

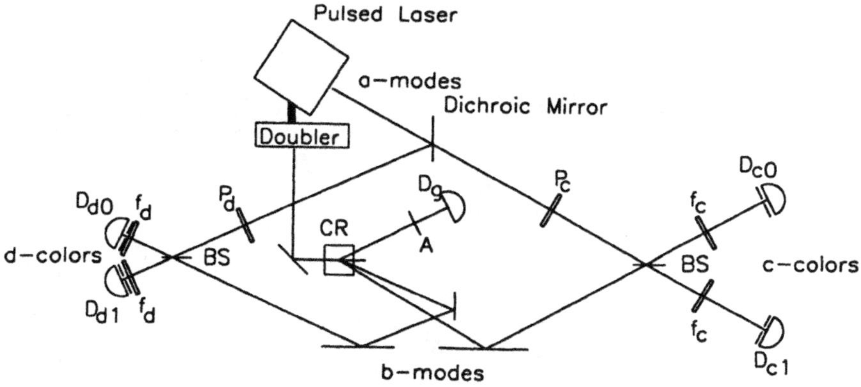

Fig. 7.a. Proposed apparatus to demonstrate a violation of Bell's inequality using heterodyne measurements of a single photon. The gating detector views a narrow band of wavelengths selected by aperture A. There remains an uncertainty in the wavelength of the gated single photons due to the bandwidth of the pump pulse. We select two of these wavelengths using filters f_c and f_d from both the gated b−modes and the undoubled pulsed laser (a−mode) and recombine these at separated beamsplitters. Measuring triple coincidences between gate, c− and d− detectors selects an entangled state from the initial product state.

minus is again replaced by a plus. The phase adjustments can be performed remote from the sources and we essentially have an apparatus that can be used to demonstrate violations of Bell's inequality [10]. A similar experiment was first analysed by Tan and coworkers [11] where the effect was described as a 'Bell inequality of a single photon'. Even when extremely attenuated, the laser can be thought of as a classical field (with a well defined phase). This experiment can then be thought of as a non-local measurement of the phase of a single photon by heterodyning with a classical field.

4. CONCLUSION AND PERSPECTIVE

The mixing and entanglement of initially separate particles is a basic requirement in the building of a generalised quantum computer. Such a machine would code all possible inputs into an N-fold superposition state and manipulate that state using coherent/interferometric logic devices. However we note the limited efficiency of existing photon counting detectors and the extremely low efficiency of parametric sources of single photon states. We expect to be able to extend our experiments to three photon interference effects but suspect better sources need to be found before we progress beyond this.

ACKNOWLEDGEMENTS

The authors would like to thank Drs A.Ekert and A.Barenco for clarifying discussions during the course of this work. This work is supported in part by the EC Framework 4 programme ACQUIRE grant number 20029.

REFERENCES

1. R.Ghosh and L.Mandel, *Phys. Rev. Lett.* **59**, 1903 (1987).
2. Fearn H. and R.Loudon, *Opt. Commun.* **64**, 485-490 (1987).
3. Z.Y.Ou, C.K.Hong, and L.Mandel *Opt. Commun.* **63**, 118-122 (1987).
4. C.K.Hong, Z.Y.Ou, and L.Mandel, *Phys. Rev. Lett.* **59**, 2044 (1987).
5. J.G.Rarity and P.R.Tapster, in *Photons and Quantum Flutuations,* E R Pike and H Walther, eds. (Adam Hilger, Bristol, 1988), p. 122; *J. Opt. Soc. Am. B* **6**, 1221 (1989).
6. J.G. Rarity and P.R.Tapster, *Phys. Rev. Lett.* **64**, 2495 (1990).
7. J.G. Rarity, P.R.Tapster, and E.Jakeman *Opt. Commun.* **62**, 201 (1987).
8. J.G.Rarity, in *Fundamental Problems in Quantum Theory* (New York Academy of Sciences, 1995); *Ann. N. Y. Acad. Sci.* **755**, 624 (1995).
9. J.G. Rarity and P.R.Tapster, submitted to *Phys. Rev. Lett.*
10. J.S.Bell, *Physics* **1**, 19 (1964)
11. S.M.Tan, D.F.Walls, and M.J.Collett, *Phys. Rev. Lett.* **66**, 252-255 (1989).

BELL'S INEQUALITY FOR A PARTICLE

M. Revzen and A. Mann

Physics Department, Technion
Haifa 32000, Israel

A systematic attempt to relate Bell's inequality to operators' noncommutativity is outlined.

Key words: Bell's inequality, uncertainty principle.

1. INTRODUCTION

Our general aim is to understand the relation of commutators of operators, which are local properties, to Bell's inequality violations (BIQV) which are nonlocal attributes, i.e. the aim of this study is to delineate notions of physical realism versus locality as these are interwoven in the customary analyses of Bell's inequality. To illustrate what we try to get as a general result we consider a special case available in the literature for some time, viz the analysis of CHSH [1] inequality. Thus the Bell's inequality introduced by CHSH [1] locality ala Bell allows 2 as an upper limit for the expectation value of

$$\hat{B} = \hat{A}_1\hat{B}_1 + \hat{A}_1\hat{B}_2 + \hat{A}_2\hat{B}_1 - \hat{A}_2\hat{B}_2, \tag{1}$$

i.e., Bell's inequality (BIQ) is, for this case,

$$|< \hat{B} >| \leq 2. \tag{2}$$

Here

$$\hat{A}_i^2 = \hat{B}_j^2 = I \quad i, j = 1, 2. \tag{3}$$

Our interest is focused on Landau's [2] derivation of the quantal upper limit [3] of the expecation value for $\hat{B}$, viz.,

$$|< \hat{B} >| \leq 2\sqrt{2} \quad \text{(quantum mechanics [3])}. \tag{4}$$

M. Ferrero and A. van der Merwe (eds.),
New Developments on Fundamental Problems in Quantum Physics, 347–354.

Thus the "mathematical" limit for the expecation value for $\hat{B}$ is 4 (recalling equation 3). The "classical" limit for "local realistic" theories is 2 [1]. The quantal limit, $2\sqrt{2}$ is gotten by noting [2] that

$$\hat{B}^2 = 4\hat{I} - [\hat{A}_1, \hat{A}_2][\hat{B}_1, \hat{B}_2]. \tag{5}$$

Now

$$| < [\hat{A}_1, \hat{A}_2][\hat{B}_1, \hat{B}_2] > | \leq 4 . \tag{6}$$

Thus [2,3] Eq. (4) follows. Here we see that BIQV is possible only if both commutators $[\hat{A}_1, \hat{A}_2] \neq 0$ and $[\hat{B}_1, \hat{B}_2] \neq 0$. Furthermore the extent of the violation is fixed by the product of the commutators. Thus, in this special case, we have demonstrated what we are looking for in general: noncommutativity determines BIQV.

A relation based on operators' commutators should not be confined to two particles correlations. We search for BIQV's fingerprints in one particle states. This because the uncertainty relation which is implied by operator noncommutativity [4] can be most simply formulated for the one particle states. We recall, in fact, that the original EPR [5] study (the root of it all...) actually relates to a single particle. The possible measurements (& results) of the "distant" particle merely underscore the dilemma facing the quantum mechanical attributes of one particle which seems, ala EPR, to have both momentum and position as "elements of reality" violating the basic tenants of quantum mechanics [5]. Furthermore the "consistent histories" interpretation of quantum mechanics maintains that all the difficulties in the EPR like cases are in the one particle quantum mechanics [6]. However, and this is a big however, Bell himself "coup de graced" all this fine argument by constructing a "pure hidden variable" one particle spin state for all to see. This should, so it seems, put an end to all attempts to construct anything for a single paricle - since local hidden variables don't mix with BIQV. Our approach, herewith reported, is based on D.Fivel's [7] observation that there are two ways (at least) to build a one particle state. In one of these, to be described shortly, the particle state is built "non invasively" with a physical picture very close to that of the original EPR. Bell's hidden variable model was not for such one particle state is our contention.

2. TWO WAYS FOR PREPARING ONE PARTICLE'S STATE

We consider a two-dimensional Hilbert space. Therein projectors onto states are defined: E(p), E(q),... Each refering to as filters for the propositions p, q,..., i.e. the particle is in the state $|p>$ if, with certainty, it will pass the E(p) filter, e.g. Stern Gerlach magnets along $\boldsymbol{p}$ with the filter picking the "up". For two particles we have a four dimensional Hilbert space. We now present Fivel's [7] two ways of constructing one particle state $|r>$.

Method a. Pass the particle through the filter E(r). This is an invasive method. Once passed through the filter (e.g. went up in a Stern-Gerlach magnet oriented along $\boldsymbol{r}$) we may caculate $\mathcal{P}_Q(s|r)$ - the probability (quantum mechanical, QM) that it will pass E(s). Formally,

$$\mathcal{P}_Q(s|r) = ||E(r)|r>||^2 = TrE(r)E(s)/TrE(r) = |<s|r>|^2. \quad (7)$$

Method b. "Ease" our particle into $|r>$ noninvasingly, by using correlation [7]. Thus prepare a two particle state $|S(1,1')>$ such that (i) each particle has equal chance to pass any filter and $|S(1,1')>$ is spherically symmetric. (ii) if $1'$ passes through E(r) our particle, 1, is known to be in $|r'>$ (E(r') = I - E(r)). We shall take $1'$ and 1 widely separated so invasing 1' can't affect 1. Thus the QM description of "easing" our particle to the state r' is for $|S(1,1')>$ to satisfy:

$$<S|E_i(r)|S>= 1/2, \quad \forall r, s \ldots, \quad (8)$$

$$U|S>=|S>, \quad \text{U denoting rotation}, \quad (9)$$

$$\begin{aligned}\mathcal{P}_Q(s|r') &=< S(1,1')|E_{1'}(r')E_1(s)E_{1'}(r')|S(1,1')> / <S|E_i|S> \\ &= Tr\{E_1(r')E_1(s)E_1(r')\}/Tr\{E_1(r')\}.\end{aligned} \quad (10)$$

The second expression involves one particle only (particle 1). It equals (7). It is easy to see that an alternative expression for the above is

$$\mathcal{P}_Q(s \mid r') = Tr\{(E(r') - E(s))^2\}. \quad (11)$$

Now local hidden variables (HV) theory means: There is a set, Λ, with elements λ, with the property that to each filter, E(p), we may associate a subset Λ(p). These have the property that

$$\begin{aligned} &\lambda \in \Lambda(p), \quad \text{particle will pass the E(p) filter} \\ &\lambda \notin \Lambda(p), \quad \text{particle will not pass the E(p) filter.}\end{aligned} \quad (12)$$

We also have a probability measure, $\mu[\Lambda(p)]$ that gives the probability for $\Lambda(p)$. When the particles are widely separated (like 1 and $1'$) invasing $1'$ cannot affect the $\rho(\lambda)$ - the local HV state of particle 1. The description of our particle being in the state built by method b allows an HV underpinning if locality is assumed because of the above mentioned noninvasiveness:

$$\mu[\Lambda(r)] = 1/2, \quad \forall r, s, \quad (13)$$

$$\Longleftrightarrow \int \rho(\lambda)A(r,\lambda)d\lambda = 0, \quad \forall r, s. \quad (14)$$

Here A is defined by

$$A(r,\lambda) = \begin{cases} +1, & \lambda \in \Lambda(r), \quad (15) \\ -1, & \lambda \notin \Lambda(r). \quad (16) \end{cases}$$

Let δ be the angle between r and s for the particle 1 where $1'$ was subjected to the filter E(r) and passed. We then write

$$\mathcal{P}_H(s \mid r') = n(\delta) = \frac{\int d\lambda \rho(\lambda) \frac{1+A(r',\lambda)}{2} \frac{1+A(s,\lambda)}{2}}{\int d\lambda \rho(\lambda) \frac{1+A(r',\lambda)}{2}}. \quad (17)$$

This expression, (17), produces faithfully our specifications. Upon integration over λ we get a contribution (=1) only when both A(r',λ) and A(s,λ) = 1. Thus the corresponding probability is

$$\mathcal{P}_H(s \mid r') = \frac{\mu[\Lambda(s) \cap \Lambda(r')]}{\mu[\Lambda(r')]}. \quad (18)$$

The subscript H signifies HV expression. We briefly summarize this section by: Noninvasive construction of a single particle state allows HV underpinning in as much as we are able to reproduce all that is required for the probabilities. Quantum mechanically the noninvasive construction appears to be a shrewd scheme for allowing "measuring" two observables pertaining to the propositions s and r simultaneously, Eqs. (10) and (11). From this vantage point for $\mathcal{P}_H$ to mimic $\mathcal{P}_Q$ is put squarely on the properties of these functions - both relating to local attributes.

3. BELL'S INEQUALITY

We defined n(δ), Eq.(17), to mean the relative number of times that our particle, 1, passes the E(s) filter when it was in the r' state:

$$\mathcal{P}_H(s \mid r') = \frac{\mu[\Lambda(s) \cap \Lambda(r')]}{\mu[\Lambda(r')]} = n(\delta). \quad (19)$$

Now it is readily shown [7] that $\mathcal{P}_H(s \mid r')$ is a metric - it is the cord joining the tips of unit vectors $\boldsymbol{r}$ and $\boldsymbol{s}$. Thus that n(0) = 0 iff $\boldsymbol{r} = \boldsymbol{s}$, and n($\delta$) $\geq$ 0 is obvious. That n(δ) is symmetric in r and s follows readily using Venn diagrams noting Eq. (18) and

$$\mu[\Lambda(s') \cap \Lambda(r)] = \mu[\Lambda(r') \cap \Lambda(s)]. \quad (20)$$

The same technique can be used to show that n(δ) obeys the triangular inequality:

$$n(\delta_1 + \delta_2) \leq n(\delta_1) + n(\delta_2) \Rightarrow n(2\delta) \leq 2n(\delta). \tag{21}$$

This is our Bell's inequality. (Note that this follows automatically if n"(0)<0.) That $\mathcal{P}_Q(s \mid r')$, Eq.(10), is mimicked by it requires that $\mathcal{P}_Q(s \mid r')$ too be a metric. Disposing of E henceforth, i.e., E(r) $\rightarrow$ R, E(s) $\rightarrow$ S etc., we require

$$\mathcal{P}_Q(s \mid r') = TrR'SR'/TrR' = \sin^2(\delta/2) =: n_Q(\delta) \tag{22}$$

to be a metric. Clearly this isn't so:

$$\left.\frac{d^2 n}{d\delta^2}\right|_{\delta=0} = \cos(0)/2 \quad > 0. \tag{23}$$

The essential point is that the violation, viz. Eq.(23), relates to local variables - i.e., at our particle's location.

4. OPERATOR NONCOMMUTATIVITY AND BIQV

We saw that $\mathcal{P}_H(s \mid r')$ is a metric while $\mathcal{P}_Q(s \mid r')$ is not. We shall now argue on quite general grounds that $\mathcal{P}_Q(s \mid r')$ can't be a metric when S and R are noncommuting projectors while it is when they do commute. In this section we consider an N (N$\geq$ 2)-dimensional Hilbert space wherein the projectors R,S,U,V are defined:

$$R^2 = R, \;\; S^2 = S, \;\; etc. \tag{24}$$

We consider the cases with

$$TrR = TrS = TrV = \ldots. \tag{25}$$

That R is a projector allows the interpretation of

$$TrR/N = \mathcal{P}(R) \tag{26}$$

as a probability of the system being in the state R, etc. Being cavalier about the normalization, we write

$$\mathcal{P}_Q(V' \mid R) = TrRV'/TrR \tag{27}$$

to mean - the probability of our system (= a particle) **not** in V given that it would pass the R filter. We first consider the commuting case. Thus for the subset of commuting projectors,

$$[R, S] = [V, S] = \ldots = 0, \tag{28}$$

$$\mathcal{P}_Q(V' \mid S) = TrSV'/TrS = \frac{TrS - TrSV}{TrS}. \quad (29)$$

SV in this case is a projector

$$(SV)^2 = SV. \quad (30)$$

Thus,

$$\frac{TrSV}{TrS} = \mathcal{P}_Q(S \cap V) = \text{Prob. of S and V.} \quad (31)$$

It can be readily verified that

$$\mathcal{P}_Q(U \cap V') \leq \mathcal{P}_Q(U \cap W') + \mathcal{P}_Q(W \cap V'). \quad (32)$$

Thus, in this case, all projectors can be simultaneously diagonalized thereby allowing an easy verification of the triangular inequality for each matrix element along the diagonal separately, thence (32) follows. The triangular inequality holds: $\mathcal{P}_Q(U \cap V')$ is a metric and BIQ is not violated. We now turn to the noncommuting case. We shall consider projectors onto pure states only. A metric for these was obtained by Fivel [7]. Thus,

$$Tr(R - V)^2 = 2(TrR - TrRV) = 2TrRV', \quad (33)$$

$$Tr(R - V)^2 = \sum_{i,j=1}^{N} \mid < i|R|j > - < i|V|j > |^2 = \sum_{k=1}^{N^2} |R_k - V_k|^2. \quad (34)$$

The last expression is the square of the "distance" between (complex) vectors in N^2 dimensional unit sphere. Thence $\sqrt{Tr(R-V)^2}$ is a (Euclidean) metric, as is

$$\sqrt{Tr(R - V)^2/2TrR} = \sqrt{TrRV'/TrR} = \sqrt{\mathcal{P}_Q(R \cap V')}. \quad (35)$$

This quantity obeys the triangular inequality and is a metric, i.e.,

$$\sqrt{\mathcal{P}_Q(R \cap V')} \leq \sqrt{\mathcal{P}_Q(R \cap W')} + \sqrt{\mathcal{P}_Q(W \cap V')}. \quad (36)$$

For the two-dimensional case, we take

$$\mathcal{P}_Q(R \cap W') = \mathcal{P}_Q(W \cap V') = \sin^2(\delta/2) \quad (37)$$

and

$$\mathcal{P}_Q(R \cap V') = \sin^2(\delta). \quad (38)$$

The inequality Eq. (36) is obeyed, but the Herbert inequality, relating to the unsquarerooted expressions, is not. In Fivel's notation (ignoring the normalizers, cf. Eq. (33)),

$$\sqrt{Tr(R-S)^2} = d(R,S), \tag{39}$$

$$Tr(R-S)^2 = D(R,S). \tag{40}$$

Bell's inequality is the triangular inequality for D(R,S). It is obeyed by commuting projectors.

5. SUMMARY AND CONCLUDING REMARKS

We have attempted to legitimize our view that one particle, $1'$, of an EPR pair is a "measuring apparatus" for the particle of interest, 1, in the sense that it (i.e., $1'$) "projects" particle 1 into a specific state. Thus the "two particles face" of the EPR pair is viewed as a shrewd technique seemingly allowing simultaneous measurements of two **non**commuting filters (projectors). E.g., spins along R and S directions that differ by an angle δ.

These "simultaneous" measurements are not for commuting observables - the latter would have a joint probability and allow hidden variables underpinning - leading to a particular metric on the projectors. The triangular inequality thereof is the Bell's inequality. At this juncture we argued that noncommutativity $[R,S] \neq 0$, implies, in a theory based on properties of a Hilbert space, a different metric, the triangular inequality of which is less strigent than the one holding for the commuting case and hence allowing violation of Bell's inequality. The bottom line of such an analysis is that Bell's inequality violation does not reflect, necessarily, nonlocality but rather the noncommutativity of operators which is a local property. Preliminary results of this work were presented at the workshop honouring Prof. Max Jammer's 80th birthday, Bar Ilan University, 1995.

This research was supported, in part, by the Fund for Promotion of Research at the Technion and by Technion VPR Fund. MR acknowledges with thanks the help of Dr. Erik Stengler.

6. REFERENCES

1. J.F. Clauser, M.A. Horne, A. Shimony, and R.A. Holt, *Phys. Rev. Lett.* **23**, 880 (1969).
2. L.J. Landau, *Phys. Lett. A* **120**, 54 (1987).
3. B.S. Cirelson, *Lett. Math. Phys.* **4**, 93 (1980).
4. H.P. Robertson, *Phys. Rev.* **34**, 163 (1929).

5. A. Einstein, B. Podolsky, and N. Rosen, *Phys. Rev.* **47**, 777 (1935).
6. R.B. Griffiths, *Am. J. Phys.* **55**, 11 (1987).
7. D. Fivel, in *Fundamental Problems in Quantum Theory*, D.M. Greenberger and A. Zeilinger, eds., *Ann. N.Y. Acad. Sci.* V 755(1995). D. Fivel, preprint.

QUANTUM GRAVITY CORRECTIONS FOR THE INITIAL STATE OF GRAVITONS

J. L. Rosales

Fakultät für Physik, Universität Freiburg
Hermann-Herder-Str. 3
D-79104 Freiburg i. Brg., Germany
rosales@phyq1.physik.uni-freiburg.de

The actual theory of quantum gravity is not yet available, nonetheless, we can cast the canonical formulation of general relativity in order to formally quantise it upon bringing into action the standard procedure consisting on replacing canonical variables by quantum operators (acting on the wave function of the universe). Moreover, the semiclassical approach would yield to obtaining, for a given complex solution of Wheeler-DeWitt equation, the Schrödinger formalism for the wave function of metric perturbations or gravitons plus quantum gravity correcting terms in pure gravity; thus, in the inflationary scenario, we should expect correcting effects in the relic graviton (Zel'dovich) spectrum from quantum gravity. These, on the other hand, could be measured in a future experiment.

Key words: quantum gravity, relic gravitons.

1. INTRODUCTION

In spite of the fact that there still lacks a full *quantum gravity* theory, a predictive semiclassical *quantum cosmology* does already exist from selecting the initial quantum state of the universe. The latter may be developed upon quantising canonical general relativity for the metric tensor corresponding to perturbations about FRW models [1–3]. In such an approximation scheme, the parameter representing the relevant energy for the initial expansion of the universe, the value of the effective cosmological constant, λ, is much lower than the Planck mass [4], $\lambda/m_{Pl}^2 \leq 10^{-9}$.

The wave function of the universe obeys a kind of Schrödinger equation with energy zero or Wheeler-DeWitt equation:

M. Ferrero and A. van der Merwe (eds.),
New Developments on Fundamental Problems in Quantum Physics, 355–360.

$$H\Psi = 0 \tag{1}$$

where H denotes the total Hamiltonian for gravity and matter. Thus, the first thing should be to derive from Eq. (1) the limit of quantum field theory in a fixed gravitational background. This has been achieved by Banks and others [5] upon using a Born-Oppenhiemer approximation scheme (that takes into account the above mentioned relative small value of matter energy with respect to the Planck mass). *Quantum gravity* corrections also arise [6] and, at least in principle, they should show how effects of quantum gravity might even be measured in the context of some inflationary scenarios. The aim of this paper is precisely to show that such effects do exist, in this case, upon introducing a shift of quantum gravity origin in the spectrum of the relic gravitons being originated during the early inflationary stages of the universe [7]. Moreover, a certain quantum gravity modification could be possibly measured in the next future by some experimental devices [8].

2. SEMICLASSICAL WAVE FUNCTION IN DE SITTER SCENARIO

The simplest model for the early stage of the Universe consists of a massive scalar field with matter potential $U(\phi)$ in a FRW spacetime described by a single degree of freedom, the scale factor, $a(t)$. After the Hartle-Hawking prescription [9], the wave function will depend on the value $U(\phi(\Sigma))$ (Σ being the boundary of the compact euclidean 4 - sphere). This is just de Sitter case with $U(\phi) \approx \lambda$ playing the role of an effective cosmological constant.

Moreover, the general approach requires a description for the Schrödinger evolution of metric perturbations or gravitons, i.e., the tensor harmonics of the three sphere d_n. Now, upon defining $\alpha = \log(a)$ as the gravitational variable, the metric perturbation Hamiltonian is

$$H_m = \sum_n \frac{-1}{2} \frac{\delta^2}{\delta d_n^2} + (n^2 - 1)e^{4\alpha} d_n^2, \tag{2}$$

where, as a consequence of its perturbative character, $d_n^2/\lambda \ll 1$. Wheeler-DeWitt equation reads [3]

$$\left[\frac{1}{2m_{Pl}^2}\frac{\partial^2}{\partial\alpha^2} + \frac{m_{Pl}^2}{2}(-e^{4\alpha} + \lambda e^{6\alpha}) + H_m\right]\Psi = 0. \tag{3}$$

In the semiclassical approach we write the wave functional $\Psi[\alpha, \{d_n\}]$ as

$$\Psi = e^{iS} \tag{4}$$

and then expand S in the form $S = m_{Pl}^2 S_0 + S_1 + m_{Pl}^{-2} S_2 + \ldots$.

We now insert this expansion in Eq. 3) and compare expressions with the same order in m_{Pl}. The highest order yields

$$\sum_n \left(\frac{\delta S_0}{\delta d_n}\right)^2 = 0. \tag{5}$$

Thus, S_0 depends only on α (the three-metric).

The next order yields to the Hamilton-Jacobi equation for gravity alone:

$$-\left(\frac{\delta S_0}{\delta \alpha}\right)^2 + \lambda e^{6\alpha} - e^{4\alpha} = 0. \tag{6}$$

If we define a functional ψ_0 according to ($S_1 := \sum_n S_{1n}$)

$$\psi_0 := D(\alpha) e^{i(\sum_n S_{1n})} = \Pi_n \psi_{0n}, \tag{7}$$

where the Van Vleck determinant is defined by $D(\alpha)^2 = (\frac{\partial S_0}{\partial \alpha}) := S_0'$, order m_{Pl}^0 will imply that ψ_0 is the solution of the functional Schrödinger equation

$$-i \frac{\delta S_0}{\delta \alpha} \frac{\delta \psi_0}{\delta \alpha} \equiv i \frac{\delta \psi_0}{\delta t} = H_m \psi_0. \tag{8}$$

3. QUANTUM GRAVITY CORRECTIONS

Upon defining $S_2 = \sigma(\alpha) + \sum_n \eta_n(d_n, \alpha)$, we obtain the second-order correction for the wave function

$$\Psi^{(2)} = \frac{1}{D} e^{i m_{Pl}^2 S_0 + i\sigma/m_{Pl}^2} \Pi_n \psi_{n0} e^{i\eta_n/m_{Pl}^2}. \tag{9}$$

Herein σ and η_n, representing quantum gravity correcting terms to the WKB wave function [6] satisfy, after some algebra,

$$-S_0'\eta_n' = \frac{i}{2}\frac{\partial^2 \eta_n}{\partial d_n^2} + \frac{i}{\psi_{0n}}\frac{\partial \eta_n}{\partial d_n}\frac{\partial \psi_{0n}}{\partial d_n} + \frac{1}{2S_0'\psi_{0n}}\psi_{0n}' S_0'' - \frac{1}{2\psi_{0n}}\psi_{0n}'', \tag{10}$$

$$\sigma' = -\frac{S_0'''}{4S_0'^2} + \frac{3}{8}\frac{S_0''^2}{S_0'^3}. \tag{11}$$

We have moreover made use of the adiabatic approximation for ψ_{0n}, i.e, we picked up the ground state [2,3]:

$$\psi_{0n}(a, d_n) = N_n(a) e^{-\frac{1}{2} n a^2 d_n^2}. \tag{12}$$

If we restrict our calculation to the region where λ is really a constant, i.e., the early stages of the universe; we can opperate the expressions a little further. To this aim, let us make our predictions for the particular semiclassical evolution parameter given by

$$\tau = \lambda D(\alpha) = \lambda e^{2\alpha}(\lambda e^{2\alpha} - 1)^{1/2}. \tag{13}$$

Thus, for $\tau \ll 1$, upon factorizing Van Vleck determinant (so that $N_n(a) = \tau(a)$) and separating the pure time dependent part of η_n in the form $\eta_j(\tau, d_j) = \eta_{j1}(\tau, d_j) + \eta_2(\tau)\delta_{nj}$, we finally obtain, after Eq. (12),

$$\frac{\partial}{\partial \tau}[\sigma(\tau) + \eta_2(\tau)] = 0, \tag{14}$$

$$\frac{i}{\lambda}\frac{\partial \eta_{n1}}{\partial \tau} = \frac{1}{2}\frac{\partial^2 \eta_{n1}}{\partial d_n^2} - d_n n a^2(\tau)\frac{\partial \eta_{n1}}{\partial d_n}, \tag{15}$$

Therefore, in $\Psi^{(2)}$, the only physically relevant quantum gravity correcting phase factor is $\eta_{n1}(\tau, d_n)$.

The initial state of the universe is taken on the three-sphere $a^2 \to 1/\lambda$. In this case we can obtain an exact solution of Eq. (15) upon imposing normalizability of the corrected wave function

$$\eta_{n1}(\tau, d_n) = \frac{(2)^{1/2} g_1}{2} e^{in\tau/2}\left[2\left(\frac{n}{\lambda}\right)^{1/2} d_n \lambda\right] + g_2 e^{in\tau}\left[\left(\frac{4nd_n^2}{\lambda} - 2\right)\lambda\right]. \tag{16}$$

We notice that $g_1 \neq 0$ in Eq. (16) (i.e., a linear term in the phase of the n-mode wave fuction of gravitons) means that the expected initial number of gravitons is different from zero and it is given by

$$N = \lim_{a^2 \to 1/\lambda}\left|\frac{2g_1\lambda(n/\lambda)^{1/2} i}{m_{Pl}^2(2na^2)^{1/2}}\right|^2 = \left|\frac{g_1\lambda}{m_{Pl}^2}\right|^2. \tag{17}$$

4. THE SPECTRUM OF RELIC GRAVITONS

Since the initial number of gravitons is different from zero, it will also change the measurable properties of the resulting spectrum corresponding to the statistic of gravitons in inflationary models.

In the first approximation, the graviton spectrum produced by an inflationary stage is enterely independent of the mechanism that produces the inflation. The only inputs which are needed to find the graviton spectrum are the classical metric of space-time, and the initial quantum state of the gravitational perturbations.

Thus, taking into account that the initial quantum state was given by a N-number coherent state, it is found that the spectrum should be given by

$$P(\omega) = \frac{1}{4\pi^2}\frac{\lambda^2}{\omega}\left[\frac{a(t_1)}{a(t_0)}\right]^4 \epsilon(\omega), \tag{18}$$

where $a(t_1)/a(t_0)$ is the ratio of scale factor values corresponding to the begining and the end of the inflationary period.

On the other hand, $\epsilon(\omega)$ differs from Zel'dovich's spectrum due to the presence of quantum gravity effects:

$$\epsilon(\omega) \equiv 1-2(\lambda/m_{Pl}^2)\left\{2Re[g_1]\left[\frac{a(t_1)}{a(t_0)}\right]\frac{\omega}{\lambda^{1/2}} + Im[g_1]\left(1-2\left[\frac{a(t_1)}{a(t_0)}\right]^2\frac{\omega^2}{\lambda}\right)\right\} \tag{19}$$

If we take into account some phenomenological values of $[\frac{a(t_1)}{a(t_0)}] \sim 10^{28}$ (see e.g., [10]), we conclude that such effects would already be present for those frequencies of the order of $\omega \sim 10^{13}$ Hz.

The result in Eq. (18) may also be expressed in terms of the effective Hubble parameter during inflation, $H = \lambda^{1/2}$, and the recombination density $\rho_R = \rho_0[\frac{a(t_1)}{a(t_0)}]^4$ (recall $\frac{8\pi G}{3}\rho_0 = \lambda$),

$$\Omega_{graviton} \equiv \frac{\omega}{\rho_R}\frac{d\rho_g}{d\omega} = \frac{2}{3\pi}\left(\frac{H}{m_{Pl}}\right)^2 \epsilon(\omega), \tag{20}$$

where $\epsilon(\omega)$ in Eq. (20) represents a perturbation to the predicted Zel'dovich plateau.

5. CONCLUDING REMARKS

Cosmic background radiation is used in order to test phenomenological models of the early universe. In these scenarios, we have just shown that, quantum gravity corrections for the spectrum of gravitons, would possibly lie on the range of frequencies technically accessible in some future experimental devices measuring CBR anisotropy [8].

Semiclassical gravity, represented by $\Psi^{(2)}$, is in this framework a predictive and testable theory of initial conditions. On the other hand, we have also seen that Hartle-Hawking prescription for the wave function of the universe, leading the ground state for the wave function of gravitons, can be thought as being technically correct since, in this case, there would not exist divergencies for the resulting quantum gravity corrections on the initial three-sphere.

6. ACKNOWLEDGEMENTS

The author whishes to thank Claus Kiefer and José Luis Sánchez-Gómez for many useful comments. I also want to express my thanks to the Organizing Committee of this second Oviedo Conference on the Foundations of Quantum Physics. This work is supported by the grant EX95-08960718 from the Spanish Ministry of Education and Culture and the project C.I.C.y T. PB 94-0194.

REFERENCES

1. S. W. Hawking, "The quantum state of the universe," *Nucl. Phys. B* **239**, 257 (1984).
2. J. J. Halliwell, and S. W. Hawking, "Origin of structure of the universe," *Phys. Rev.* **31**, 1777 (1985).
3. S. Wada, "Quantum cosmological perturbations in pure gravity," *Nucl. Phys. B* **276**, 729 (1986).
4. V. A. Rubakov, M. V. Sazhin, and A. V. Veryaskin, "Graviton creation in the inflationary universe and the grand unification scale," *Phys. Lett. B* **115**, 189 (1983).
5. T. Banks, "Quantum gravity, the cosmological constant and all that," *Nucl. Phys. B* **249**, 332 (1985).
6. C. Kiefer, and T. P. Singh, "Quantum gravitational corrections to the functional Schrödinger equation," *Phys. Rev. D* **44**, 1067 (1991).
7. P. J. Steinhardt, "Cosmology confronts the cosmic microwave background," *Int. J. Mod. Phys. A* **10**, 1091 (1995).
8. N. Mandolesi, et al., "COBRAS/SAMBA: the ESA medium-size mission for measurements of CBR anisotropy," *Planet. Space Sci.* **43**, 1459 (1995).
9. J. B. Hartle, and S. W. Hawking, "Wave function of the universe," *Phys. Rev. D* **28**, 2960 (1983).
10. B. Allen, "Stochastic gravity-wave background in inflationary-universe models," *Phys. Rev. D* **37**, 2078 (1988).

QUANTUM PRIVACY AMPLIFICATION: A SECURE METHOD FOR CRYPTOGRAPHY

A. Sanpera and C. Macchiavello

Clarendon Laboratory, Oxford University
Parks Road, OX1 3PU Oxford, United Kingdom

In the present contribution we describe a quantum protocol, originally developed by Deutsch et al.[1] that allows to solve the problem of security of quantum cryptography over noisy channels. In contrast to all previous methods, where quantum mechanical effects were used only to detect eavesdropping and privacy amplification was performed in a classical way, our method exploits quantum phenomena both for detecting eavesdropping and for privacy amplification.

Key words: quantum cryptography, entanglement, purification.

1. QUANTUM CRYPTOGRAPHY IN NOISY CHANNELS

The basic idea of quantum cryptography is to exploit quantum effects, such as the Heisenberg uncertainty relations or the quantum correlations between two separate systems, to establish a common secret key between two parties, traditionally known as Alice and Bob [2-4]. To establish a secure random cryptographic key, Alice and Bob must have access to both a quantum communication channel (presumably noisy) and a public classical channel that allows them to exchange classical messages. These messages can be intercepted by an eavesdropper (Eve) but not altered. Under the existing protocols Alice and Bob can detect the presence of an eavesdropper (Eve) in the quantum channel by performing quantum measurements on subensembles of the set of transmitted particles to determine if there is some entanglement with an external system. To overcome the problem of eavesdropping, the existing quantum protocols have adopted classical privacy amplification techniques [5] that allow them to distill from the classical bits that Alice and Bob share a shorter but safer sequence of random bits that can then be used as the key to transmit the secret message. However, existing proofs of

M. Ferrero and A. van der Merwe (eds.),
New Developments on Fundamental Problems in Quantum Physics, 361–367.

the security of classical privacy amplification apply only to classical communication channels and classical eavesdropping. This means that the security of quantum cryptography has only been proved in the idealised case of a noiseless quantum channel in the absence of eavesdropping. In this contribution we present a new quantum protocol which is secure in the presence of noise and an eavesdropper. A more exhaustive description of the scheme can be found in reference [1].

2. DESCRIPTION OF THE PROTOCOL

The protocol that we propose is based on the distribution of pairs of entangled particles between Alice and Bob, as in the original entanglement based quantum cryptographic scheme [3], but the particles are not immediately measured to establish the key. The measurements are now performed as the last step of the protocol after Alice and Bob have applied the quantum privacy amplification technique to the previously stored particles. The aim of such procedure is to disentangle the distributed pairs from any external system (including an eavesdropper). In this way the external system will have no information at all about the outcomes of the measurements that Alice and Bob will later perform in order to establish the key. The QPA step is therefore the quantum analogue of classical privacy amplification procedures in the sense that it allows to distill a smaller fraction of "safer" pairs from an ensemble of corrupted pairs, but the ultimate security it provides comes from the fact that it is performed in a quantum mechanical fashion. Our protocol is composed of four basic steps:

(i) Distribution of entangled pairs.
(ii) Evaluation of the maximum information that an eavesdropper could obtain on the state of the distributed pairs.
(iii) Quantum privacy amplification.
(iv) Measurements to establish the secret key.

We shall mainly describe the second and third steps, which contain the new ingredients needed to guarantee a perfect security of the shared key. To allow for complete generality we analyse the scenario which is the most favorable for an eavesdropper, *i.e.*, where the eavesdropper herself is allowed to prepare all the qubit pairs that Alice and Bob will receive and they will subsequently use for cryptography. Any realistic situation involves environmental noise that is not under Eve's control, but this may be treated as a special case in which Eve is not using the full information available to her. The most general state resulting from Eve's preparation can be written as

$$|\Psi\rangle = c_0\,|00\rangle\,|R_0\rangle + c_1\,|01\rangle\,|R_1\rangle + c_2\,|10\rangle\,|R_2\rangle + c_3\,|11\rangle\,|R_3\rangle\,. \quad (1)$$

Here the states $|0\rangle$ and $|1\rangle$ of each particle of the pair distributed to Alice and Bob denote, for example, spin up and spin down states along the z axis, and the four states of Eve's system $|R_i\rangle$ are normalised but, in general, not orthogonal to each other. By preparing the state $|\Psi\rangle$, Eve predetermines the joint probability distribution of the four possible outcomes of the spin measurements along the z-axis and therefore she will have some a priori information about the actual bit values registered by Alice and Bob. For example, she knows that Alice will register bit value "0" with probability $|c_0|^2 + |c_1|^2$ and bit value "1" with probability $|c_2|^2 + |c_3|^2$ which gives her $1 - H(|c_0|^2 + |c_1|^2)$ bits of a priori information about Alice's result ($H(p) = -p\log p - (1-p)\log(1-p)$). Similarly she has $1 - H(|c_0|^2 + |c_2|^2)$ bits of a priori information about Bob's result. Eve can also measure her ancilla after Alice's and Bob's measurements and try to determine its relative state $|R_i\rangle$. In general Eve cannot reliably distinguish between those states because they are not orthogonal to each other and an upper bound on the amount of information she can acquire is given by Holevo's theorem [6]:

$$I_{\mathrm{acq}} = S(\varrho_R) = -\mathrm{Tr}\varrho_R \log \varrho_R\,, \quad (2)$$

where

$$\varrho_R = \sum_{i=0}^{3} |c_i|^2\,|R_i\rangle\langle R_i|\,. \quad (3)$$

Since the global state (1) is pure, for any set of relative states $\{|R_i\rangle\}$ the von Neumann entropy $S(\varrho_R)$ is equal to the von Neumann entropy $S(\rho)$ where ρ is the reduced density operator describing the state of the two particles $\rho = \mathrm{Tr}_{\mathrm{ancilla}}\,|\Psi\rangle\langle\Psi|$. Thus Eve's acquired information about the outcomes registered by Alice and Bob is bounded by

$$I_{\mathrm{acq}} = S(\rho). \quad (4)$$

The total information available to Eve is then bounded by the sum of the a priori and the acquired information. Alice and Bob can evaluate such bound by choosing at random a subensemble of their pairs and estimating ρ by evaluating spin correlations on this subensemble. If the resulting value of the total information exceeds a prefixed threshold and the resulting key cannot be considered secure, the QPA step is then applied. The way to estimate the maximum amount of information available to Eve that we have just described may not be the only or the most general one, but as we will now show the QPA procedure that Alice and Bob perform allows to reduce any bound to any arbitrarily small value. Therefore, the security of the protocol is completely independent of the criterion adopted to evaluate the degree of eavesdropping at step (ii).

For the sake of simplicity, let us now see how the QPA works in the case of an ensemble of pairs of spin 1/2 particles all described by the following density matrix

$$\rho = a\,|\,\phi^+\rangle\,\langle\phi^+| + b\,|\,\psi^-\rangle\,\langle\psi^-| + c\,|\,\psi^+\rangle\,\langle\psi^+| + d\,|\,\phi^-\rangle\,\langle\phi^-|, \tag{5}$$

namely a diagonal matrix in the Bell basis representation ($|\,\phi^\pm\rangle = \frac{1}{\sqrt{2}}(|\,00\rangle \pm |\,11\rangle)$ and $|\,\psi^\pm\rangle = \frac{1}{\sqrt{2}}(|\,01\rangle \pm |\,10\rangle)$). Alice and Bob have first to divide the set of particles they receive into groups of two particles each and perform the following simple local operations that represent the basic step of our QPA algorithm. Alice performs a rotation by $\pi/2$ along the x axis on both her particles and Bob performs a rotation in the opposite direction $R_{-\pi/2}$ along the x-axis on his two particles. Then both Alice and Bob perform a quantum Controlled-NOT operation [7] where one of the pairs gives the two control qubits and the other one the two target qubits. Such operation acts on two qubits and gives

$$|\,x\rangle\ |\,y\rangle \longrightarrow |\,x\rangle|\,x \oplus y\rangle, \qquad (x,y) \in \{0,1\}, \tag{6}$$

where the first qubit is the control and the second the target and $\oplus$ is addition modulo 2. Subsequently Alice and Bob measure the spin z-component of the target qubits and publicly exchange the results they have obtained. If the outcomes coincide they regard the operation as successful and keep the control pair for the second round. If the outcomes do not coincide they agree to discard the two pairs.

We have now described a single step of the QPA procedure. If it is successful the remaining pairs will still be described by a density matrix of the form (5), still diagonal in the Bell basis representation, with diagonal elements $\{a', b', c', d'\}$ given by

$$a' = (a^2 + d^2)/p, \tag{7}$$
$$b' = 2bc/p, \tag{8}$$
$$c' = (c^2 + b^2)/p, \tag{9}$$
$$d' = 2ad/p. \tag{10}$$

Here $p = (a+d)^2 + (b+c)^2$ is the probability of success, that is that Alice and Bob obtain coinciding outcomes in the measurement on the target pair. If Alice and Bob have sufficiently many pairs at their disposal the procedure just described can be repeated with those pairs which survived the first filtering. The aim of the procedure is to drive the surviving pairs as close

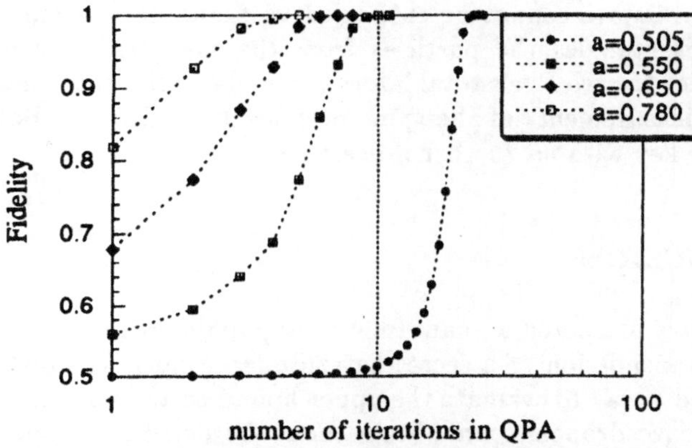

Fig. 1. Average fidelity as a function of the number of iterations for four different values of the initial fidelity a_0.

as possible to the original uncorrupted state $|\phi^+\rangle$, at the cost of sacrificing some of the distributed pairs (the target pairs and the control pairs that are discarded after the measurements). We take the fidelity

$$F = \langle \phi^+ | \rho | \phi^+ \rangle \tag{11}$$

as a measurement of the closeness of the state ρ to the uncorrupted state $|\phi^+\rangle\langle\phi^+|$ [8]. Such quantity is given by the first diagonal element a (or a') of the density matrix in the Bell basis representation. If the initial fidelity a_0 is greater than 0.5 then the average fidelity increases in the process and asymptotically goes to one [1,8].

In order to check that such condition is fulfilled before starting the QPA procedure Alice and Bob can estimate the initial average fidelity a_0 by measuring spin correlations on a randomly selected set of distributed pairs. The fidelity is given by

$$a = \frac{1}{4}(1 + \langle \sigma_x \sigma_x \rangle - \langle \sigma_y \sigma_y \rangle + \langle \sigma_z \sigma_z \rangle). \tag{12}$$

In Fig. 1 we plot the fidelity as a function of the number of iterations for four different initial states of the form (5) with $b = c = d$. As we can see, the convergence to the unit value is very fast and few iterations of the procedure are needed in practice.

When the fidelity approaches 1 the average density operator of the remaining pairs necessarily approaches the pure state $|\phi^+\rangle\langle\phi^+|$ and the joint state of Eve's ancilla and the particles necessarily approaches a product state of the form

$$\left|\tilde{\Psi}\right\rangle = 1/\sqrt{2}(|00\rangle + |11\rangle)\,|R\rangle, \tag{13}$$

because a pure state cannot be entangled with another system. The QPA procedure disentangles the particles from the eavesdropper's ancilla and clearly reduces to zero Eve's total information about the outcomes of Alice's and Bob's measurements of the spin components. Alice and Bob can now establish the key without further processing.

3. CONCLUSION

We have presented a quantum cryptographic protocol that allows for the secure transmission of a secret key through a noisy channel. We have first discussed a way to evaluate the upper bound on the amount of information that an eavesdropper could obtain about the distributed signals between the two legitimate users. We have then presented a "quantum privacy amplification" procedure to be applied on the distributed pairs before the key is established. Although the procedure has been explicitly discussed in the simplest case of an ensemble of distributed pairs all described by the same density matrix (diagonal in the Bell basis), the procedure can be straightforwardly generalized for more complex initial preparations [1]. The method guarantees a complete secrecy of the key that is subsequently established.

We would like to thank David Deutsch, Artur Ekert, Richard Jozsa and Sandu Popescu for a joint work on quantum privacy amplification, which we reviewed here. C.M. is sponsored by the European Union HCM Programme. A.S. is financially supported by the U.K. Engineering and Physical Science Research Council. This contribution has also been presented in the IV Central European Workshop on Quantum Optics held in Budmerice, Slovaquia, in June 1996.

REFERENCES

1. D. Deutsch, A. Ekert, R. Jozsa, C. Macchiavello, S. Popescu, and A. Sanpera, *Phys. Rev. Lett.* **77**, 2818 (1996).
2. C.H. Bennett and G. Brassard, in *Proceedings, IEEE International Conference on Computers, Systems and Signal Processing*, Bangalore, India (IEEE, 1984), p.175.
3. A.K. Ekert, *Phys. Rev. Lett.* **68**, 661 (1991).
4. C.H. Bennet, *Phys. Rev. Lett.* **68**, 3121 (1992).
5. C.H. Bennett, G. Brassard, and J.-M.Robert, *SIAM J. Comp* **17**, 210 (1988).
6. A.S. Holevo, *Problems of Information Transmission* **9**, 177 (1973).
7. A. Barenco, D. Deutsh, A. Ekert, and R. Jozsa *Phys. Rev. Lett.* **74**, 4083

(1995).
8. C.H. Bennett, G. Brassard, S. Popescu, B. Schumacher, J. Smolin, and W.K. Wootters, *Phys. Rev. Lett.* **76**, 722 (1996).

THE PROGRAM OF LOCAL HIDDEN VARIABLES

Emilio Santos

Departamento de Física, Universidad de Cantabria
Santander, Spain

Arguments are given supporting the ensemble (or statistical) interpretation of quantum mechanics and locality. A critical analysis is made of the performed empirical tests of Bell's inequalities, and it is stressed that no real violation has been shown, so that the possibility of local hidden variables is still open. A restriction on the postulates of quantum mechanics is proposed which might make the theory compatible with local hidden variables.

Key words: hidden variables, locality, Bell's theorem, quantum noise.

1. THE ENSEMBLE INTERPRETATION OF QUANTUM MECHANICS

1.1. Space-Time Description in the Various Interpretations

Seventy years after the discovery of quantum mechanics the debate about the interpretation is still alive. The variety of interpretations may be illustrated with the different views about the space-time description of quantum processes. Let us consider, for instance, the alpha decay of an atom in a $J = 0$ state, with detectors all around in a 4π geometry. The state of the alpha particle before detection should be represented by a Ψ-function in the form of a spherical wave. At the moment of detection, the Ψ-function suddenly "collapses" and becomes localized near the detection point.

The most obvious (but naïve?) interpretation of the Ψ-function is that it represents just our information, i.e., it has the same role as the probability $p = 1/6$ that we attach to every possible outcome in playing dice. Thus the collapse of Ψ-function is just a change of our information as a consequence of an observation. This view of quantum mechanics has been called "statistical" [1] or "ensemble" interpretation and it was supported by

M. Ferrero and A. van der Merwe (eds.),
New Developments on Fundamental Problems in Quantum Physics, 369–379.

Einstein [2] amongst others. However, the ensemble interpretation is not free from difficulties, and it is a fact that nobody has been able to develop it in a consistent and complete form. The main purpose of the present article is to present, and defend the possibility of, a program aimed at a full development of the ensemble interpretation.

The difficulties of the ensemble interpretation have caused that several alternatives have become more popular. The main part of the scientific community supported, until recently, the Copenhagen interpretation proposed by N. Bohr. According to this interpretation, which is pragmatic and positivistic, quantum mechanics does not provide a space-time picture of the microscopic world, but only a set of rules for the prediction of the results of observations or experiments. The referent of the theory is not the microscopic system, but the conjunction of the system and an external oberver. However when the physical system is the whole universe there is no external observer. Consequently, the increasing interest for quantum cosmology has made the Copenhagen interpretation obsolete in the opinion of people working in foundations, although it is still supported by most of the physicists using quantum mechanics as a tool.

The most fashionable interpretation in the last few years, at least among physicists interested in fundamental problems, is the popularly named "many worlds" interpretation, which essentially consists of assuming the unrestricted validity of quantum mechanics even for the universe as a whole. In the modern form, this interpretation is supplemented with the decoherence approach [3], which takes into account the fact that no physical system (except the universe itself) is fully isolated. The idealization of perfect isolation may be a good approximation for microscopic systems, but certainly not for large and complex systems. Thus, even if the universe is adequately represented by a Ψ-function or, equivalently, by an idempotent density matrix, the density matrix appropriate for a laboratory system should be obtained by taking the partial trace of the (idempotent) density matrix of the universe with respect to the environement (that is, everything except the system at hand). In this form we get, for macroscopic systems, a density matrix interpretable as a mixture. As is well known [3] this solves, at least in practice, the problems put by the collapse of the Ψ-function. With this interpretation the quantum infinite-dimensional Hilbert space is, in some sense, more fundamental than ordinary three-dimensional space. The latter emerges at the macroscopic limit as a consequence of the decoherence induced by the environment.

1.2. Ensemble Interpretation and Local Hidden Variables

Returning to the ensemble interpretation, it is my opinion that it requires hidden variables. Indeed if the Ψ-function refers to an ensemble, it is necessary to characterize the different elements of the ensemble by means of suitable parameters, and these are the hidden variables. For instance, in the example of the atomic decay, above mentioned, each possible decaying alpha particle may be described by the time of emission and a unit vector fixing the direction of motion and these will be the hidden variables (assuming a straight line motion of the particle after decay, with velocity fixed by conservation of energy and momentum). In summary, hidden variables are a natural consequence of the ensemble interpretation, although it might be difficult, or even impossible, to introduce them explicitely.

But I also think that the hidden variables should be local. The arguments for locality were very clearly expressed by Einstein. I shall quote just a sentence from his famous authobiography:

> On one supposition we should, in my opinion, hold absolutely fast: the real factual situation of the system S_2 is independent on what is made with the system S_1 which is spatially separated from the former. (Ref. 2, p. 85.)

This locality condition was very well captured by John Bell [4], and I shall always speak about local hidden variables (LHV) in the sense of Bell.

Any reader of the present article knows that there are two very strongs objections against the possibility of LHV theories, namely

(i) They have been empirically refuted, via the violation of Bell's inequalities, in the experiments using entangled photon pairs produced in atomic cascades [5] or, more recently, parametric down-conversion [6,7].

(ii) Bell's theorem forbids the existence of LHV theories [4,8].

In the following I argue against these difficulties.

2. CRITIQUE OF THE EXPERIMENTAL TESTS OF BELL'S INEQUALITIES

2.1. What Bell's Inequaltities Have Been Violated?

In all experimental tests of Bell's inequalities there is a source where particle pairs (most frequently photons) are produced. Each particle, which we shall label 1 or 2, may arrive at a detector after crossing some appropriate device (e.g., a polarization analyzer) with a parameter, ϕ_1 or ϕ_2, which may be set at will by the experimenter. These parameters may be, for instance, the angles of the polarizing planes of the analyzers. The recorded data are single detection rates, R_1 and R_2, and coincidence rates R_{12}. In practice the former are independent of ϕ_1 and ϕ_2 and the latter are given by

$$R_{12} = R_{ave}\left[1 + V\cos(\phi_1 - \phi_2)\right], \tag{1}$$

where R_{ave} is the coincidence rate averaged over the paramteters ϕ_1 and ϕ_2, and V a constant measuring the polarization correlation. We shall consider also the production rate, R_O,usually not measurable.

In order to see whether LHV theories have been refuted by the experiments, we should check whether it is impossible to find functions $(\rho(\lambda), P_1(\lambda,\phi_1)$ and $P_2(\lambda,\phi_2)$, fulfilling Bell's local hidden variables assumptions

$$\rho(\lambda) \geq 0, \quad \int \rho(\lambda)d\lambda = 1, \quad 0 \leq P_j(\lambda,\phi_j) \leq 1, \quad j = 1,2, \tag{2}$$

and such that the measured single and coincidence rates may be obtained from the integrals

$$R_j = R_O \int \rho(\lambda)P_j(\lambda,\phi_j)d\lambda, \quad R_{12} = R_O \int \rho(\lambda)P_1(\lambda,\phi_1)P_2(\lambda,\phi_2)d\lambda. \tag{3}$$

As is well known, such functions will not exist, and LHV models will be impossible, if a Bell inequality is violated. We shall consider the inequality given by Clauser and Horne [8] (see their Eq. (4)), which reads

$$\eta(1 + \sqrt{2}V) \leq 2, \eta = 4R_{ave}/(R_1 + R_2). \tag{4}$$

It is a fact that this inequality has never been violated because, although values of V greater than 0.98 have been obtained, the parameter η has never surpassed a few per cent. Why it is then claimed that LHV theories have been refuted? The truth is that a genuine Bell inequality like (4) (which is necessarily inhomogeneous, involving both single and coincidence rates [5]) has never been put to a test. The tested inequalities have been homogeneous (i.e., involving only coincidence rates), most frequently the following (easily derivable from Eq. (11) of Ref. 8, using our Eq. (1)):

$$1 + \sqrt{2}V \leq 2 \Rightarrow V < 0.71. \tag{5}$$

It is well known, but rarely mentioned, that the derivation of an inequality like (5) requires additional assumptions, like "no-enhancement" [8]

$$P_j(\lambda,\phi_j) \leq P_j(\lambda,\infty), \tag{6}$$

where $P_j(\lambda,\infty)$ is the detection probability when the corresponding polarizer has been removed. We shall conclude, therefore, that only the restricted family of LHV theories fulfilling the additional assumptions have been refuted, but other kinds of LHV theories are still possible.

At this moment it is worth mentioning that a genuine Bell inequality should always involve the measurement of two observables, one of them being the position. In fact, without position measurements locality cannot be tested. In the inequality (4) the position observables are apparent through the detection rates, but in the inequality (5) any reference to position has been lost. This shows that the polarization correlation measured in the tests of (5) refers to a subensemble of the photon pairs produced (the subensemble where both photons were detected). But it is known that the violation of Bell's inequalities in subensembles (post-) selected at the time of measurement is compatible with LHV models [5]. In summary, a genuine Bell inequality only forbids that the position correlation and the polarization correlation are simultaneously high, whilst the empirically violated inequality (5) gives a much stronger constraint.

2.2. Loopholes and Conspiracies of Nature

The fact that performed experiments have not refuted a (genuine) Bell inequality is well known to the experts, but there is the widespread belief that it is only due to the nonideality of the measuring devices (in particular the low efficiency of available photon counters). Furthermore, the loophole is considered a minor practical problem that surely will be solved in the future. This belief has been expressed by many people, including John Bell:

> It is hard for me to believe that quantum mechanics works so nicely for inefficient practical setups and is yet going to fail badly when sufficient refinements are made. (Ref. 4, Chap. 2.)

We see that, according to a common belief, the empirical results so far obtained are sufficient to disprove all LHV theories, except if there is a kind of conspiracy in nature. There are two general objections to that belief. Firstly, ideal preparations and measurements do not exist in physics, even less in quantum physics where noise is an essential ingredient, as we shall stress at the end of this article. Second, in the history of physics there are examples of believed "conspiracies of nature" that turned out to be the consequence of fundamental principles. For instance, the impossibility of the *perpetuum mobile* has the roots in the principle of conservation of energy, and the impossibility to measure the absolute velocity of Earth in the principle of relativity. I conjecture that an equally important principle—local realism—may lie at the roots of the difficulty (impossibility?) to violate a genuine

Bell inequality, as shown by the failure of repeated efforts during the last 30 years.

But there are even more clear arguments for the relevance of the loopholes in the performed experiments. In fact, there exist LHV models in agreement with the quantum predictions, *even with ideal measuring devices* (polarizers, detectors, etc.) for all performed tests of Bell's inequalities [5,7]. For instance, in atomic cascade experiments the ideal quantum predictions are

$$R_1 = R_2 = 1/2(\Omega/4\pi)R_0, \quad R_{12} = 1/4(\Omega/4\pi)^2 R_0 \left[1 + F\cos 2(\phi_1 - \phi_2)\right], \tag{7}$$

where Ω is the solid angle covered by the apertures of the lens system and F a depolarization factor. It is a fact that Ω and F are related in such a way that F decreases when Ω increases and the following inequality always hold [5]:

$$\Omega F < 1.2 \quad \Rightarrow \quad \eta V < 0.2, \tag{8}$$

As a consequence the Bell inequality (4) is fulfilled even if the polarizers and detectors are ideal, and LHV models can be constructed for these experiments [5].

From another point of view we may recall that the state of the photon pair produced in the source is

$$|\Psi\rangle = \Sigma c\,(p, k_1, \varepsilon_1, k_2, \varepsilon_2)\,|p\rangle|k_1, \varepsilon_1\rangle|k_2, \varepsilon_2\rangle, \tag{9}$$

where $k_1(k_2)$ is the wavevector and $\varepsilon_1(\varepsilon_2)$ the polarization of the first (second) photon in the pair and p is the momentum of the recoil atom. If the atom were detected, as we have proposed recently [9], the two-photon state associated with a detected atom might violate a Bell inequality, but if the atom is not detected, as in the performed experiments, the two-photon state should be represented by a density matrix obtained from (9) after taking the parcial trace with respect to the (not measured) atom variables. That density matrix does not violate any (genuine) Bell inequality.

It might be argued that the apertures produce a collapse of the state to an entangled two-photon state, but this is also not true. Actually the collapsed state is a mixture of vacuum state, plus single photon states, plus entangled two-photon states, and this mixture does not violate Bell's inequalities either. It should be stressed that the "ensemble of photon pairs both passing the apertures", frequently mentioned as a justification for the claim that LHV theories have been refuted, is not a quantum state. The assumption that such ensemble exists derives from a naïve corpuscular model of the photon. It is the corpuscular model of the photon which is refuted by the performed experiments.

The situation is not better with parametric down-conversion experiments [6,7]. In this case even the ideal quantum prediction (the homologous to (7)) has not yet been calculated. Only the limits $V \rightarrow 1, \eta \rightarrow 0$ and $V < 1/2, \eta \rightarrow 1$ are known.

3. RESTRICTED QUANTUM THEORY

We go to the second difficulty for LHV theories, namely the incompatibility with quantum mechanics stated by Bell's theorem. For a few people, like the present author, it is unbelievable that local realism does not hold in nature. But for most authors what is unbelievable is a failure of quantum mechanics. I also think that a conflict between quantum mechanics and future experiments, if not unbelievable, is quite improbable. Therefore, I shall make every effort to discover loopholes in the proof of Bell's theorem.

In order to prove the incompatibility of quantum mechanics with LHV theories it is necessary to define rigorously the two terms of the comparison. A clear definition of LHV theories has been given by Bell, as said above, but strangely enough the definition of quantum mechanics is not so clear. Quantum mechanics is defined by several postulates and they will be the subject of our attention in what follows.

3.1. The Quantum Postulates Are Too Strong

The standard postulates of QMy w may be grouped in three classes:

(i) Equations of motion (Schrödinger, Dirac, etc.).

(ii) Correspondence between vectors and states and between self-adjoint operators and observables.

(iii) Postulates of the measurement (e.g., the assumption that the measurement of the observable A gives eigenvalue a_j with probability $|\langle\Psi|\Phi_j\rangle|^2$)

But we argue that QM does not loose its (truly spectacular) predictive power if we restrict the set of postulates according to the rule "Retain (i), restrict (ii), reject (iii)," as explained in the following. We propose the name "restricted quantum theory" for the theory obtained in that way.

Of course "reject (iii)" does not mean that the measurement "postulates" should be considered radically false. They are certaily correct to a good approximation, but maybe not exactly. And, in any case, they should not be postulated but derived as theorems. Indeed, if quantum mechanics is a fundamental theory of nature, measurement should be explained like any other physical process, without specific postulates. On these grounds the existence of measurement postulates has been critiziced by philosophers like Popper or Bunge and by leading scientists like Einstein, Schrödinger (remember his cat paradox) and, in more recent times, John Bell. As a result it is widely accepted today that such postulates are neither necessary nor convenient. What is not so widely realized is that the postulates of measurement are required for the standard proofs of Bell's theorem. A correct proof, not involving the measurement postulates, should consist of the design of a (loophole-free) experiment truly able, in principle, to exhibit a contradiction between LHV theories and quantum mechanics. Consequently, the best proof of Bell's theorem is, at the moment, the experiment proposed by Fry et al. [10].

Going to postulates of class (ii), it is usually assumed that the corresponcences states-vectors and observables-operators both are one to one, but this assumption is unreasonably strong. In fact, in any theory of physics it may be appropriate to demand that there exists an element of the theory corresponding to every element of reality, in order that the theory is complete, but not viceversa.

3.2. Two Examples from Classical Physics Against One-One Correspondences

Two examples taken from classical physics may illustrate the argument against the one to one correspondences above commented. In classical mechanics each possible state of a system is represented by a point in phase space, but obviously not all points may be realized in practice. For instance, it is beyond experimental capabilities to prepare a system in a state corresponding to a point with irrational coordinates, because any measurement gives a rational number.

The second example is the diffusion equation in one dimension,

$$\partial f/\partial t = \partial^2 f/\partial x^2, \tag{10}$$

$f(x,t)$ being the density of the diffusing matter. With boundary condition $f(0,t) = f(\pi,t) = 0$, the solution can be always written

$$f(x,t) = \Sigma c_n \sin(nx) \exp(-n^2 t), \tag{11}$$

where the coefficients c_n may be obtained from the initial condition $f(x,0)$. We need all the $f_n = \sin(nx)$ with integer n, in order to be able to express the solution (11) of the problem in the form of a Fourier series (this requirement is called completeness of the basis), but it is obvious that not all series of the form (11) may represent physcal states. In particular, the functions $\sin(nx)$ themselves are not positive definite (except for $n = 1$) and cannot represent densities. Similarly, we should assume that some vectors in Hilbert space (Ψ-functions) may not correspond to physically realizable states although all these vectors are necessary in order to represent every possible physical state as a linear combination of them. In particular we might conjecture that Ψ-functions violating a genuine Bell inequality are not physically realizable.

If the one to one correspondences above discussed are not true, even the proposal of a loophole-free experimental, like Ref. 10, cannot be considered a rigorous proof of Bell's theorem. Actually "Bell's theorem" should not be considered a theorem but a working hypothesis, extremely usefull because it *may* allow the empirical test of LHV theories.

3.3. The Interpretation of Restricted Quantum Theory

In the ensemble interpretation there is no fundamental distinction between vectors in Hilbert space (or Ψ-functions) and density operators (ρ), because both may represent mixtures. Consequently the wavefunction Ψ loses its privileged role. In order to make possible an ensemble interpretation we should postulate that states are represented by density operators, without distinguishing pure states (represented by Ψ-functions or idempotent density operators) from mixed states (represented by non-idempotent density operators). On the other hand, as explained in the previous subsection, we should not postulate that all density operators represent states. Well known restrictions are the superselection rules. Constraints of the form $Trf(\rho) > 0, f$ being some appropriate function, are also derived from the decoherence approach for open systems. But additional restrictions might also exist. Searching for them should be an essential part of a LHV research program.

In summary, I conjecture that restricted quantum theory as defined above may be compatible with LHV theories. This conjecture would be refuted by the violation of a genuine Bell inequality in future experiments (e.g., Ref. 10). But our experience with the performed tests dictates that a very carefull analysis of the results of the experiments will be required before all the loopholes are truly blocked.

4. THE RELEVANCE OF NOISE

Quantum theory, interpreted as a LHV theory within the ensemble interpretation, is a stochastic theory. The essential ingredient of any stochastic theory is noise, but in quantum mechanics noise is not apparent. We shall try to understand this difficulty.

Let us consider a classical oscillator with equation of motion

$$d^2x/dt^2 + 2\gamma dx/dt + \omega_o^2 x = f(t), \tag{12}$$

$f(t)$ being some colored (non-white) noise. A possible solution is a "probability packet" with center moving according to the equation

$$x_o(t) = A\cos(\omega_\gamma t + \phi)\exp(-\gamma t), \quad \omega_\gamma = \sqrt{(\omega_o^2 - \gamma^2)}. \tag{13}$$

We may compare this solution with the coherent state of a quantum oscillator. It is a "wavepacket" with center moving according to

$$x_o(t) = A\cos(\omega_o t + \phi). \tag{14}$$

A comparison of (13) with (14) shows that in the classical case there is fluctuation (spreading of the "probability" packet) and dissipation (giving rise to the exponentially decaying term exp $(-\gamma t)$). In contrast, in the quantum case we have fluctuation without dissipation! It seems that there is a sharp constrast between quantum noise (non-dissipative) and classical noise (dissipative).

However, if we go from quantum *mechanics* to quantum *field theory* the difference between classical and quantum noise is not so sharp. In fact, quantum electrodynamics of the (charged) harmonic oscillator predicts for the coherent state a wavepacket with center moving like [11]

$$x_o(t) = A\cos(\omega_\gamma t + \phi)\exp(-\gamma t). \tag{15}$$

The similarity with the classical oscillator is now remarkable. In quantum field theory noise is accompanied by dissipation!. It may be argued that (15) requires to treat the system as open, allowing fot the "photons" emitted by the oscillator to escape to infinity. But also classical dissipation is associated to open systems. The moral of this example is the following: we should not try to understand quantum mechanics, but quantum field theory (involving infinitely many degrees of freedom).

We have argued that noise is essential for a LHV interpretation of quantum theory, but noise is also what makes difficult the violation of the Bell inequalities. In the first place we should take into account that, in the microscopic domain, measurements do not consists of just revealing the values of observables. They are usually rather indirect procedures to relate an input (e.g., a light beam) with an output (e.g., a pulse of electric current in a photomultiplier). The relation is obviously contaminated by noise. There are false negatives (we say that photodetectors are less than 100% efficient) and false positives (e.g., dark counts). It is not realistic to hope that the efficiency may be increased without increasing, at the same time, the dark counts. The argument may be extended to any preparation or measurement. This is why I conjecture that the combination of all sources of noise, usually not taken into account in standard quantum calculations, may prevent the violation of the Bell inequalities, if no noise subtraction is made.

I am aware that most of the readers will not take seriously the conjecture but for me the principles of realism and locality are so fundamental that I cannot believe that nature does not respect them.

Acknowledgement.

I acknowledge financial support of DGICYT Project No. PB95-0594 (Madrid).

REFERENCES

1. L. E. Ballentine, "The statistical interpretation of quantum mechanics," *Rev. Mod. Phys.* **42**, 358-381 (1970).
2. A. Einstein, "Autobiographical Notes," in *Albert Einstein, Philosopher Scientist*, P. A. Schilpp, ed. (Open Court, Evanston, 1949).
3. W. H. Zurek, "Decoherence and the transition from quantum to classical," *Phys. Today* **44** (10), 36 (1991).
4. J. S. Bell, *Speakable and Unspeakable in Quantum Mechanics* (Cambridge University Press, Cambridge, 1987).
5. E. Santos, "Critical analysis of the empirical tests of local hidden-variables theories," *Phys. Rev. A* **46**, 3646-3656 (1992).
6. R. Y Chiao, P. G. Kwiat, and A. M. Steinberg, "Quantum nonlocality in two-photon experiments at Berkeley," *Quantum Semiclass. Opt.* **7**, 259-278 (1995).
7. E. Santos, "Unreliability of performed tests of Bell's inequality using parametric down-converted photons," *Phys. Lett. A* **212**, 10-14 (1996).
8. J. F. Clauser and M. A. Horne, "Experimental consequences of objective local theories," *Phys. Rev. D* **10** 526-535 (1974).
9. S. F. Huelga, M. Ferrero, and E. Santos, "Loophole-free test of the Bell inequality," *Phys. Rev. A* **51**, 5008-5011 (1995).
10. E. S. Fry, T. Walther, and S. Li, "Proposal for a loophole-free test of the Bell inequalities," *Phys. Rev. A* **52**, 4381-4395 (1995).
11. E. Santos, "The harmonic oscillator in stocahstic electrodynamics," *Nuovo Cimento* **19B**, 57-89 (1974).

NONINVARIANT VELOCITY OF LIGHT AND LOCALLY EQUIVALENT REFERENCE FRAMES

F. Selleri

Università di Bari, Dipartimento di Fisica
INFN - Sezione di Bari
I-70126 Bari, Italy

The large lifetime increase of muons observed in storage ring experiments is independent of muon acceleration. We assume this to be a manifestation of a general principle: only velocities matter for length contraction and time dilation phenomena. Therefore *local* measurements of space, time and velocity in an accelerated reference frame must give the same results as those in a "locally equivalent" inertial frame having the same instantaneous velocity. This principle is used for deducing the velocity of light both along the rim of a rotating platform and relative to inertial reference frames, and leads to noninvariant values. A set of transformations of space and time different from the Lorentz transformations can be deduced.

Key words: special relativity, velocity of light, conventionality.

1. TIME TRANSFORMATION TO A ROTATING DISK

Lifetimes of positive and negative muons were measured in the CERN Storage-Ring experiment [1] for muon speed 0.9994 c, corresponding to a γ factor of 29.33. Muons circulated in a 14 m diameter ring with an acceleration of 10^{18}g. Excellent agreement was found with the formula

$$\tau_0 = \frac{\tau_{\text{rest}}}{\sqrt{1-\beta^2}}, \tag{1}$$

where τ_0 is the observed muon lifetime, τ_{rest} is the lifetime of muons at rest, and $\beta = v/c$, v being the laboratory speed of the muon on its circular orbit.

Consider an ideal platform rotating with the same angular velocity as the muon in the e.m. field (with respect to such a platform the muon is at rest). Consider also four different observers:

M. Ferrero and A. van der Merwe (eds.),
New Developments on Fundamental Problems in Quantum Physics, 381–394.

1. O_0 is the observer in the laboratory reference frame S_0, assumed to be an inertial frame. Thus O_0 be the CERN storage ring experimenter.

2. $\tilde{O}_a$ is an observer standing on the platform. He knows everything about the platform (the accelerated frame S_a) through which he can move freely;

3. O_a is a second observer on the platform S_a. He stands very near the muon which, to him, appears to be at rest; O_a has a local knowledge of the platform and of its physical properties extending only to the immediate surroundings of his position;

4. O_T observes from an inertial frame S_T, with respect to which O_a is instantaneously at rest. S_T will be called the "tangent" inertial frame.

We next consider the muon lifetime according to these four observers. According to O_a the muon lifetime τ_0 is greatly enhanced with respect to that (τ_{rest}) of muons at rest in S_0 and is given by Eq. (1). According to the accelerated observer $\tilde{O}_a$ the clocks on the platform have a pace dependent on their distance from the centre, the fastest one being at the centre; in agreement with the equivalence principle he attributes this phenomenon to the presence of a position-dependent radial gravitational field of cosmic origin. He can check that the lifetime of muons near the rim of the platform is either τ_0 or τ_{rest} depending on the clock chosen (in the centre or near the rim, respectively) for measuring it. Therefore he explains the value τ_{rest} found by O_a to be a consequence of a cosmic gravitational field which delays in the same way muon decay and the clock used by O_a for lifetime measurements. According to O_a, who knows only the time shown by his local clock, the muon lifetime is τ_{rest}. Of course O_a is under the action of a large acceleration (10^{18}g), but nevertheless his lifetime measurements give just τ_{rest}, as for muons at rest in S_0 observed by O_0. Finally, according to the observer O_T belonging to the tangent inertial frame S_T the lifetime is τ_{rest} (measured of course on muons at rest in his frame).

The first conclusion to be drawn concerns the transformation of time given by Eq. (1) which can be extended to all kinds of processes. The laboratory time interval Δt_0 between any two events taking place at a given spot on the rotating disk (muon injection and decay, in the last example) is seen dilated by the well-known factor, compared with the corresponding time interval Δt measured by the accelerated observer O_a:

$$\Delta t_0 = \frac{\Delta t}{\sqrt{1-\beta^2}}. \tag{2}$$

For example, the pace of a clock C_a placed on the rim of the disk would be slowed down according to Eq. (2), if compared with clocks at rest in S_0. And a light signal propagating along the rim (e.g., on the internal surface of a cylindrical mirror) from a point P near C_a and back to P on a circular path, would require a time interval measurable both by the clock C_a itself and by the laboratory clocks, thus again satisfying Eq. (2). Notice that Eq. (2) holds unmodified, irrespective of muon position on the circular orbit: in other words time is transformed from the laboratory to the rim of the disk without a space-dependent time-lag term.

The second conclusion is that the observers O_a and O_T agree on the results of measurements of space and time, for example on the decay rate of muons at rest, even though O_a feels the presence of a radial gravitational field (of cosmic origin) while O_T does not. Of course this conclusion is not new, even though it was never used systematically. Already Einstein in his 1911 article [2] assumed that the acceleration of a clock C_a relative to an inertial system has no influence on the rate of C_a, and that the increase in proper time of C_a at any time is the same as that of the standard clocks of the inertial system in which C_a is momentarily at rest. A similar conclusion clearly holds for length: a measuring unit rod placed parallel to the circumference of the disk undergoes a contraction dependent only on the rotational velocity and indifferent to the centrifugal acceleration. Therefore its length is the same as that of another unit rod at rest in S_T set parallel to the velocity of S_T with respect to S_0. The identical conclusions of O_a and O_T concerning their measurements of time and space imply that the one-way speed of light found locally along the rim of the rotating disk should be the same as that observed in the "tangent" inertial frame. But in Einstein's special theory of relativity (STR) the latter speed is always c and we are so brought to conclude that the speed of light along the rim of the rotating disk should also be c. However this conclusion is not acceptable, as shown in the third section.

2. IS CLOCK SYNCHRONISATION CONVENTIONAL?

When Einstein [3] formulated his theory of relativity he *postulated* that the velocity of light has always the same value c. There seems to be agreement among many physicists on the idea that the constancy of the one-way velocity of light is a useful convention and that it must not be considered as something dictated by objective properties of the natural world. E.g. see: Mansouri and Sexl [4] Grünbaum [5], Jammer [6], Sjödin [7], Cavalleri [8], Ungar [9], and Vetharaniam and Stedman [10]. The idea is however rather old.

Already in 1898 Poincarè [11], discussing the independence of the speed of light of its propagation direction stated: "This is a postulate without which it would be impossible to start any measurement of this velocity. It will always be impossible to verify experimentally the said postulate." Similarly in 1916 Einstein [12] wrote: "that light requires the same time to traverse the path AM ... as the path BM [M being the midpoint of the line AB] is in reality *neither a supposition nor a hypothesis* about the physical nature of light, but a *stipulation* which I can make of my own free will." In 1925 Reichenbach [13] considered the following situation: in an inertial system S a flash of light leaves point A at time t_1, is reflected back in point B at time t_2, and arrives again in A at time t_3. The problem is how to synchronise the clock near point B with the clock near point A. In the theory of relativity it is *assumed* that the one-way velocity of light has the same value from A to B as from B to A, so that $t_3 - t_2 = t_2 - t_1$, whence the clock-B time t_2 can be written in terms of the two clock-A times t_1 and t_3 as follows:

$$t_2 = t_1 + \frac{1}{2}(t_3 - t_1) \tag{3}$$

Reichenbach commented [13]:

> "This definition is essential for the special theory of relativity, but it is not epistemologically necessary. If we were to follow an arbitrary rule restricted only to the form
>
> $$t_2 = t_1 + \varepsilon(t_3 - t_1), \qquad 0 < \varepsilon < 1 \tag{4}$$
>
> it would likewise be adequate and could not be called false. If the special theory of relativity prefers the first definition, i. e., sets ε equal to 1/2, it does so on the ground that this definition leads to simpler relations."

Clearly, different values of ε correspond to different values of the one-way speed of light. In fact, one can write

$$t_2 - t_1 = \frac{L}{\tilde{c}(\theta)} \quad \text{and} \quad t_3 - t_2 = \frac{L}{\tilde{c}(\theta + \pi)}, \tag{5}$$

where L is the $A - B$ distance, $\tilde{c}(\theta)$ is the one-way speed of light from A to B, θ being the angle between the light propagation direction and the absolute velocity of the considered inertial frame. Similarly $\tilde{c}(\theta + \pi)$ is the one-way speed from B to A. By adding the previous relations one gets

$$t_3 - t_1 = \frac{L}{\tilde{c}(\theta)} + \frac{L}{\tilde{c}(\theta + \pi)} = \frac{2L}{c}, \tag{6}$$

the last step being a necessary assumption, because the *two-way* speed of light has been measured with great precision and always found to be c. From (4), (5) and (6) one easily gets

$$\varepsilon = \frac{t_2 - t_1}{t_3 - t_1} = \frac{c}{2\tilde{c}(\theta)} \tag{7}$$

Therefore freedom of choice of ε would mean freedom of choice of the one-way speed of light.

In the following section we will show that $\tilde{c}(\theta)$ is not really a free parameter of relativistic physics because there is only one choice which leads to a consistent theory allowing one to pass smoothly from an accelerated reference frame to an inertial reference frame istantaneously at rest with respect to one another. However the standard assumption $\tilde{c}(\theta) = c$ leads to unsurmountable difficulties, as will be shown next.

3. TIME MEASUREMENTS ON A ROTATING PLATFORM

The laboratory is assumed to be an inertial frame in which clocks have been synchronised with the standard method, that is by assuming that the one-way speed of light is c in all directions.

We consider a clock C_Σ that is on the rim of a uniformly rotating platform having radius R and angular velocity ω. We assume this clock to be set as follows: When a clock of the laboratory momentarily very near C_Σ shows time $t_0 = 0$ then also C_Σ is set at time $t = 0$. Given that C_Σ keeps a constant position on the platform we can apply Eq. (2) with $\Delta t_0 = t_0 - 0 = t_0$ and $\Delta t = t - 0 = t$ and conclude that any observer at rest in the laboratory near the rim of the platform whose clock marks the time t_0 will see the clock C_Σ passing by in that very moment marking the time

$$t = t_0\sqrt{1-\beta^2} \tag{8}$$

with $\beta = \omega R/c$.

On the rim of the platform besides the clock C_Σ there is also a light source Σ placed in a fixed position very near C_Σ. Two light flashes leave Σ at time t_1, such that the description of light propagation given by the laboratory observers is the following: Two light flashes leave Σ at time t_{01}. The first one propagates circularly in the sense opposite to the platform rotation and comes back to Σ after a 2π rotation at time t_{02}. The second one propagates circularly in the same rotational sense of the platform and comes back to Σ after a 2π rotation at time t_{03}. These laboratory times, all relative to events taking place in a fixed point of the platform very near C_Σ, are related to the corresponding platform times via (8):

$$t_{0i} = \frac{t_i}{\sqrt{1-\beta^2}}, \quad i = 1, 2, 3. \tag{9}$$

If L_0 is the disk circumference length measured in the laboratory, light propagating in the direction opposite to the disk rotation must cover a distance smaller than L_0 by a quantity $x = \omega R(t_{02} - t_{01})$ equalling the shift of Σ during the time $t_{02} - t_{01}$ taken by light to reach Σ. Therefore,

$$L_0 - x = c(t_{02} - t_{01}), \quad x = \omega R(t_{02} - t_{01}) \,. \tag{10}$$

From these equations, one gets

$$t_{02} - t_{01} = \frac{L_0}{c(1+\beta)}. \tag{11}$$

Light propagating in the rotational direction of the disk must instead cover a distance larger than the disk circumference length L_0 by a quantity $y = \omega R(t_{03} - t_{01})$ equalling the shift of Σ during the time $t_{03} - t_{01}$ taken by light to reach Σ. Therefore,

$$L_0 + y = c(t_{03} - t_{01}), \quad y = \omega R(t_{03} - t_{01}) \tag{12}$$

One now gets

$$t_{03} - t_{01} = \frac{L_0}{c(1-\beta)}. \tag{13}$$

By taking the difference between (13) and (11) we see that the time delay between the arrivals of the two light flashes back in Σ is observed in the laboratory to be

$$t_{03} - t_{02} = \frac{2L_0\beta}{c(1-\beta^2)}. \tag{14}$$

This is the well known time delay for the Sagnac effect calculated in the laboratory.

We show next that these relations fix the time intervals on the disk. In fact, (9) applied to (11) and (13) gives

$$t_2 - t_1 = \frac{L_0\sqrt{1-\beta^2}}{c(1+\beta)}, \quad t_3 - t_1 = \frac{L_0\sqrt{1-\beta^2}}{c(1-\beta)} \tag{15}$$

Of course one has

$$L_0 = L\sqrt{1-\beta^2}, \tag{16}$$

because the rotating disk circumference length appears contracted in the laboratory. Therefore the velocities of the two light flashes relative to the disk are measured to be

$$\tilde{c}(\pi) = \frac{L}{t_2 - t_1} = \frac{c}{1-\beta} \tag{17}$$

for the propagation in the direction of the disk rotation, and

$$\tilde{c}(0) = \frac{L}{t_3 - t_1} = \frac{c}{1+\beta} \tag{18}$$

for the propagation in the opposite direction. Equations (17)-(18) are particular cases of the formula $\tilde{c}(\theta) = c/(1+\beta\cos\theta)$, shown in Ref. [14] to be compatible with the experimental evidence at the special relativistic level.

Starting from (17)-(18), it is a simple matter to show that the two-way velocity of light along the rim is just c. But every line element of the circumference of the rotating disk is instantaneously at rest in some inertial reference frame (the "tangent" frame). Since in inertial frames the *two-way* velocity of light is experimentally well established to be c, the local equivalence with the accelerated frame requires this. Nothing different would be acceptable.

Notice also that Eq. (9) applied to (14) gives

$$t_3 - t_2 = \frac{2L\beta}{c}, \tag{19}$$

where also (16) was used. This is again the Sagnac formula, this time calculated on the rim of the disk.

By accepting (17)-(18), we can overcome the long-standing "mystery" of the rotating platform and find a perfectly rational description of the Sagnac effect [15], essentially the observation of a rotation dependent phase shift between two coherent beams travelling on opposite paths in an interferometer placed on a rotating disk. Many authors deduce the Sagnac formula in the laboratory (our Eq. (14) above), but do not try to describe the phenomenon on the rotating platform. Exceptions are Langevin [16], Anandan [17], Dieks and Nienhuis [18], and Post [19], but none of these treatments is free of ambiguities. For example, Langevin's approach leads to difficulties at the general relativistic level [in his 1937 paper he even recognised the possibility of a non-standard velocity of light on the rotating platform and gave formulae agreeing to first order with our results (17)-(18)]. As a second example, Post's "relativistic" formula is limited to the arbitrary case where the origin of the "tangent" inertial frame coincides with the centre of the rotating disk.

If, instead of (17)-(18), one had $\tilde{c}(0) = \tilde{c}(\pi) = c$ it would obviously follow that the two light pulses require the same time to go around the disk in opposite directions, or that $t_2 = t_3$. The prediction for the Sagnac effect would become $t_3 - t_2 = 0$, totally unacceptable because contradicting the theoretical prediction in the laboratory (14) which agrees with experimental evidence. A speed of light along the rim of the circular disk equal to c is thus seen to lead to a logical contradiction within the theory. But such a value is required by the principle of local equivalence between the accelerated frame and the "tangent" inertial frames *if in the latter frames clocks are synchronised with Einstein's method*, as required by the assumed validity of the Lorentz transformations for all inertial frames. This is the reason why something fundamental has to be modified.

4. SPEED OF LIGHT RELATIVE TO INERTIAL FRAMES

One can always choose Cartesian coordinate systems in two inertial reference frames S and S_0 and assume:

(1) that space is homogeneous and isotropic, and that time is also homogeneous;

(2) that relative to S_0 the velocity of light is the same in all directions, so that Einstein's synchronisation can be used in this frame and the velocity v of S relative to S_0 can be measured;

(3) that the origins of S and S_0 coincide at $t = t_0 = 0$;

(4) that planes (x_0, y_0) and (x, y) coincide at all times t_0 ; that also planes (x_0, z_0) and (x, z) coincide at all times t_0; but that planes (y_0, z_0) and (y, z) coincide at time $t_0 = 0$ only.

It then follows [14] that the transformation law from S_0 to S is necessarily

$$\begin{cases} x = f_1(x_0 - vt_0), \\ y = g_2 y_0, \\ z = g_2 z_0, \\ t = e_1 x_0 + e_4 t_0, \end{cases} \tag{20}$$

where the coefficients f_1, g_2, e_4, and e_1 can depend on v. If at this point one assumes the validity of the relativity principle (including invariance of light speed) these transformations reduce *necessarily* to the Lorentz ones. It was shown in [14] that the most general transformations of the general type (20) satisfying the conditions of constant two-way velocity of light and of time dilation according to the usual relativistic factor are a particular case of (20) with

$$f_1 = \frac{1}{R(\beta)}, \quad g_2 = 1, \quad e_4 = R(\beta) - e_1 \beta c,$$

where $\beta = v/c$, and

$$R(\beta) = \sqrt{1 - \beta^2}, \tag{21}$$

so that

$$\begin{cases} x = \dfrac{x_0 - \beta c t_0}{R(\beta)}, \\ y = y_0, \\ z = z_0, \\ t = R(\beta) t_0 + e_1 (x_0 - \beta c t_0). \end{cases} \tag{22}$$

In (22) e_1 is the only unknown function of v. Length contraction by the factor $R(\beta)$ is also a consequence of (22). The inverse speed of light compatible with (22) was shown to be

$$\frac{1}{\tilde{c}(\theta)} = \frac{1}{c} + \left[\frac{\beta}{c} + e_1 R(\beta)\right] \cos\theta, \tag{23}$$

where θ is the angle between the direction of propagation of light and the absolute velocity $\boldsymbol{v}$ of S. The transformations (22) represent the complete set of theories "equivalent" to STR: If e_1 is varied, different elements of this set are obtained, which, according to Reichenbach's conventionality thesis, should all be equivalent for explaining experimental results. Different values of e_1 are obtained from different clock-synchronisation conventions. In all cases but that of STR such values exclude the validity of the relativity principle, and imply the existence of a privileged frame [14]. The Lorentz transformation is recovered as a particular case with $e_1 = -\beta / c R(\beta)$.

In the previous sections we found values of the one-way speed of light along the rim of the disk, and relative to the disk itself, different from c and given by Eqs. (17) and (18). Our principle of local equivalence between the rim of the disk and the "tangent" inertial frame requires (17) and (18) to apply in the latter frame as well. Equation (23) applied to the cases $\theta = 0$ and $\theta = \pi$ gives

$$\frac{1}{\tilde{c}(0)} = \frac{1}{c} + \left[\frac{\beta}{c} + e_1 R(\beta)\right], \quad \frac{1}{\tilde{c}(\pi)} = \frac{1}{c} - \left[\frac{\beta}{c} + e_1 R(\beta)\right]. \tag{24}$$

This agrees with (17) and (18) if and only if

$$e_1 = 0. \tag{25}$$

The space dependent term in the transformation of time is thus seen to disappear from (22).

5. THE INERTIAL TRANSFORMATIONS

In the previous section we showed that the condition $e_1 = 0$ has necessarily to be used. This generates a transformation different from the Lorentz one, but nevertheless particularly nice [14]:

$$\begin{cases} x = \dfrac{x_0 - \beta c t_0}{R(\beta)}, \\ y = y_0, \\ z = z_0, \\ t = R(\beta) t_0. \end{cases} \tag{26}$$

The speed of light relevant to a theory based on (26) can easily be found by taking $e_1 = 0$ in (23):

$$\frac{1}{\tilde{c}(\theta)} = \frac{1 + \beta \cos \theta}{c}. \tag{27}$$

The transformation (26) can be inverted and gives

$$\begin{cases} x_0 = R(\beta) \left[x + \dfrac{\beta c}{R^2(\beta)} t \right], \\ y_0 = y, \\ z_0 = z, \\ t_0 = \dfrac{1}{R(\beta)} t. \end{cases} \tag{28}$$

Note the formal difference between (26) and (28). The latter implies, for example, that the origin of $S_0 (x_0 = y_0 = z_0 = 0)$ in S is described by $y = z = 0$ and by

$$x = -\frac{\beta c}{1 - \beta^2} t.$$

This origin is thus seen to move with speed $\beta c/(1 - \beta^2)$, which can exceed c, but cannot be superluminal. In fact a light pulse seen from S to propagate in the same direction as S_0 has $\theta = \pi$ and thus [using (26)] has speed $\tilde{c}(\pi) = c/(1 - \beta)$, which can easily be checked to satisfy

$$\frac{c}{1 - \beta} \geq \frac{c\beta}{1 - \beta^2}.$$

One of the typical features of these transformations is of course the presence of relative speeds exceeding c. Absolute speeds are instead always smaller than c [14]. In STR one is used to relative velocities that are always equal and opposite, but this symmetry is a consequence of the particular synchronisation used and cannot be expected to hold more generally [14].

Consider now a third inertial system S' moving with velocity $\beta' c$ and its transformation from S_0 , which of course is

$$\begin{cases} x' = \dfrac{x_0 - \beta' c t_0}{R(\beta')}, \\ y' = y_0, \\ z' = z_0, \\ t' = R(\beta') t_0, \end{cases} \tag{29}$$

where $R(\beta')$ is given by (7) with β' replacing β. By eliminating the S_0 variables from (29) and (28) one obtains the transformation between the two moving systems S and S':

$$\begin{cases} x' = \dfrac{R(\beta)}{R(\beta')} \left[x - \dfrac{\beta' - \beta}{R^2(\beta)} ct \right], \\ y' = y, \\ z' = z, \\ t' = \dfrac{R(\beta')}{R(\beta)} t. \end{cases} \tag{30}$$

A possible name for (26)-(28)-(30) is "inertial transformations". In its most general form (30) the inertial transformation depends on two velocities (v and v'). When one of them is zero, either S or S' coincide with the privileged system S_0 and the transformation (30) becomes either (26) or (28).

A feature characterising the transformations (26)-(28)-(30) is ***absolute simultaneity***: two events taking place in different points of S but at the same t are judged to be simultaneous also in S' (and viceversa). Of course the existence of absolute simultaneity does not imply that time is absolute: on the contrary, the β-dependent factor in the transformation of time gives rise to time-dilation phenomena similar to those of STR. *Time dilation* in another sense is however also absolute: a clock at rest in S is seen from S_0 to run slower, but a clock at rest in S_0 is seen from S to run *faster* so that both observers will agree that motion relative to S_0 slows down the pace of clocks. The difference with respect to STR is however more apparent than real: a meaningful comparison of rates implies that a clock T_0 at rest in S_0 must be compared with clocks at rest in different points of S, and the result is therefore dependent on the convention adopted for synchronising the latter clocks.

Absolute length contraction can also be deduced from (26)-(28). A rod at rest on the x axis of S between the points with co-ordinates x_2 and

x_1 is seen in S_0 to have end points x_{02} and x_{01} at a common time t_0, where, from (26),

$$x_2 = \frac{x_{02} - vt_0}{\sqrt{1-\beta^2}}, \quad x_1 = \frac{x_{01} - vt_0}{\sqrt{1-\beta^2}}. \tag{31}$$

From this one obtains

$$x_2 - x_1 = \frac{1}{\sqrt{1-\beta^2}}(x_{02} - x_{01}). \tag{32}$$

The reasoning can be inverted by considering the rod at rest in S and observed from S_0, and using the transformation (28). One gets then, after a few simple steps:

$$x_{02} - x_{01} = \sqrt{1-\beta^2}(x_2 - x_1), \tag{33}$$

which could also be obtained by inverting (32). The two results are thus mathematically equivalent and lead to the conclusion (with which both observers agree) that motion relative to S_0 leads to contraction. Again the discrepancy with the STR is due to the different conventions concerning clock synchronisation: the length of a moving rod can only be obtained by marking the *simultaneous* positions of its end points, and therefore depends on the very definition of simultaneity of distant events.

6. CONCLUSIONS

Our choice of synchronisation (called "absolute" by Mansouri and Sexl [4]) has been made by considering accelerations. It is worth stressing that the normally accepted relationship between STR and accelerated systems is far from negligible. Accelerations are essential in the so-called twin paradox, which is a prediction of STR made before the general theory was even conceived. Very large accelerations enter in the muon experiment with the CERN Storage Ring, considered the most accurate test of time dilation. The Earth acceleration is essential for perceiving the retardation of the eclipses of Jupiter's satellites. The same can be said about the detection of stellar aberration. And so on. When they seem to generate difficulties within the existing theory accelerations are instead avoided. We have tried to do better.

There remain at least two problems to solve, before a theory based on the inertial transformations can be considered reasonably complete:

(1) Maxwell's equations must be reformulated. They will take a more general form, dependent on the absolute velocity of the inertial frame with respect to which they are considered, and will assume the usual form only in the privileged frame. This work is underway and should generate agreement

with experiments, given the results already obtained[14] concerning optical interference experiments.

(2) Also the general theory needs modifications because the ds^2 cannot be considered invariant anymore, given that the inertial transformations predict a frame-dependent one-way speed of light. It will be necessary to show that the modified theory makes the same successful predictions as the general theory of relativity. We know already that the new theory will have an advantage over the old one in the continuity principle between accelerated frames and locally tangent inertial frames.

Kinematics of high energy particle interactions was studied in Ref. [14] showing that complete equivalence exists between our predictions and those of STR. In fact energy and momentum are defined in such a way as to coincide *numerically* (not analytically) with those of STR for all particles and in all inertial frames, once they coincide in the fundamental frame. Therefore the kinematics of high energy processes, the determination of particle masses, and so on, do not require a different analysis from the one successfully carried out up to the present time.

REFERENCES

1. J. Bailey et al., *Nature* **268**, 301 (1977) .
2. A. Einstein, *Ann. Phys.* **35**, 898 (1911).
3. A. Einstein, *Ann. Phys.* **17**, 891 (1905).
4. R. Mansouri and R. Sexl, *Gen.Relat. Gravit.* **8**, 497 (1977).
5. A. Grünbaum, *Philosophical Problems of Space and Time* (Reidel, Dordrecht, 1973).
6. M. Jammer, in *Problems in the Foundations of Physics*, G. Toraldo di Francia, ed. (North Holland, Amsterdam, 1979).
7. T. Sjödin, *Nuovo Cim.* **51B**, 229 (1979).
8. G. Cavalleri, *Nuovo Cim.* **104B**, 545 (1989).
9. A. A. Ungar, *Found. Phys.* **21**, 691 (1991).
10. I. Vetharaniam and G.E. Stedman, *Phys. Lett.* **A183**, 349 (1993).
11. Quoted by A.A. Tyapkin, *Relatività speciale* (Jaca Book, Milano, 1993), p. 67.
12. A.Einstein, *Relativity, The Special, The General Theory* (Chicago, 1951).
13. H. Reichenbach, *The Philosophy of Space & Time* (Dover, New York, 1958).
14. F. Selleri, *Found. Phys.* **26**, 641 (1996); *Found. Phys. Lett.* **9**, 43 (1996).
15. G. Sagnac, *Compt. Rend.* **157**, 708 (1913); *ibid.* 1410.
16. P. Langevin, *Compt. Rend.* **173**, 831 (1921); **205**, 304 (1937).
17. J. Anandan, *Phys. Rev. D* **24**, 338 (1981).

18. D. Dieks and G. Nienhuis, *Am. J. Phys.* **58**, 650 (1990).
19. E. J. Post, *Rev. Mod. Phys.* **39**, 475 (1967).

QUANTUM INTERFEROMETRY, MEASUREMENT AND OBJECTIVITY: SOME BASIC FEATURES REVISITED

Markus Simonius

Institut für Teilchenphysik, Eidgenössische Technische Hochschule CH-8093 Zürich, Switzerland

The reduction paradigm of quantum interferometry and the objectivation problem in quantum measurements are reanalyzed. Both are shown to be amenable to straightforward mathematical treatment within "every-users" simple-minded quantum mechanics without reduction postulate etc., using only its probabilistic content.

Key words: quantum interferometry, state reduction, measurement, objective events.

1. INTRODUCTION: POSTULATES AND INTERPRETATION

Every-users quantum mechanics, based only on it's minimal probabilistic interpretation, is a concise and extremely successful theory without any intrinsic contradictions. Difficulties usually arise only when additional interpretational assumptions are introduced in order to bridge missing links in its understanding.

It is the object of this paper to demonstrate that once the basic axioms expressing the probabilistic content of quantum theory are laid down problems can be solved by careful definitions and simple but rigorous mathematics, without invoking any additional ad hoc ingredients.

Only the minimal probabilistic postulates I and II of quantum mechanics listed below will be used. No state reduction and no axiom of measurement etc.! Postulate III is included for completeness. It is not used but not violated either.

I States of a physical system are represented by elements X of the convex set $\mathcal{S}$ of density matrices on a Hilbert space $\mathcal{H}$ characterized by

M. Ferrero and A. van der Merwe (eds.),
New Developments on Fundamental Problems in Quantum Physics, 395–401.

$X \geq 0$ and $\mathrm{Tr}(X) = 1$.

II The probability for any given kind of event for a system in a state represented by X is expressed by an expectation value $(A, X) \equiv \mathrm{Tr}[AX]$ where A is a hermitian operator on $\mathcal{H}$, called an observable, which obviously must obey $0 \leq A \leq 1$ or, equivalently, $0 \leq \langle\varphi|A|\varphi\rangle \leq 1$ for all $\varphi \in \mathcal{H}$, $\|\varphi\| = 1$. The set of all such operators is denoted by $\widehat{\mathcal{O}}$.

III The evolution of an isolated system is reversible on $\mathcal{S}$ and thus unitary on $\mathcal{H}$.

Note the important fact that (A, X) is convex linear on $\mathcal{S}$ under mixing, $(A, |c_1|^2 X_1 + |c_2|^2 X_2) = |c_1|^2 (A, X_1) + |c_2|^2 (A, X_2)$, as required by its interpretation as a probability and that any such function can be represented by some $A \in \widehat{\mathcal{O}}$.

For a pure state $X = |\varphi\rangle\langle\varphi|$ with $\varphi \in \mathcal{H}$, $\|\varphi\| = 1$ and $(A, X) = \langle\varphi|A|\varphi\rangle$. Pure states play no conceptually distinguished role, and in particular it is *not* assumed that a system is always in some, perhaps unknown, pure state.

In II above "probability" is just a number between zero and one expressible by a formula within the theory. This number is *interpreted* as the probability for the occurrence of something named an *event* $e \in \{1, 0\}$ where $e = 1$ means that the event occurs in a given trial and $e = 0$ that it does not. There is no corresponding formula for events, however, and actually none is needed for any practical application. Nevertheless, it will be shown that events *can* be described in quantum mechanics using only I and II above and without violating III, where an event is *defined* by the following objectivity criterion [1]:

The basic property defining an event and guaranteeing its objectivity is that it can be observed or read in at least two independent, mutually non-interfering ways with necessarily agreeing results (zero probability for disagreement).

2. SINGLE QUANTUM INTERFEROMETRY [2]

Consider a typical interferometer arrangement as sketched in Fig. 1. Its properties are well known and often discussed: even if only one quantum (photon or neutron etc.) is inside the arrangement at a given time, the configuration of Fig. 1a can reveal interference between the states passing the two arms I and II provided none of them is blocked. Correspondingly, a pure single-particle state within the interferometer is represented by a normed wave function of the form

$$\varphi = c_1 \varphi_1 + c_2 \varphi_2, \quad \|\varphi\| = \|\varphi_1\| = \|\varphi_2\| = 1, \tag{1}$$

where, φ_1 and φ_2 represent states passing completely along one of the two

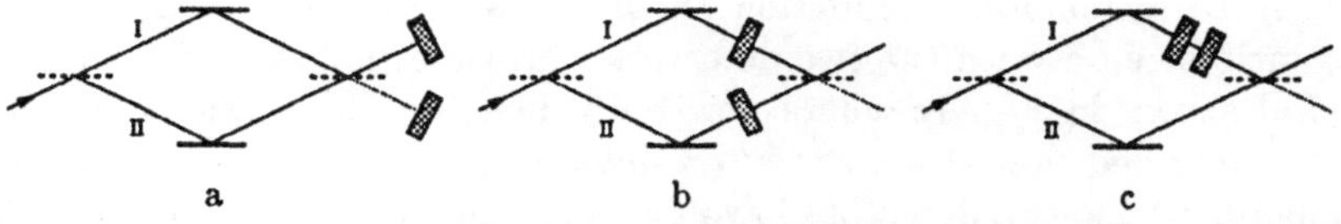

Fig. 1. Sketch of typical interferometer arrangements with differently placed detectors.

paths I or II, respectively, in the interferometer and have zero component in the other (and thus are mutually orthogonal). For ideal 50:50 beam-splitting $|c_1|^2 = |c_2|^2 = \frac{1}{2}$.

Now consider instead coincidences between two detectors inserted into the two paths of the interferometer as shown in Fig. 1b. From the appearance of the wave function (1) one might suspect that such coincidences should occur. But quanta (particles or photons) are supposed to be *indivisible entities* and thus a *single quantum* is not able to trigger both detectors in the arrangement of Fig. 1b (nor in Fig. 1a, of course). Though all this is much discussed textbook wisdom, one looks in vain in the literature for a satisfying *derivation* of this fact which uses only the basic structure of single particle quantum theory and does not amount to just stating the fact as postulate in one way or another. But clearly it should be possible to deduce such basic properties mathematically once the general premises of the theory are laid down [3]. And this, in fact, turns out to be the case.

The analysis is based on the following "silly" theorem which is a simple consequence of the positivity postulated in II:

Theorem 1: *Let A be a positive operator on $\mathcal{H}$ such that $\langle\psi_1|A|\psi_1\rangle = \langle\psi_2|A|\psi_2\rangle = 0$ for given $\psi_1, \psi_2 \in \mathcal{H}$. Then $\langle\psi|A|\psi\rangle = 0$ for all superpositions $\psi = c_1\psi_1 + c_2\psi_2$ between them.*

In order to apply this theorem to Fig. 1b with the two detectors in coincidence one only has to remark that one of the fundamental requirements for a coincidence between the two detectors is that the probability for a coincidence event be zero for any state which has zero component in one of the two arms of the interferometer, i.e., if either $c_1 = 0$ or $c_2=0$ in Eq. (1). (This is actually what careful experimenters check in order to verify that their arrangement does not produce spurious coincidence events!) Thus $\langle\varphi_1|A|\varphi_1\rangle = \langle\varphi_2|A|\varphi_2\rangle = 0$ where A is the operator describing the probability of coincidence events. It then follows from the above theorem that $\langle\varphi|A|\varphi\rangle = 0$ also for arbitrary $\varphi = c_1\varphi_1 + c_2\varphi_2$ and thus that *also for arbitrary superpositions between the states in the two arms of the interferometer the two detectors in Fig. 1b have zero probability to produce a coincidence event.*

A corresponding "reduction theorem" is obtained similarly (using anticoincidence between the two detectors, this time) for an arrangement of the kind shown in Fig. 1c where now both detectors are in the same path and it is assumed that at least the first detector transmits (does not absorb) the quanta: If the two detectors have unit efficiency either both of them fire or none. Of course two-slit or Stern-Gerlach arrangements can be analyzed correspondingly.

Thus "what must be" is graciously born out by the mathematical analysis in spite of appearance of the wave function in Eq. (1), "with nothing left to the discretion of the theoretical physicist" [3] (except to formulate the problems properly). Astonishingly, in spite of its simplicity, no such treatment is found in the literature, let alone in textbooks where it would belong.

The structure of theorem 1 and of the corresponding physical statements is emphasized: Conclusions for arbitrary superpositions between two states φ_i are obtained from conditions imposed only for the two states φ_i themselves. There is no need to *postulate* what happens in the case of superpositions.

3. MEASUREMENTS AND OBJECTIVE EVENTS [1]

The demonstration of the emergence of objective events is conceptually more involved and only a rather condensed exposition of the concepts and main result can be given here. For a more general and detailed discussion refer to Ref. [1].

The following operational definitions of discrimination and superpositions [1], formulated for density operators and probabilities directly, should be visualized with the help of the different arrangements of Fig. 1. and related to the more familiar formulations in terms of Hilbert space elements in case of pure states.

An observable $A \in \widehat{\mathcal{O}}$ *discriminates* between two states X_1, X_2 if $(A, X_1) = 1$ and $(A, X_2) = 0$ or vice versa. Such A exists provided X_1 and X_2 are *orthogonal.*

A state X is a *(general) superposition* between two orthogonal states X_1 and X_2 with some fixed weights $|c_1|^2$ and $|c_2|^2$ such that $|c_1|^2 + |c_2|^2 = 1$ if

$$\begin{aligned} (A, X) &= |c_2|^2 (A, X_2) \ \forall A \in \widehat{\mathcal{O}} : (A, X_1) = 0 \\ (A, X) &= |c_1|^2 (A, X_1) \ \forall A \in \widehat{\mathcal{O}} : (A, X_2) = 0. \end{aligned} \tag{2}$$

The set of all states X with this property is denoted by $\mathcal{S}(|c_1|^2 X_1, |c_2|^2 X_2)$. It is obviously convex and contains the incoherent mixture $|c_1|^2 X_1 + |c_2|^2 X_2$.

An observable A is *sensitive to interference* between X_1 and X_2 if $(A, X) \neq (A, |c_1|^2 X_1 + |c_2|^2 X_2)$ for some $X \in \mathcal{S}(|c_1|^2 X_1, |c_2|^2 X_2)$.

An immediate consequence of this definition of superpositions is the following generalization of theorem 1 (and thus its implications!) to non-pure states:

Theorem 1′: *If in an arrangement* $(A, X_1) = (A, X_2) = 0$ *for two orthogonal states* X_1, X_2*, then* $(A, X) = 0$ *also for any superposition* X *between these two states.*

A central role plays the careful analysis and definition of the concept of measurement: In a measurement information on the state X of an object is distributed onto different, at least two, separated channels (identified by greek upper indices) consisting of different systems on which mutually non-disturbing observations, using channel observables or *readings* $A^\mu \in \mathcal{O}^\mu$, $\mu = 1, 2, ...$, may be performed.

A given measurement is characterized by a function $m(\{A^\mu,\ \mu \in M\}; X) : \otimes_{\mu \in M} \mathcal{O}^\mu \otimes \mathcal{S} \to [0,1]$ where M is the subset of channels which are read. It will be denoted also $m(A^\mu; X)$ in case only one channel μ is read or $m(A^\mu, A^\nu; X)$ in case two channels μ and ν are read simultaneously etc.. It is to be interpreted as the probability for *coincidence events* $\prod e^\mu$, $\mu \in M$, i.e $m(\{A^\mu, \mu \in M\}; X) = Prob(\prod_{\mu \in M} e^\mu = 1)$ where e^μ is the event obtained by the reading A^μ.

The function m must be (convex) linear in $\mathcal{S}$ and all $\mathcal{O}^\mu$ in analogy to the corresponding properties of (A, X) appropriate for probabilities as discussed. In addition it has to fulfill the following important requirement expressing *separability or mutual non-disturbance* of readings of different channels:

$$m(A^\mu, A^\nu; X) + m(A^\mu, \bar{A}^\nu; X) = m(A^\mu; X) \tag{3}$$

$(\mu \neq \nu)$ for all $A^{\mu,\nu} \in \mathcal{O}^{\mu,\nu}$ independent of A^ν, and correspondingly for an arbitrary number of channels. Here the *complement* $\bar{A}$ of an observable A is defined by $(\bar{A}, X) = 1 - (A, X)$ for arbitrary state X which can be thought of as being the observable obtained from A by the replacement of the corresponding event e by its negation $\bar{e} = 1 - e$ (i. e., by inverting the output). The l.s.h of Eq. (3) then means that the event corresponding to A^ν is ignored though it has been obtained, and the r.h.s that no observation of channel ν is performed at all.

A channel μ of a measurement *discriminates* between X_1 and X_2 if reading of that channel alone, without further observation, allows one to discriminate between X_1 and X_2. A corresponding reading A^μ *discriminates* X_1 *against* X_2 if $m(A^\mu; X_1) = 1$ and $m(A^\mu; X_2) = 0$. This defines what a measurement is good for.

The fact that for arbitrary fixed readings A^μ, $\mu \in M$ a measurement realizes some observable $A \in \widehat{\mathcal{O}}$ such that $(A, X) = m(\{A^\mu,\ \mu \in M\}; X)$ together with the defining Eq. (2) of superpositions then leads to the central result:

Theorem 2 (Probability and Objectivity): *Consider a measurement with readings A^μ, A^ν for two different channels which both discriminate X_1 against X_2 and assume that $X \in \mathcal{S}(|c_1|^2 X_1, |c_2|^2 X_2)$ is any superposition between the X_i. Then*

$$m(A^\mu, A^\nu; X) = m(A^\mu; X) = m(A^\nu; X) = \|c_1\|^2 \tag{4}$$

and

$$m(A^\mu, \bar{A}^\nu; X) = m(\bar{A}^\mu, A^\nu; X) = 0, \tag{5}$$

implying that for all $X \in \mathcal{S}(|c_1|^2 X_1, |c_2|^2 X_2)$ two such discriminating readings performed on separate channels have zero probability to disagree. The last statement is exactly the defining property of an objective event as given in Sec. 1, which leads to the following:

Corollary (Objective Events): *A measurement which discriminates between two states X_1 and X_2 according to the condition of theorem 2, performed on any superposition X between them, leads for each trial to an (objective) event for an observable A which discriminates between X_1 and X_2.*

The *value* of the event is not predicted in general, only its probability according to Eq. (4). However, under the condition of theorem 2, the probabilistic element is completely eliminated in the correlation between two discriminating channels. Reading one of them one can or could with certainty predict the outcome of an eventual reading of the other: "The moon is *objectively* there even if nobody looks."

The mutual independence of the readings of different channels required by the definition of objectivity is contained in the important separability condition (3). Besides this, the central ingredient of the proof of Eq. (5) is (again) theorem 1′: The condition of theorem 2 together with the separability condition (3) which yields $m(A^\mu, \bar{A}^\nu; X) \leq m(A^\mu; X)$ etc. implies $m(A^\mu, \bar{A}^\nu; X_i) = m(\bar{A}^\mu, A^\nu; X_i) = 0$ which then must hold for any superposition X between X_1 and X_2 by theorem 1′.

Thus objective events emerge automatically from the definition of measurements. It is emphasized that *only the mathematical properties of the functions $m(\cdot\,;\cdot)$ and $(\cdot\,,\cdot)$ are actually used in the proof without reference to their interpretation in terms of events. Moreover, only the response of each channel separately to the two states X_i is presupposed and, as in section 2, nothing about superpositions between them nor about correlations among different channels.*

It can be shown [1] that the definition and emergence of objective events does not violate any principle of quantum mechanics. The rational behind this is quite simple: *One just cannot violate the defining equations (2) of superpositions however complicated one chooses the arrangement for an observation to be.*

The crucial feature for objectivity and consistency turns out to be the distribution of the information onto different channels. In the classical domain this happens continuously due to the abundance of channels consisting of light.

REFERENCES

1. M. Simonius, *Helv. Phys. Acta* **66**, 721 (1993); **65**, 884 (1992).
2. M. Simonius, *Helv. Phys. Acta* **69** Separanda, 15 (1996).
3. J. S. Bell, *Physics World* **3**(8), August 1990, p. 33.

EFFECT OF QED FLUCTUATIONS ON THE DYNAMICS OF THE MACROSCOPIC PHASE

Fernando Sols and Ivar Zapata

Departamento de Física Teórica de la Materia Condensada
Universidad Autónoma de Madrid
E-28049 Madrid, Spain

The dynamics of the relative phase between two separate superconductors or superfluids is studied theoretically, with particular emphasis on the effect of the quantum radiation field. We show that QED fluctuations are inefficient as a source of decoherence for the macroscopic phase. In particular, we find that the recurrent dynamics predicted for the relative phase between two isolated but previously connected superconductors is preserved essentially intact.

Key words: macroscopic quantum mechanics, Josephson junction, quantum measurement, quantum electrodynamics, quantum dissipation.

1. INTRODUCTION

The effect of a complex environment with many degrees of freedom on the dynamics of a physical system has been the object of intense study during the last decade [1]. Quantum dissipation techniques have been applied to a wide range of physical problems. In this context, however, relatively little attention has been given to the electromagnetic field as a dissipative environment. Barone and Caldeira [2], as well as and Joos and Zeh [3], have studied the coherence properties of charged particles in the presence of the background radiation field. Here, we address the question of dissipation by the photon field from a different viewpoint. Our aim is to investigate the decoherent effect of the radiation field on macroscopic quantum systems. We focus on the most elementary case of a Josephson junction and look at how the coherence properties of the relative phase are modified by the presence of the background radiation field. We show that the dominant term in the effective low-energy Hamiltonian is given by the coupling of the electromag-

M. Ferrero and A. van der Merwe (eds.),
New Developments on Fundamental Problems in Quantum Physics, 403–413.

netic (EM) field to the electric dipole, which is directly proportional to the number of transferred Cooper pairs. We also analyse the phase dynamics *after* the two Josephson junctions have been separated. It has been shown [4] that, in the absence of coupling to the quantum EM field, the phase undergoes an intrinsically recurrent dynamics. We have addressed here the question of whether the dissipative effect of the radiation field could, in the long term, destroy such a recurrence. We have found that, upon separation, the EM field does not add any decoherence to that existing before separation. Surprisingly, this result holds even at finite temperatures and for arbitrarily long times.

2. THE JOSEPHSON EFFECT

The low-energy Hamiltonian for a Josephson junction is

$$H_J = E_J(1 - \cos\varphi) + E_C N^2/2, \tag{1}$$

where φ is the phase difference, N is the number of transferred Cooper pairs, and E_J and E_C are the Josephson coupling and capacitance energies, respectively. N and φ are quantum canonically conjugated variables satisfying $[\varphi, N] = i$. For two superconductors, $E_C = (2e)^2/2C$, where $2e$ is the Cooper pair charge and C is the electrostatic capacitance. In the case of two baths of superfluid ^{3}He, $E_C = gM_0^2/\rho A$ if the capacitance is due to gravitational forces (g is the acceleration of gravity, M_0 is twice the mass of the ^{3}He atom, ρ is the mass density, and A is the superfluid area), and $E_C = M_0^2/\rho^2 V\kappa_T$ if the volume V is held fixed, κ_T being the compressibility. In a typical Josephson junction, $E_C \ll E_J$, and (1) can be approximated as

$$H \simeq E_J\varphi^2/2 + E_C N^2/2, \tag{2}$$

as long as $k_B T \ll E_J$. In equilibrium, the phase is narrowly peaked around $\varphi = 0$. In a semiclassical description (C large), φ is a c-number and the energy is given by the first term of Eq. (1).

The operators phase and particle number obey the uncertainty relation $\Delta N \Delta\varphi \geq 1/2$, which we write here as if φ were an extended variable. In a standard Josephson junction, E_J/E_C is large and the phase uncertainty is very small, and there are large coherent fluctuations in the relative particle number. On the contrary, two separate superconductors that have never been connected, and which therefore have not developed any phase correlations, will have, at sufficient low temperatures, a well-defined particle number and the relative phase will be completely ill-defined. A question that can be raised is what happens when the two linked superconductors are suddenly separated. What is the density matrix that describes the system after separation? One may expect that, after a sufficiently long time, the density matrix becomes diagonal in the particle number representation, and, assuming that is the case, one may wonder about the details of the evolution. We will see that the evolution from a density matrix that is quasi-diagonal in the phase representation to one that is diagonal in the number representation is by no means guaranteed. In particular, we show that the quantum radiation field destroys the coherence between N eigenstates only weakly, and thus cannot "measure" the relative charge between two superconductors.

3. RANDOMIZATION OF THE PHASE WITHOUT THE RADIATION FIELD

The randomization of the phase has been previously analysed in the absence of dissipation [4,5] and in the presence of a normal metal reservoir [4,6]. For completeness, we review here some of the main results. Within the harmonic oscillator approximation, the equilibrium average energy is $E_i = (\hbar\omega_0/2k_BT)\coth(\hbar\omega_0/k_BT)$, where $\omega_0 = (E_C E_J)^{1/2}/\hbar$ is the oscillator (Josephson plasma) frequency. The initial phase uncertainty is $\Delta\varphi(0) = (E_i/E_J)^{1/2} \ll 1$. If, by some mechanism, the Josephson coupling energy E_J is suddenly made zero at time $t = 0$, the initially narrow phase distribution tends to broaden at $t > 0$. If the phase φ is formally treated as an extended variable, and in the absence of any other relevant degrees of freedom, the phase uncertainty evolves ballistically and, at long times, it can be written

$$\Delta\varphi(t) \simeq \sqrt{E_i E_C}\,t/\hbar. \tag{3}$$

Thus $\Delta\varphi$ becomes of order 2π at time

$$\tau_B = h/\sqrt{E_i E_C}, \tag{4}$$

which may be called the *ballistic randomization time*. Typical values for a superconductor are $E_J \sim 5$ eV and $C \sim 1$ nF (a compromise value between a self-capacitance and and a plate capacitance), which leads, with $E_i \simeq \hbar\omega_0/2$, to $\tau_B \sim 30$ ns. For superfluid ^{3}He, $E_J \sim 1$ K. If the dimensions of the baths are taken of order 1 cm (implying $E_i \simeq k_BT$), one obtains [5] $\tau_B \sim 4$ hours, if capacitance is caused by gravitational forces, and $\tau_B \sim 1$ minute, if it is dominated by compressibility forces. Thus, in contrast with the superconductor case, the intrinsic randomization times of superfluids are remarkably long.

The randomization time for the relative phase between two superconductors is altered drastically if these are not isolated but coupled to a large normal metal reservoir. The incoherent exchange of particles with the normal reservoir is similar to the random exchange of momentum between a Brownian particle and its surrounding viscous medium. Drawing on this analogy, we note that the phase plays the role of the position, and the capacitance is analogous to the mass. The particle number N is equivalent to the momentum p. The chemical potential difference $\mu \propto \dot{\varphi}$ is the the analogue of the particle velocity v. The friction coefficient η is defined as $\dot{p} = -\eta v$, with $\langle v \rangle = 0$. Analogously, a coefficient λ may be defined relating $\dot{N}$ to a spontaneous fluctuation in the chemical potential of the superconductor with respect to the normal reservoir:

$$\dot{N} = -\lambda \mu, \tag{5}$$

with $\langle \mu \rangle = 0$. Dissipation affects the particle dynamics in a time scale $\gamma^{-1} = m/\eta$. In the superconductor case, $\gamma = \lambda E_C$, and γ^{-1} turns out to be the $R_N C$ time, where R_N is the normal state resistance between our system and the reservoir. Taking a contact area of 1 cm^2 and an average transmission probability at the Fermi level of $T_0 \sim 0.01$, we arrive at the rather small time $\gamma^{-1} \sim 7 \cdot 10^{-19}$ s. The inverse temperature time is $\hbar/k_B T \sim 7$ ps at $T = 1$ K.

Once the Brownian particle analogy has been established, we may write for the evolution of the phase uncertainty [7]

$$\Delta\varphi^2(t) = \frac{2k_B T}{\hbar^2 \lambda}[t - \frac{1}{\gamma}(1 - e^{-\gamma t})]. \tag{6}$$

At short times ($\gamma t \ll 1$), we reproduce the ballistic randomization given by (3). At long times ($\gamma t \gg 1$), the randomization becomes diffusive,

$$\Delta\varphi(t) \simeq \sqrt{2k_B T t/\hbar^2 \lambda}. \tag{7}$$

Equation (7) is a classical result. However, from the work of Ref. [8], we know that (7) applies at long times $t \gg \hbar/k_B T \gg \gamma^{-1}$, provided $\Delta\varphi(0) \ll 1$

The *diffusive randomization time* is defined as

$$\tau_D = \hbar^2 \lambda / 2k_B T. \tag{8}$$

With the parameters given above, we find $\tau_D \sim 12$ minutes, a surprisingly long time if we compare it with our estimates of the ballistic time τ_B. The large disparity between the two time scales may be understood by noting that, while the fast ballistic randomization is due to the small value of the electrostatic capacitance, the diffusive dynamics (which takes over in a time scale $\gamma^{-1} \ll \tau_B$) is essentially independent of it and is determined instead by the coupling to the normal reservoir.

Going back to the dissipation free case, we note that the phase is not an extended variable. It is actually an angular variable for which 0 and 2π

are to be identified, and N has only integer eigenvalues. When $E_J = 0$, the wave function evolves as

$$\Psi(\varphi,t) =< \varphi|\Psi(t)>= \sum_N a_N\psi_N(\varphi)\exp(-iE_CN^2t/2\hbar), \tag{9}$$

where $\psi_N(\varphi) =< \varphi|N>$ and $a_N =< N|\Psi(0)>$. From (9), we infer the existence of a *recurrence time*

$$\tau_R = 2h/E_C, \tag{10}$$

after which the wave packet recombines, $\Psi(\varphi,\tau_R) = \Psi(\varphi,0)$. The property $E_C \ll E_J$ implies $\tau_R \gg \tau_B$, so that the two time scales are well separated.

The resulting picture for the conservative evolution of a particle on a circle is the following: An initially narrow wave packet begins to spread according to the law given in (3). When t becomes of order $0.5\tau_B$, the two tails of the wave packet begin to overlap and interference on a fine scale begins to develop. By the time $t \sim 2\tau_B$, the wave packet is almost uniformly spread around the circle, appart from very fast (both in time and in angle) oscillations. At a much later time τ_R, the wave function recombines and broadens again in a time scale τ_B. With the parameters given above, we find $\tau_R \sim 13\ \mu s$ for the superconductor and astronomically large values for the superfluid. The superconductor value of τ_R falls entirely within experimental reach, although it must be noted that recurrence would only be observed in quite ideal conditions (very low noise, good time resolution, etc.), perhaps not easily realizable. One may wonder whether, for two superconductors ideally isolated in free space, the quantum fluctuations of the free photon field would be sufficient to destroy the recurrence and, in particular, to act as a measurement apparatus which measures the relative charge. This question is addressed in the following two sections.

4. COUPLING TO THE ELECTROMAGNETIC FIELD

In this section, we study the equilibrium dynamics of a connected Josephson junction in the presence of quantum electromagnetic fluctuations. Our starting point is a Hamiltonian for a Josephson junction that includes charging effects and coupling to the transverse photon field. In the Coulomb gauge, we may write [9]

$$H = H_L + H_R + H_T + H_Q + H_{\mathrm{rad}}. \tag{11}$$

Here $H_{L(R)}$ is the BCS Hamiltonian [9,10] for the left (right) superconductor. It includes the effect of the transverse electromagnetic field through minimal coupling to the quantum vector potential (in this section $c = \hbar = 1$)

$$\mathbf{A}(\mathbf{r}) = \sum_l \mathbf{A}_l(\mathbf{r})a_l + \mathrm{H.c.}, \tag{12}$$

where a_l anihilates a photon in mode l. The charging term H_Q contains the effect of the longitudinal electromagnetic field, which in its simplest form may be written

$$H_Q = (Q_L - Q_R)^2/8C, \tag{13}$$

where $Q_{L(R)}$ is the charge accumulated on the left (right) superconductor and C is the geometrical capacitance. H_T is the tunneling Hamiltonian between the left and right superconductors, which also depends on $\mathbf{A}(\mathbf{r})$. Finally, $H_{\mathbf{rad}}$ describes the unperturbed transverse photon field,

$$H_{\mathbf{rad}} = \sum_l \omega_l a_l^\dagger a_l. \tag{14}$$

In free vacuum $l \equiv (\mathbf{k}, \lambda)$, where $\mathbf{k}$ and λ are the photon momentum and polarization.

In order to proceed further, we introduce the *electric dipole approximation*

$$\mathbf{A}(\mathbf{r}) \simeq \mathbf{A}(0). \tag{15}$$

We may write the electric dipole operator as

$$\mathbf{P} \equiv e \int \mathbf{r}\rho(\mathbf{r})d^3r \simeq 2eN\mathbf{R}, \tag{16}$$

where ρ is the electron density operator and $\mathbf{R}$ is the vector linking the centers of the electronic clouds within each superconductor. Its approximate amplitude is $R \simeq d + 2\lambda_{\mathrm{TF}}$, where d is the junction thickness and λ_{TF} is the Thomas-Fermi screening length. If we introduce the canonical transformation

$$U_{\mathbf{P}} = \exp\{-i\mathbf{P} \cdot \mathbf{A}(0)\}, \tag{17}$$

the Hamiltonian (11) becomes

$$H = H_L^0 + H_R^0 + H_T + H_{\mathbf{rad}} + H_{\mathrm{int}}, \tag{18}$$

where $H_{L(R)}^0$ are the BCS Hamiltonians with $\mathbf{A} = 0$, and the electron-photon interaction takes the form

$$H_{\mathrm{int}} = -\mathbf{P} \cdot \mathbf{D}(0) \simeq \sum_l g_l \left(a_l - a_l^\dagger + \frac{g_l N}{\omega_l}\right) N, \tag{19}$$

$\mathbf{D}(0)$ being the electric displacement field at position $\mathbf{r} = 0$. Hence, within the approximation (15), the effective spectral density for coupling to free photons is

$$J(\omega) \equiv \pi \sum_l g_l^2 \delta(\omega - \omega_l) = \frac{8e^2R^2}{3}\omega^3. \tag{20}$$

Being super-Ohmic [1], we expect the transverse photon field to be relatively inefficient as a source of decoherence. We note in this respect that coupling to the electric quadrupole moment yields a ω^5 law for the spectral density, which is even more weakly dissipative. We conclude that coupling to the electric dipole moment must give the dominant contribution to decoherence.

We may adopt a phase representation by writing [11]

$$|\varphi>= \frac{1}{\sqrt{2\pi}} \sum_N e^{-iN\varphi}|N>, \tag{21}$$

where $|N>$ is the factorised ground state of the right and left superconductors with N transferred Cooper pairs. At low enough temperatures, quasiparticles can be neglected and the effective Hamiltonian becomes

$$H = E_J(1-\cos\varphi) - \frac{E_C}{2}\frac{d^2}{d\varphi^2} + \sum_l \omega_l(a_l^\dagger - i\frac{g_l}{\omega_l}\frac{d}{d\varphi})(a_l - i\frac{g_l}{\omega_l}\frac{d}{d\varphi}). \tag{22}$$

Equation (22) has the structure of a system-plus-reservoir Hamiltonian [1]. The interesting degree of freedom is the phase φ, or its canonical conjugate $N \equiv -id/d\varphi$, and the bath of harmonic oscillators is formed by the transverse photons, which couple to N with the spectral density (20).

The problem posed by the model Hamiltonian (22) cannot be solved exactly and one has to resort to approximations. Here we employ a *variational approximation* to treat the coupling of the phase to the radiation field. For that purpose, we introduce the canonical transformation

$$U = \exp\{i\frac{d}{d\varphi}\sum_l f_l\omega_l(a_l - a_l^\dagger)\}, \tag{23}$$

where f_l will be determined variationally. The resulting Hamiltonian is

$$\tilde{H} = \tilde{H}_0 + \tilde{V}, \tag{24}$$

where $\tilde{H}_0$ describes an effectively decoupled system

$$\begin{aligned}
\tilde{H}_0 &= \tilde{E}_J(1-\cos\varphi) - \frac{\tilde{E}_C}{2}\frac{d^2}{d\varphi^2} + \sum_l \omega_l a_l^\dagger a_l, \\
\tilde{E}_C &= E_C + \sum_l \frac{(g_l - f_l)^2}{\omega_l}, \\
\tilde{E}_J &= E_J e^{-\gamma}, \quad \gamma = \sum_l \frac{f_l^2}{2\omega_l}\coth(\frac{\beta\omega_l}{2}),
\end{aligned} \tag{25}$$

and $\tilde{V}$ is a residual interaction of zero average $[\mathrm{Tr}(e^{-\beta\tilde{H}_0}\tilde{V}) = 0]$. We make us of the Bogoliubov-Peierls inequality [12] and adjust f_l to minimize the free energy for the decoupled Hamiltonian H_0. The result is

$$f_l = \frac{g_l}{1 + \tilde{\omega}/\omega_l}, \tag{26}$$

where $\tilde{\omega}$ is a complicated function of the renormalized parameters $\tilde{E}_J$ and $\tilde{E}_C$ involving the eigenvalues of the Mathieu equation. Eq. (26) tells us that photons with frecuency $\omega_l \ll \tilde{\omega}$ are essentially unaltered by the presence of the Josephson junction ($f_l \simeq 0$), while photons with frequency $\omega_l \gg \tilde{\omega}$ follow the quantum dynamics of the phase adiabatically ($f_l \simeq g_l$). Therefore, the relevant time scale is given by $\tilde{\omega}^{-1}$. In the limit $E_C \ll E_J$, we find $\tilde{\omega} \simeq \omega_0$.

In terms of the total density matrix $\bar{\rho}$, the variational approximation can be stated as

$$\bar{\rho} \simeq U^\dagger e^{-\beta\tilde{H}_0} U / \mathrm{Tr}(e^{-\beta\tilde{H}_0}). \tag{27}$$

This is the approximate equilibrium density matrix which we assume describes a Josephson junction during the connection period $t \leq 0$.

5. RANDOMIZATION WITH QED FLUCTUATIONS

If, at $t = 0$, we make $E_J = 0$, the system is described again by the Hamiltonian (22) with the Josephson coupling term removed. The subsequent evolution of the reduced density matrix $\rho(t) \equiv \mathrm{Tr}_{\mathrm{ph}}\{\bar{\rho}(t)\}$ can be calculated without much difficulty. In the long time limit, we obtain

$$< N|\rho(t)|N' > \simeq \Psi_N^* \Psi_{N'} e^{-\gamma(N-N')^2} e^{-i(E'_C/2\hbar)(N^2-N'^2)t}, \tag{28}$$

where $\Psi_N = < N|\Psi >$ is the wave function of the ground state. The term $\Psi_N^*\Psi_{N'}$ is the pure state component of the density matrix. The γ-dependent term contains the effect of decoherence due to the electromagnetic field. It must be noted, however, that this term was already present before the separation. Finally, Eq. (28) contains an oscillatory term describing the recurrent dynamics characteristic of zero coupling. E'_C is the capacitance energy in the disconnected configuration. Surprisingly, recurrence is preserved in the presence of a fluctuating EM field, and this result holds at arbitrary temperatures.

The decoherence factor $\exp\{-\gamma(N - N')^2\}$ has a disturbing dependence on the upper frequency cutoff Ω which must be introduced in (22). At zero temperature, one obtains

$$\gamma = \frac{2\alpha}{3\pi}\left(\frac{R\Omega}{c}\right)^2. \tag{29}$$

where $\alpha \simeq 1/137$ is the fine structure constant. The cutoff dependence is the price we pay for the use of the simplifying dipole approximation. One usually identifies Ω with the lowest frequency above which there is important physics not described by the formalism. In our case, this frequency is given by the superconducting gap $\Delta/\hbar$, since, for $\omega > \Delta/\hbar$, the excitation of quasiparticles within the superconductor cannot be neglected [9]. Typically, $\Delta \sim 1$ meV, while, for a standard tunnel junction or point contact, $R \sim 10 - 20$ Å. This makes $\gamma \sim 10^{-14}$. According to this estimate, the EM field would induce decoherence between number eigenstates with differences $|N - N'| > 10^7$.

It is interesting to think of a hypothetical situation in which the superconducting gap were so large that the effect of quasiparticles could be completely neglected. Then, the natural cutoff to introduce in (22) would be $\Omega \sim c/R$, and, from (29) γ would acquire the universal value

$$\gamma \simeq \frac{2\alpha}{3\pi}, \tag{30}$$

where $\alpha \simeq 1/137$ is the fine structure constant. We conclude that, due to the effect of quantum electromagnetic noise, the phase uncertainty would have a minimum, universal value

$$\Delta\varphi_{\rm min} \simeq \sqrt{2\gamma} \simeq 5.6 \times 10^{-2}, \tag{31}$$

which would be relevant for junctions with $E_C/E_J < 10^{-5}$, because for them the quantum uncertainty of the phase due to the charging energy is smaller than that caused by the zero point noise of the EM field.

6. SUMMARY AND DISCUSSION

We have studied some aspects of the dynamics of the macroscopic phase between superconductors, with emphasis on the effect of the quantum electromagnetic field. We have analysed the randomization of the relative phase after suppression of the Josephson coupling. In the presence of a common normal metal reservoir, the relative phase between two disconnected superconductors randomizes diffusively with time scales that can be of the order of minutes. In the absence of a dissipative environment, and because of the intense Coulomb forces, the phase between two isolated superconductors has a strong quantum character, with randomization times of order 10-100 ns. However, because of the absence of Josephson coupling, this phase broadening is not irreversible but displays a recurrent dynamics with a period of the order of 10 μs.

By contrast, the relative phase between disconnected superfluids yields ballistic randomization times of the order of minutes or hours, indicating that the phase of a superfluid behaves essentially like a classical variable. It is interesting to note in this context that, due to the extremely weak interatomic interaction, the effective capacitances between the recently discovered Bose-Einstein condensates [13] can be extremely large, and in that case the phase may display a much more classical character than its superfluid counterpart.

For a connected junction, the EM field destroys the coherence between largely differing eigenstates. This causes an intrinsic randomness in the relatively phase which is, however, much smaller than that due to the charging energy. The effect is weak because of the super-Ohmic spectrum with which the EM field couples to the electric dipole operator. We note here that, in a parallel plate geometry in which the two superconductors are separated by a dielectric slab, the low-frequency transverse photons which really act on the phase dynamics are strongly confined around the plane of of the slab, due both to the Meissner effect and to the poor matching between the dielectric and the vacuum [14]. Quasi-bound photons have a discrete spectrum that makes photon-induced decoherence even smaller, and essentially zero.

When the two superconductors are separated, one might expect that, due to the lack of competition from the Josephson coupling term (which tends to favor the definition of the phase), the quantum EM field would destroy the coherence between number eigenstates more efficiently. The surprising result is that photon-induced decoherence does not increase upon separation, and this remains true for arbitrary temperature.

Other authors [2,3] have analysed the role of the quantum electromagnetic field as a dissipative environment. Barone and Caldeira [2] have noticed that the transverse component of the electromagnetic field is quite inefficient in destroying the coherence properties of an electron in a typical interference experiment. Connection with our work can be established by noting that the position of an electron is akin to a fluctuating electric dipole. Thus, our conclusions are in qualitative agreement with those of Ref. [2]. Joos and Zeh [3] have noted, however, that weak scattering of photons by dust particles in free space is sufficient to prevent the indefinite coherent spreading of the wave packet which would occur in the absence of radiation. The analogous effect in our problem would be described by the weakly decoherent term shown by the reduced density matrix in the number representation (see Eq. 28).

In our opinion, important questions still remain on the decoherent role of the QED field. For one thing, it would be desirable to make predictions of physical quantities (such as the interference pattern displayed by a charged particle) that were not cutoff dependent. On the other hand, one would not like to abandon the simplifying and physically meaningful (at least in the context here considered) electric dipole approximation. In a reply to Santos [15], Diosi [16] has argued that the zero-point EM field yields not real decoherent effect, since any decoherence should be absorbed in an effective pure state contribution. We hope that future work will clarify these questions further.

ACKNOWLEDGMENTS

We would like to thank A. J. Leggett and J. L. Sánchez-Gómez for stimulating discussions. This work has been supported by DGCICyT (Project no. PB93-1248) and by the HCM Programme of the European Union.

REFERENCES

1. U. Weiss, *Quantum Dissipative Systems* (World Scientific, Singapore, 1993).
2. P.M.V.B. Barone and A.O. Caldeira, *Phys. Rev. A* **43**, 57 (1991).
3. E. Joos and H.D. Zeh, *Z. Phys. B* **59**, 223 (1985).
4. F. Sols, *Physica B* **194-196**, 1389 (1994).
5. A. J. Leggett and F. Sols, *Found. Phys.* **21**, 353 (1991).
6. F. Sols and R.A. Hegstrom, in *Fundamental Problems in Quantum Physics*, M. Ferrero and A. van der Merwe, eds. (Kluwer, Dordrecht, 1995).
7. F. Reif, *Foundations of Statistical and Thermal Physics* (McGraw-Hill, New York, 1965).
8. V. Hakim and V. Ambegaokar, *Phys. Rev. B* **32**, 423 (1984).
9. U. Eckern, G. Schön, and V. Ambegaokar, *Phys. Rev. B* **30**, 6419 (1984).
10. M. Tinkham, *Introduction to Superconductivity* (McGraw-Hill, New York, 1996).
11. P. G. de Gennes, *Superconductivity of Metals and Alloys* (Addison-Wesley, New York, 1989).
12. M. D. Girardeau and R. M. Mazo, *Adv. Chem. Phys.* **24**, 187 (1973).
13. M. H. Anderson, J. R. Ensher, M. R. Matthews, C. E. Wieman, and E. A. Cornell, *Science* **269**, 198 (1995).
14. D. D. Coon and M. D. Fiske, *Phys. Rev.* **138**, A744 (1965); I. O. Kulik, *JETP Lett.* **1**, 84 (1965).
15. E. Santos, *Phys. Lett. A* **188**, 198 (1994); **197**, 185 (1995).
16. L. Diósi, *Phys. Lett. A* **197**, 183 (1995).

WHAT WHEN GLEASON'S THEOREM FAILS?

Rolf Tarrach

Department d'ECM, Facultat de Física
Universitat de Barcelona
E-08028 Barcelona, Spain

Gleason's theorem, which associates a density matrix to any probability distribution, does not hold for two dimensional (two state) Hilbert spaces. Thus probability distributions might in principle exist which would correspond to a state which cannot be described by a density matrix. We show why this cannot happen, and that when it apparently happens it is due to a wrong measuring apparatus - selfadjoint operator association.

Key words: Gleason, density matrix, probability distributions.

1. INTRODUCTION

Consider a Hilbert space of dimension larger than 2. Consider any orthonormal basis in it, $\mid i>$. Let E be any projector in it, and $E_i \equiv \mid i >< i \mid$. Let the probability of obtaining 1 when measuring E be $P(E)$. Then

$$P(I) = 1, \quad 0 \leq P(E) \leq 1, \quad P(0) = 0, \tag{1}$$

$$P(E_1 + E_2 + E_3 + ...) = P(E_1) + P(E_2) + P(E_3) + \ldots, \qquad E_i E_j = \delta_{ij} E_i.$$

A state s is determined by the function $P(E)$ which satisfies Eqs. (1). Gleason's theorem [1] states that, for any $P(E)$ which satisfies Eqs. (1), there exists a density matrix ρ such that

$$P(E) = Tr(\rho E). \tag{2}$$

In other words, s is described by the density matrix ρ.

Gleason's theorem does not hold in two-dimensional Hilbert spaces. Consider a two-dimensional Hilbert space $\mathcal{H}$. With the notation

$$\sigma.\hat{n} \mid \hat{n} >= \mid \hat{n} >, \quad \hat{n} = (\sin\theta\cos\varphi, \sin\theta\sin\varphi, \cos\theta), \tag{3}$$

M. Ferrero and A. van der Merwe (eds.),
New Developments on Fundamental Problems in Quantum Physics, 415–419.

$$0 \leq \theta \leq \pi, \quad 0 \leq \varphi < 2\pi, \quad | \hat{n} >= \begin{pmatrix} \cos\frac{\theta}{2} \\ \sin\frac{\theta}{2} e^{i\varphi} \end{pmatrix},$$

the projector onto $| \hat{n} >$ is given by

$$E_{\hat{n}} \equiv | \hat{n} >< \hat{n} | = \begin{pmatrix} \cos^2\frac{\theta}{2} & \frac{1}{2}\sin\theta\ e^{-i\varphi} \\ \frac{1}{2}\sin\theta\ e^{i\varphi} & \sin^2\frac{\theta}{2} \end{pmatrix} = E(\theta, \varphi). \quad (4)$$

Equations (1) still hold, the last being

$$P(E_{\hat{n}} + E_{-\hat{n}}) = P(E_{\hat{n}}) + P(E_{-\hat{n}}) = P(I) = 1, \quad E_{\hat{n}} E_{-\hat{n}} = 0. \quad (5)$$

Now, in two-dimensional Hilbert spaces it is not difficult to find probability distribution functions

$$P(E_{\hat{n}}) \equiv P(\theta, \varphi) \quad (6)$$

such that no density matrix ρ exists which satisfies Eq. (2). How do we then describe in $\mathcal{H}$ the state s characterized by $P(\theta, \varphi)$?

Usually one dismisses the problem altogether by stating that no such probability distribution will ever appear. And one justifies the statement saying that any two-dimensional Hilbert space always corresponds eventually to a reduction of a higher-dimensional Hilbert space and that well-tested rules exist which allow one to get the two-dimensional density matrix from the higher-dimensional one. Then, by construction, a density matrix exists in the two-dimensional Hilbert space. But, can one not envisage genuine two-dimensional Hilbert spaces? Is there an inconsistency in two-dimensional Hilbert spaces? We will now show that no Gleason exceptional probability distributions can actually appear in $d = 2$ Hilbert spaces, and we will do so without refering to higher dimensional Hilbert spaces.

2. EXAMPLES

Let us first recall a couple of examples of such a distribution. The most general density matrix in 2 dimensions is

$$\rho = \begin{pmatrix} p & \sqrt{p - p^2}\ \cos q\ e^{\ ir} \\ \sqrt{p - p^2}\ \cos q\ e^{-ir} & 1 - p \end{pmatrix}, \quad (7)$$

$$0 \leq p \leq 1, \qquad 0 \leq q \leq \frac{\pi}{2}, \qquad 0 \leq r < 2\pi,$$

so that the most general probability distribution which satisfies Gleason's theorem is

$$\begin{aligned} P_\rho(\theta, \varphi) &= Tr \rho E_{\hat{n}} \\ &= \frac{1}{2} + (p - \frac{1}{2})\ \cos\theta + \sqrt{p - p^2}\ \cos q\ \sin\theta\ \cos(r + \varphi) \quad (8) \\ &\equiv P(\theta, \varphi, p, q, r). \end{aligned}$$

Notice that from the dependence on θ and φ one obtains p, q and r, and thus ρ. This is how the state is determined. Mutatis mutandis, from the dependence on p, q and r one obtains θ and φ and thus $E(\theta,\varphi)$. This is how the observable is determined.

An example of a Gleason-exceptional distribution is [2]

$$P_e(\theta,\varphi) = \frac{1}{2} + \frac{cos^3\theta}{2}, \tag{9}$$

which satisfies

$$0 \le P_e(\theta,\varphi) \le 1, \quad P_e(\theta,\varphi) + P_e(\pi-\theta,\varphi\pm\pi) = 1 \tag{10}$$

and is therefore a correct probability distribution.

A more radical example is the one of dispersion free states [3]

$$\begin{aligned} P_e(\theta,\varphi) &= 1, \quad \text{if } 0 \le \varphi < \pi \quad \text{or } \theta = 0, \\ P_e(\theta,\varphi) &= 0, \quad \text{if } \pi \le \varphi < 2\pi \quad \text{or } \theta = \pi. \end{aligned} \tag{11}$$

For no values of p, q and r, which satisfy Eq. (7) does Eq. (8) reproduce Eq. (9) or Eqs. (11). What then if an experiment gives Eq. (9) or Eqs. (11)?

3. ANALYSIS

Let us analyse a bit more what a probability distribution means. One first prepares the system with a well defined procedure. The system is then in a state s. Let us call the universal apparatus used in preparing s $B(P,Q,R)$, where P, Q and R are the dials which by varying them allow to prepare all possible states s, one-to-one. One then performs measurements on s with another universal apparatus $A(\Theta,\Phi)$ which depends on the positions of the dials Θ and Φ in such a way that there is a one-to-one correspondence between $A(\Theta,\Phi)$ and all possible yes-no observables. All we know about s comes from these measurements, which give the probability distribution $\mathcal{P}_s(\Theta,\Phi)$. Notice that we are using the fact that one does not have to deal with all observables, it is enough to deal with the ones corresponding to projectors. The function $\mathcal{P}_s(\Theta,\Phi)$ determines s and thus P, Q and R. If one knows the relations $\Theta(\theta,\varphi)$ and $\Phi(\theta,\varphi)$ then $\mathcal{P}_s(\Theta(\theta,\varphi),\ \Phi(\theta,\varphi)) \equiv P_s$. If then $P_s = P_e$ we are in trouble, as we have no means of describing s in $\mathcal{H}$. How do we know $\Theta(\theta,\varphi)$ and $\Phi(\theta,\varphi)$?

Both the correspondence by which an apparatus A acts as a projector in $\mathcal{H}$ and the one by which an apparatus B determines the density matrix which represent s in $\mathcal{H}$ are obtained from the comparision of P_ρ and $\mathcal{P}_s$, i.e.,

$$P_\rho(\theta,\varphi) = \mathcal{P}_s\ (\Theta,\Phi). \tag{12}$$

Thus, a unique correspondence

$$\begin{aligned} &\theta = \theta(\Theta, \Phi), \quad \varphi = \varphi(\Theta, \Phi), \\ &p = p(P, Q, R), \quad q = q(P, Q, R), \quad r = r(P, Q, R) \end{aligned} \tag{13}$$

should emerge from Eq. (12); notice the independence of the two groups of parameters, or, in other words, that preparation and measurement have independent, uncorrelated, representations in $\mathcal{H}$. Thus, going from Θ and Φ to θ and φ is only possible if a ρ exists which represents s, so that actually no Gleason exceptional probability distribution is possible. Where did then P_e in Eq. (9) or Eqs. (11) come from?

Suppose the experimentalist has found a probability distribution

$$\mathcal{P}_s(\Theta, \Phi) = \frac{1}{2} + \frac{\cos^3 \Theta}{2}, \tag{14}$$

and suppose the theorist *assumes* $\Theta = \theta$, $\Phi = \varphi$. Then

$$P_s(\theta, \varphi) = \frac{1}{2} + \frac{\cos^3 \theta}{2} = P_e \tag{15}$$

is Gleason exceptional. Well, since the theorist's measuring apparatus-projector correspondence is an assumption, which, leading to a state $s \neq \rho$ cannot be correct, he better changes his assumption. This he will have to do until, necessarily, $s = \rho$ and the distribution function will be Gleason consistent. Indeed, let the theorist now assume $\cos^3 \Theta = \cos \theta$, $\Phi = \varphi$. Then

$$P_s(\theta, \varphi) = \frac{1}{2} + \frac{\cos \theta}{2}, \tag{16}$$

and $s = \rho(p = 1)$. Now the theorist can go on checking his assumption with the help of Eq. (12), i.e., checking if his measuring apparatus-projector correspondence holds for all preparations. If it does, then all s are represented by ρ, and no Gleason exceptional situation exists. If it does not, he will have to change again his assumption to a new one. If the process ends, no Gleason exceptional probability distribution exists; if not, then no representation of the physics exists in a two dimensional Hilbert space, and one has to go to higher dimensional ones. Again, and now because Gleason's theorem always applies, no Gleason exceptional probability distribution exists.

A similar analysis, with some modifications, applies to Eqs. (11). Indeed, one would get it from

$$\begin{aligned} &\mathcal{P}_s(\Theta, \Phi) = 1, \quad \text{for } 0 \le \Phi < \pi \quad \text{or} \quad \Theta = 0, \\ &\mathcal{P}_s(\Theta, \Phi) = 0, \quad \text{for } \pi \le \Phi < 2\pi \quad \text{or} \quad \Theta = \pi, \end{aligned} \tag{17}$$

with $\Theta = \theta$ and $\Phi = \varphi$. One could then change to the correspondence

$$\Theta, \Phi / \mathcal{P}_s = 1 \longrightarrow \theta = 0, \quad \Theta, \Phi / \mathcal{P}_s = 0 \longrightarrow \theta = \pi, \tag{18}$$

which leads to $s = \rho(p = 1)$ and is thus not Gleason expectional anymore. Now, however, the measuring apparatus is not one-to-one anymore, it only leads to two projectors in the Hilbert space. The process continues as before, so that no Gleason exceptional probability distributions exist, but this time the measuring apparatus cannot be universal.

4. CONCLUSIONS

Gleason exceptional probability distributions correspond to a wrong association of projectors with measuring apparatuses. In $d > 2$ a wrong association of this type can, for any particular given state, be hidden by a wrong density matrix - physical state association. This is not so in $d = 2$. Also, in $d = 2$, probability distributions exist which exclude universal measuring apparatuses.

ACKNOWLEDGMENTS

I acknowledge financial support of CICYT and CIRIT through grants AEN95-0590 and GRQ93-1047, respectively.

REFERENCES

1. A.H. Gleason, "Measures on the closed subspaces of a Hilbert space," *J. Math. Mech.* **6**, 885 (1957).
2. A. Galindo and P. Pascual, *Quantum Mechanics* (Springer, Berlin, 1990).
3. J. Bell, "On the problem of hidden variables in quantum mechanics," *Rev. Mod. Phys.* **38**, 447 (1966).

QUANTUM DECOHERENCE INDUCED BY GRAVITATIONAL FLUCTUATIONS IN THE MEASUREMENT APPARATUS

J. Unturbe

Universidad Autónoma de Madrid
28049 Madrid, Spain

From the interaction of a quantum system with its own gravity, we derive bound states for the center of mass of a body with a mass greater than a critical value, which we obtain too. Also, in a more general, different approach, we obtain a master equation for the density matrix of that center of mass (with a non-unitary term responsible for decoherence). Both results are in agreement.

Key words: decoherence, theory of measurement, gravitational noise.

When a microscopic system interacts with a measurement apparatus, their wave functions will entangle. This means macroscopic superposition of the "measurement results," say, cat alive and dead. To interpret this we resort to decoherence.

In a quantum system, the quantum decoherence could be to define as the fast damping of the off-diagonal elements of the density matrix when it interacts with the environment. Decoherence takes place in a definite basis which has been chosen in order to produce consistent histories. Since these off-diagonal elements are undetectable in practice, one should average over them. As Zeh said in its contribution, this environment must be the rest of the universe, to be free of contradictions.

An investigation line maintains this decoherence to be produced by vacuum gravitational fluctuations as that environment [1]. Generally this approach only take into account the reaction of the matter to fluctuations of fundamental origin, which implies unpleasant free parameters. Instead of that Sánchez-Gómez and I have considered the reaction of the metric to the body as well [2]. In this line we present here a very simple model illustrating how a body can be decoherent when its mass goes beyond certain critical

M. Ferrero and A. van der Merwe (eds.),
New Developments on Fundamental Problems in Quantum Physics, 421–424.

value without resort free parameters.

For the sake of illustration we consider an extreme idealized pointer apparatus (i) as an isolated homogeneous ball of mass M and radius R, (ii) that produces a Newtonian gravitational field, and (iii) its center of mass (CM) as a wavepacket.

The assumptions are the following: (a) because the CM is not a point, the gravitational field fluctuates according with the CM wavefunction, and (b) these fluctuations obey the uncertainty principle

$$\Delta E_G \tau_G \sim \hbar, \tag{1}$$

where ΔE_G= gravitation energy fluctuation τ_G = gravitation time fluctuation.

Our contribution shows that *these gravitational fluctuations could make the CM wavefunction decoherent* beyond a critical mass. We choose two independent approaches with the same results, namely, "harmonic potential" and "gravitational noise."

A. The Harmonic Potential Approach

In this approach we consider only the mean value of the metric fluctuations inside the body. In the lowest approximation we must consider the Newtonian potential which is harmonic in this region. The wave function of the CM is affected by the gravitational potential, and the schrödinger equation gives us a harmonic solution, that is, bound states with an energy separation of $\hbar\omega$, where

$$\omega = (G\rho)^{1/2}; \tag{2}$$

here ρ is the matter density (normal densities $\sim$ 1g/cm^3).

The most important consequences are the following [3]:

(i) The energy of the fundamental level is $\hbar\omega/2$ ($\sim 10^{-19}$eV for normal densities).

(ii) The spreading of the CM packet in the fundamental level is

$$\overline{\Delta X} = (\hbar/M\omega)^{1/2}. \tag{3}$$

(For M = 1g, then $\overline{\Delta X} \sim 10^{-12}$cm, for normal densities.)

(iii) The condition for the existence of bound states is given by

$\hbar\omega/2$ (fundamental energy) $< GM^2/2R$ (high of well energy),

which implies $M > M_{tr}$, where the transition mass is

$$M_{tr} = (\hbar^6/\rho/G^3)^{1/10}. \tag{4}$$

($M_{tr} \sim 10^{-14}$gm for normal densities.)

(iv) The time decay from an arbitrary level of energy, say with ΔX_0 spread, is easy to obtain:

$$\tau \sim \hbar/\Delta E \sim \frac{\hbar}{\frac{1}{2}M\omega^2\Delta X_0^2 - \frac{1}{2}M\omega^2\overline{\Delta X}^2}$$

and, if $\Delta X_0 \gg \overline{\delta X}$,

$$\tau \sim \omega^{-1}(\overline{\Delta X}/\Delta X_0)^2. \tag{5}$$

(If the initial spread of the CM of a mass of 1 gm is about 1 cm, it will decay to a spread of 10^{-12}cm in a time of about 10^{-20}s.)

B. The Gravitational Noise Approach

Using the Feynman-Vernon formalism [4] and by considering the CM gravitational fluctuations as a stochastic noise, we arrive at an equation for the density matrix representing the CM evolution:

$$\dot{\rho}(\mathbf{x},\mathbf{x}';t) = -\frac{i}{\hbar}[H,\rho] - \eta(\mathbf{x},\mathbf{x}')\rho(\mathbf{x},\mathbf{x}';t),$$

where the last term of the r.h.s. side is a nonunitary term and is given by (for $M > M_{tr}$)

$$\eta(\mathbf{x},\mathbf{x}') = \tau_G(M\omega^2/\hbar)^2(\Delta X)^2(\mathbf{x}-\mathbf{x}')^2,$$

where ω is given by (2). Now for an arbitrary $\Delta X, \Delta E_G \sim M\omega^2(\Delta X)^2$ inside the body. Taking this and (1) into account, we easily obtain

$$\begin{aligned}\eta(\mathbf{x},\mathbf{x}') &\sim \tau_G\Delta E_G(M(\omega^2/\hbar^2)(\mathbf{x}-\mathbf{x}')^2 \sim (M\omega^2/\hbar)(\mathbf{x}-\mathbf{x}')^2\\ &\sim (GM\rho/\hbar)(\mathbf{x}-\mathbf{x}')^2 \text{ or } (\omega/\overline{\Delta X}^2)(\mathbf{x}-\mathbf{x}')^2,\end{aligned}$$

where ω and $\overline{\Delta X}$ are given by (2) and (3), respectively.

Hence the behavior of CM is the following: *independently of the characteristics of CM initial state* (that is, no matter how it has been prepared) *it will eventually evolve into a Gaussian wavepacket with a constant microscopic width* $\overline{\Delta x}$.

This happens after a time which is of the order of η^{-1} which can we rewriten as Eq. (5).

This kind of treatment can be also applied to microscopic systems ($M \ll M_{tr}$), in which case we find that $\eta = 0$ in a very large region, regaining the unitarian behavior [3].

C. Main Conclusions

For a *macroscopic body* ($M > M_{tr}$) we find that its gravitational field is capable of *producing decoherence* in the CM dynamic evolution, then precluding quantum superpositions of macroscopic states of CM.

This result practically *coincides with that obtained in the GRW model* [5] with the advantage that our model is parameter-free.

For masses clearly less than M_{tr}, gravitational effects are practically *negligible*, so that no decoherence is produced in such a case.

Bodies with mass about M_{tr} (10^{-14}gm for normal densities) have a very complex behavior and our approximation is not valid. The more simple model predicts fluctuations of the CM. It is not foolish to think that these bodies could have something to do with chaos and even the origin of life (because of the property of causing and undergoing reductions).

REFERENCES

1. J. L. Rosales and J. L. Sánchez-Gómez, *Phys. Lett. A* **119**, 320 (1995).
2. J. Unturbe and J. L. Sánchez-Gómez, *Nuovo Cimento B* **107**, 1992; *Phys. Essays* **6**, 580 (1993).
3. A more detailed work is in progress with Sánchez-Gómez.
4. R. P. Feynman and A. R. Hibbs, *Quantum Mechanics and Path Integrals* (McGraw-Hill, New York, 1965). M. R. Gallis, *Phys. Rev. A* **45**, 47 (1992).
5. G. C. Ghirardi, A.. Rimini, and T. Weber, *Phys. Rev. D* **34**, 470 (1986).

SCHRÖDINGER CAT STATES IN QUANTUM OPTICS

David Vitali

Dipartimento di Matematica e Fisica
Università di Camerino and
Istituto Nazionale Fisica della Materia, Italy

Two different schemes for the generation and the detection of linear superpositions of two coherent states of a radiation mode are considered. Both of them are based on apparatuses already employed for the experimental demonstration of quantum non-demolition (QND) measurements of a radiation field.

Key words: macroscopic quantum coherence, quantum non demolition measurements, optical nonlinearities.

Quantum optics offers various possibilities to study and maybe produce in the laboratory quantum superpositions of macroscopically distinguishable states of the field, the so called "Schrödinger cats". A promising way is offered by conditional measurements: several physical processes can produce correlated subsystems, in such a way that a quantum measurement done on one subsystem leaves the other one in a well-defined state, which can be a Schrödinger cat. This way was explored in detail by Brune *et al.* [1], considering the possibility to produce quantum superpositions of microwave fields inside a very high Q superconducting cavity. The cat is "prepared" in a conditional way through the detection of an atom which has crossed the cavity and got entangled with the state of the field. In this paper we investigate theoretically the possibility of using an *all-optical* preparation and detection process, which would involve two field modes rather than atoms and field.

We consider two radiation modes in a cavity, which we shall call the signal and the meter, with annihilation operators a and b, respectively. The two modes interact with each other and this interaction is described by a Hamiltonian term H_{int}. This is the typical situation considered in QND

M. Ferrero and A. van der Merwe (eds.),
New Developments on Fundamental Problems in Quantum Physics, 425–429.

measurements, in which an observable of the signal mode can be measured by performing an appropriate measurement on the meter mode only. In an ideal QND measurement, the result of the meter measurement allows to determine exactly the state of the signal without perturbing it [2]. We will explicitely use the QND property of the present scheme to develop a detection strategy which is able to preserve as much as possible the interference fringes signalling the presence of quantum coherence and which are usually quickly destroyed by the dissipative interaction with the vacuum modes outside the cavity.

We explicitely consider two different examples, both referred to apparatuses already employed for the experimental realization of optical QND measurements. The first model refers to the experiment by La Porta *et al.* [3], in which the two orthogonally polarized modes generated by a pumped KTP crystal are mixed by two $\lambda/2$ wave plates, preceding and following the parametric amplifier. The interaction Hamiltonian for the two modes can be written as $H_{int} = \hbar(K_+ a^\dagger b^\dagger + H.c.) + \hbar(K_- a^\dagger b + H.c.)$; the coupling constant K_+ incorporates the nonlinear susceptibility of the KTP crystal and the pump amplitude, while K_- gives the strength of the mode mixing at the polarizers. The QND condition is realized when the optic axis of two half-wave plates forms an appropriate angle with respect to the KTP ordinary polarization axis [3]. In this case the two coupling constants $K_\pm$ have the same modulus and phases can be chosen so that $K_+ = -K_- = i\chi/4$, so that the interaction becomes

$$H_{int} = \hbar\chi X_a Y_b, \tag{1}$$

where $X_a = (a + a^\dagger)/2$ and $Y_b = (b - b^\dagger)/2i$, showing that the apparatus becomes able to perform QND measurements of the signal quadrature X_a.

The second model is characterized by a crossed Kerr interaction of the form

$$H_{int} = -\hbar f a^\dagger a b^\dagger b \tag{2}$$

realized experimentally in [4] for the QND measurement of an optical intensity. This effective nonlinear interaction can be realized using beams of atoms quasi-resonant with the two cavity modes and such that the one-photon detuning is much larger than the two-photon detuning. In this way, the self-Kerr terms proportional to $(a^\dagger a)^2$ and $(b^\dagger b)^2$ can be neglected.

The conditional measurement scheme for the cat state generation is identical for the two models: the system evolves for a given time t_0 starting from an appropriate initial condition $|\psi_a(0)\rangle|\psi_b(0)\rangle$ and is then subject to the measurement of an appropriate meter observable. The meter is projected into an eigenstate $|\Psi_m\rangle$ and the signal is projected onto a state conditioned to the result m of the measurement, given by

$$|\psi^{aft}\rangle = \frac{\langle\Psi_m|e^{-\frac{i}{\hbar}H_{int}t_0}|\psi_a(0)\rangle|\psi_b(0)\rangle}{\sqrt{P_m}}, \tag{3}$$

where P_m is the probability of the result m. In the case of the model (1), taking the vacuum for the meter mode and a squeezed vacuum for the signal as initial states, a measurement of the meter photon number with result $b^\dagger b = m$ projects the signal in a Schrödinger cat-like state, tending to the even coherent state $|\sqrt{m}\rangle + |-\sqrt{m}\rangle$ in the limit of large m and large signal input squeezing along the quadrature orthogonal to X_a, provided that the interaction time is $t_0 = 2/\chi$ [5]. In the crossed-Kerr model, starting from a coherent state in each mode $|\psi_a(0)\rangle|\psi_b(0)\rangle = |\alpha_0\rangle|\beta_0\rangle$, the system evolves at the time $t_0 = \pi/f$, in the entangled state [6]

$$|\psi(t_0)\rangle = \frac{1}{2}\{|\alpha_0\rangle + |-\alpha_0\rangle\}|\beta_0\rangle + \frac{1}{2}\{|\alpha_0\rangle - |-\alpha_0\rangle\}|-\beta_0\rangle . \quad (4)$$

A meter measurement able to discriminate between $|\beta_0\rangle$ and $|-\beta_0\rangle$, is the homodyne measurement of the quadrature $X_b = (be^{-i\phi_b^0} + b^\dagger e^{i\phi_b^0})/2$, where $\phi_b^0 = arg(\beta_0)$. For large enough $|\beta_0|$, the overlap $\langle\beta_0| - \beta_0\rangle$ can be neglected and a measurement of X_b, even with a poor resolution, will select the even (odd) coherent state for the signal whenever the result $X_b = |\beta_0|$ ($X_b = -|\beta_0|$) is obtained [6].

Anyway, these two generation schemes are unrealistic and too "optimistic," because they completely neglect losses, i.e., the decoherence effect of the vacuum modes outside the cavity. In the realistic case one has to include damping of both modes, described by the cavity decay rates k_a and k_b. We include also the two cavity detunings δ_a and δ_b, which will be useful for the detection of the interference fringes, and consider the model defined by the following master equation for the density matrix D of the two modes

$$\dot{D} = \mathcal{L}_a D + \mathcal{L}_b D - \frac{i}{\hbar}[H_{int}, D] , \quad (5)$$

where H_{int} is given by (1) or (2), the superoperator $\mathcal{L}_a$ is given by $\mathcal{L}_a\rho = k_a(2a\rho a^\dagger - a^\dagger a\rho - \rho a^\dagger a) - i[\delta_a a^\dagger a, \rho]$ and an analogous expression holds for $\mathcal{L}_b$. The master equation (5) can be solved analytically for both models using the Wigner function representation (see [6,7]). In general, the conditioned signal Wigner function immediately after the measurement is connected to the Wigner function of the whole system before the measurement by the convolution:

$$W^{af}(\alpha) = \frac{\pi}{P_m}\int d^2\beta W_m(\beta) W^{bf}(\alpha,\beta) , \quad (6)$$

where $W_m(\beta)$ is the Wigner function of the meter state after the measurement $|\Psi_m\rangle$ and the probability to get this state is $P_m = \int d^2\alpha d^2\beta \pi W_m(\beta) W^{bf}(\alpha,\beta)$. In the ideal case of zero damping and detuning for both modes, the state of Eq. (6) coincides with the cat states discussed above. In the realistic damped case, this Schrödinger cat-like state is modified, especially in its interference terms which are notoriously fragile to the presence of dissipation. Anyway, if the two decay rates k_a and k_b are not large with respect to the QND coupling constants χ or f, the state (6) will describe a "weakly damped cat state."

After the generation of the linear superposition state, one has to perform a second measurement to detect it. Anyway, performing measurement directly on the signal mode could be too destructive, because an efficient measurement on the signal mode needs a sufficient leakage loss at the output mirror and this will necessary cause a quick destruction of the interference fringes, yielding the transformation of the quantum linear superposition into a classical mixture of two approximate coherent states. Our idea is to use the QND nature of the models and deduce the state of the signal mode from measurements performed on the meter mode only. If we choose a situation where $k_a \ll k_b$ and $k_b \sim \chi$, we have that the meter leakage loss is sufficiently large to perform homodyne measurements of a meter quadrature and that the signal decay rate can be chosen to be very small so that the cat state generated is only slightly affected by dissipation. Moreover, it can be seen [6,8] that in this adiabatic limit $k_a \ll k_b$, the meter homodyne photocurrent $I(t)$ can be directly expressed in terms of the "instantaneous" mean values of signal mode operators conditioned on the result of the measurement. For the model (1), one has [8]

$$I(t) = 2\eta\chi \left[\frac{k_b^2}{k_b^2 + \delta_b^2} \langle X_a(t) \rangle_c + \sqrt{\frac{k_b}{2\eta\chi^2}} \xi(t) \right] , \tag{7}$$

where η is the efficiency of the homodyne detection and $\xi(t)$ is a Gaussian white noise with $\langle \xi(t)\xi(t') \rangle = \delta(t - t')$, while for the model (2) one has a similar expression, with the signal photon number $a^\dagger a$ replacing the quadrature X_a [6]. Actually, for a complete experimental characterization of a Schrödinger cat state one has to verify both the presence of two distinct "classical" field states and the presence of quantum coherence between these two states, in order to distinguish the linear superposition from the corresponding statistical mixture. In the case of the model (2) this characterization can be obtained using only signal photon number probability distributions by using the same method proposed in Ref. [1], in which a coherent field in phase or $\pi/2$ out of phase with respect to the coherent states of the cat is appropriately injected into the cavity.

In the case of the model (1), this characterization can be given by the probability distributions of two orthogonal quadratures [9]. Actually, our indirect homodyne measurement of the signal quadrature measures the QND quadrature X_a only; anyway, the other signal quadrature Y_a could be measured with the same QND strategy if we could change the interaction Hamiltonian (1) soon after the meter photon number measurement. This can be done with an electronic switch, driving an electro-optic modulator which shifts the phase of the pump with respect to its initial value. In this way one can change the sign of K_+ and the interaction Hamiltonian (1) turns to $H_{int} = -\hbar\chi Y_a X_b$. Eq. (7) will now hold for the homodyne photocurrent associated with the measurement of the meter quadrature Y_b, with $\langle X_a(t)\rangle_c$ replaced by $\langle Y_a(t)\rangle_c$ [7].

The analysis of Refs. [6,7] shows that the damped cat state generated in the model (1) is much more robust against the decoherent effects of vacuum damping than that generated by the crossed Kerr model and this makes the model (1) very attractive. In our opinion, this experiment could be feasible with current technologies, expecially in the light of the recent experimental results [10], where this apparatus has been used for the realization of QND measurement of a quadrature clearly in the quantum regime. Another interesting aspect of model (1) is that it is possible to adjust the two detunings in a such a way as to significantly increase the decoherence time, i.e., the "lifetime" of the interference fringes associated to the quantum coherence. The detunings therefore can greatly facilitate the detection of the cat state and this is the reason why we have explicitely considered nonzero cavity detunings in the model.

REFERENCES

1. M. Brune, S. Haroche, J.M. Raimond, L. Davidovich, and N. Zagury, *Phys. Rev. A* **45**, 5193 (1992).
2. C.M. Caves, K.S. Thorne, R.W.P. Drever, V.D. Sandberg, and M. Zimmermann, *Rev. Mod. Phys.* **52**, 341 (1980).
3. A. La Porta, R.E. Slusher, and B. Yurke, *Phys. Rev. Lett.* **62**, 28 (1989).
4. J.P. Poizat and P. Grangier, *Phys. Rev. Lett.* **70**, 271 (1993).
5. B. Yurke, W. Schleich, and D.F. Walls, *Phys. Rev. A* **42**, 1703 (1990).
6. D. Vitali, P. Tombesi, and P. Grangier, to appear in *Appl. Phys. B.*
7. P. Tombesi and D. Vitali, *Phys. Rev. Lett.* **77**, 411 (1996).
8. H.M. Wiseman and G.J. Milburn, *Phys. Rev. A* **47**, 642 (1993).
9. B. Yurke and D. Stoler, *Phys. Rev. Lett.* **57**, 13 (1986).
10. S.F. Pereira, Z.Y. Ou, and H.J. Kimble, *Phys. Rev. Lett.* **72**, 214 (1994); K. Bencheikh, J.A. Levenson, P. Grangier, and O. Lopez, *Phys. Rev. Lett.* **75**, 3422 (1995).

AN EXPERIMENTAL REALIZATION OF BOHM'S SPIN-1/2 PARTICLE EPR *GEDANKEN* EXPERIMENT

Thomas Walther and Edward S. Fry

Department of Physics, Texas A&M University
College Station, Texas 77843-4242

The Hg isotope ^{199}Hg (nuclear spin I=1/2) is suited for an experimental realization of Bohm's spin-1/2 particle EPR *gedanken* experiment. The experiment is based on the photo dissociation of a ^{199}Hg dimer followed by the analysis of the nuclear spin components of the resulting fragments. Dissociation and analysis of spin components are carried out by pulsed lasers. The experiment would close both loopholes, which were left open by previous experiments testing Bell inequalities.

Key words: Bell-inequality, Einstein-Podolsky-Rosen experiment.

1. INTRODUCTION

One of the fundamental problems not yet completely resolved in quantum mechanics is the *gedanken* experiment that was described by Einstein, Podolsky and Rosen in 1935 and that supposedly demonstrates the incompleteness of quantum mechanics [1]. In particular, the concept of non-locality in quantum mechanics contradicts the intuitive understanding of quantum mechanics. In order to save locality, which is assumed in classical physics, local hidden variable (LHV) theories were later introduced; these supposedly lead to a complete description of reality. In these theories all interactions are local in nature. The hidden variables are introduced in such a fashion that the predictions of these LHV theories will mimic the outcome of experiments. The parameters themselves are, however, not accessible through any known experimental approaches. The crucial insight first recognized by John Bell [2] is that such parameters would also restrict the magnitude of correlations in certain phenomena and that the statistical predictions of quantum mechanics may, in fact, give strong correlations that violate that restriction. Mathematically this behavior can be expressed in form of inequalities, the

M. Ferrero and A. van der Merwe (eds.),
New Developments on Fundamental Problems in Quantum Physics, 431–439.

Bell inequalities. They are valid in the context of LHV-theories, but can be violated in quantum mechanics. Therefore, experimentally testing these inequalities against the quantum mechanical predictions can yield a decisive decision between the non-locality of quantum mechanics and the locality of LHV theories.

Because of experimental inadequacies, i.e., low detection efficiencies, the strongest form of the Bell inequalities (Bell-Clauser-Horne inequality, BCH) has never been tested [3,4]. Previous experiments, which were predominantly based on the detection of photons, tested weaker forms, which required an additional assumption [4,5,6]. Consequently, all previous experiments have left open a loophole due to their low detection efficiencies [7]. So far there has only been one attempt to enforce the locality condition [8]; this, however, was not rigorous and therefore a second loophole involving the enforcement of locality is not yet completely closed [9].

Different approaches to close the detection efficiency loophole can be undertaken [10]. Since photo-ionization is a very efficient detection method, it should be possible to make use of these high detection efficiencies by performing experiments on atoms in entangled states to close the detection efficiency loophole. Furthermore the photo dissociation of diatomic molecules can produce atom pairs which are highly spatially correlated, which is also one of the requirements to test the strongest form of the Bell inequalities. By choosing the right molecular system it is possible to generate a spin entanglement between two spin-1/2 atoms resulting from a photo dissociation process. One of the possible experiments, which is currently under way at Texas A&M University [11], will be discussed briefly in this paper. It is an exact realization of Bohm's spin-1/2 particle version of the original EPR-experiment.

2. THE ENTANGLED SINGLET STATE

The entangled singlet state of two spin-1/2 particles, which is necessary for a realization of Bohm's version of an EPR-experiment, can be written as

$$|\Psi\rangle = \frac{1}{\sqrt{2}}\Big(|\uparrow\rangle_1 |\downarrow\rangle_2 - |\downarrow\rangle_1 |\uparrow\rangle_2\Big). \qquad (1)$$

The indices refer to the particles 1 and 2, respectively and $|\uparrow\rangle$ means spin up and $|\downarrow\rangle$ spin down, respectively. The search for atoms which are suited for such a test should be restricted to those atoms which allow a particularly easy analysis of their spins. The Hg isotope ^{199}Hg with a nuclear spin I=1/2 fulfills this requirement as it has no orbital angular momentum and no electron spin in the electronic ground state. Our EPR-experiment, will require the measurement of the correlation of nuclear spins.

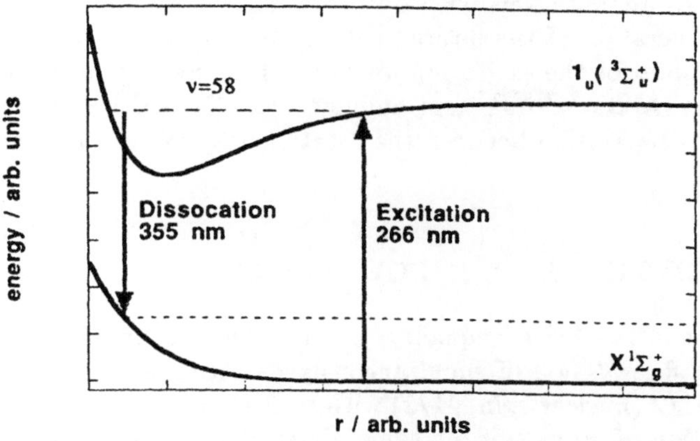

Fig. 1. Photo dissociation scheme of the dimer $^{199}Hg_2$ for the generation of a nuclear spin entangled state between the two resulting Hg atoms. First a 266 nm laser pulse excites the molecules to the $^3\Sigma_u^+$ state. Afterwards a 355 nm laser pulse will transfer the molecule onto a dissociating part of the $^1\Sigma_g^+$ ground state. The two resulting atoms have a very well defined kinetic energy, which is given by the laser energies, binding potential and exact rotational state dissociated.

An entangled state corresponding to Eq. 1 for two spatially separated ^{99}Hg atoms can be generated by a two-photon photo dissociation process of $^{199}Hg_2$ dimer (see Fig. 1). In the first excitation step the dimers are excited ɔ an electronic excited state; a second laser pulse with different wavelength ransfers the molecules in a Raman process onto a dissociating part of the round state potential energy surface. The second pulse not only defines a ime zero at which the atoms start separating, but also defines the kinetic nergy of the two fragments. The arrival time of the two fragments at the etectors is therefore well known. This is important, since the detection will lso be based on the use of pulsed lasers.

The $^{199}Hg_2$ dimer has no electron spin or angular momentum spin ı the electronic ground state. The two nuclear spin states are a singlet or riplet state. The selective generation of only the singlet entangled state equired by Eq. 1 is based on the symmetry properties of the molecular ave function. Specifically, because of the nuclear spin of the ^{199}Hg atoms, ıe overall dimer wave function has to be anti-symmetric with respect to n exchange of the two nuclei in order to conform to the Pauli principle. 'herefore the nuclear spin singlet states are only associated with rotational ;ates with even rotational quantum number; triplet states are only associated with rotational levels with odd rotational quantum number [11,12]. Thus, the generation of the desired entangled state is assured by selecting those transitions for the excitation process, which start from rotational levels with even rotational quantum number. The linewidth of the laser and the frequency separation between the rotational transitions permits this selection.

3. TEST OF THE BELL INEQUALITIES

A test of the BCH-inequality requires the measurement of the coincidence rate $R_{++}(\theta_1, \theta_2)$ of simultaneously detecting atom one of a dimer with m_F=+1/2 (nuclear spin +1/2) with respect to direction θ_1 at detector 1 and detection of atom 2 of the same dimer at detector 2 with m_F=+1/2 with respect to the direction θ_2. Furthermore the singles rates $R_{i+}(\theta_i)$ have to be determined, i.e. the detection rate of the atoms with m_F=+1/2 in direction θ_i at detector i, i=1, 2. In the context of our experiment, the BCH-inequality yields [3, 4]:

$$S(a,b,a',b') = \frac{R_{++}(a,b) - R_{++}(a,b') + R_{++}(a',b) + R_{++}(a',b')}{R_{1+}(a') + R_{2+}(b)} \leq 1, \tag{2}$$

where a,a' are two detection directions for the nuclear spin at detector 1, and b, b' are two directions at detector 2. The significant difference here is that this strong Bell inequality is a ratio of coincidence rates to singles rates, whereas the Bell inequalities tested in all previous experiments have involved a ratio of coincidence rates to coincidence rates. With the quantum mechanical predictions for these rates, the maximum violation of this inequality by quantum mechanics can be calculated. One finds for a, b, a', b' equal to 135°, 0°, 225°, and 90°, respectively,

$$S_{QM}(135°, 0°, 225°, 90°) = \frac{1}{2}\eta g \varepsilon_+ \left(1 + \sqrt{2}\left(\frac{\varepsilon_-}{\varepsilon_+}\right)^2\right), \tag{3}$$

where η is the detection efficiency, g the conditional probability that given atom 1 enters detector 1 that then atom 2 of the pair enters detector 2, and $\varepsilon_\pm$ are measures of the detector discrimination between the two spin states. The latter are defined as $\varepsilon_\pm = \varepsilon_S \pm \varepsilon_s$, where ε_S is the transmission for the spin component under consideration (ideally equal to one) and ε_s is the leakage for the other unwanted spin component (ideally 0). The parameters η and g must be as large as possible, in order to obtain the maximum violation of Eq. 2 by quantum mechanics. Furthermore, $\varepsilon_\pm$ should be as close as possible to unity.

The conditional probability g is given by the geometry of the experiment and the angular distribution of the fragments. Specifically, the value

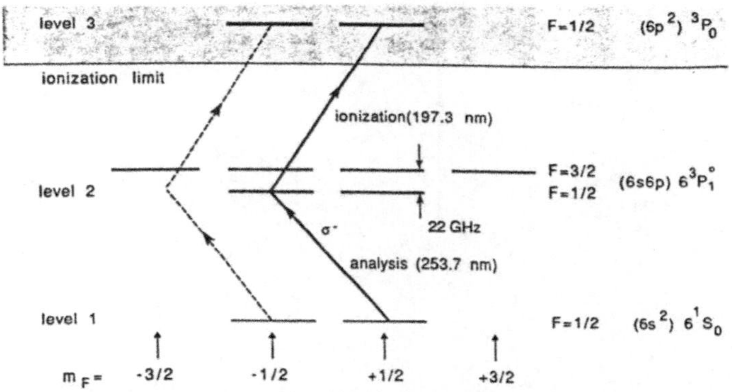

Fig. 2. The Hg energy level diagram with the levels relevant for our experiment. As an example the analysis scheme with left-hand circular polarized 253.7 nm laser light is shown. Since m_F has to decrease by one in such a transition, only those Hg atoms can be excited to the $(6s6p)\ 6^3P_1^\circ$ (F=1/2) level, which are in the $m_F=+1/2$ Zeeman sublevel of the ground state. The dashed transitions are detuned by 22 GHz and correspond to the leakage in our analysis scheme.

of g is determined by the size and position of the detector apertures, the source volume, i.e., the intersection between laser and dimer beams, as well as the velocity spread of the dimer in the beam. The excitation laser at 266 nm will be aligned copropagating with the dimer beam. Because of its very narrow linewidth, dimers with a very narrow velocity spread can be spectroscopically selected via the Doppler effect. Monte-Carlo simulations show that values of $g>0.975$ can be achieved [11].

In contrast to the suggestion in Bohm's original version, the analysis of the spin components will not be performed with Stern-Gerlach magnets - the nuclear magneton is too small. Instead it will be performed via a two photon-ionization process. The analysis is based on the polarization properties of the laser light employed and therefore on the specific selection rules between Zeeman sublevels. After the atoms enter the detectors, they pass two lasers with wavelengths 253.7 nm and 197.3 nm, respectively. The first laser is resonant to the transition between the ground state $(6s^2)\ 6^1S_0$ (F=1/2) (level 1) and the $(6s6p)\ 6^3P_1^\circ$ (F=1/2) level (level 2). The second laser drives the transition from level 2 to the auto-ionizing state $(6p^2)\ 6^3P_0$ (level 3); see Fig. 2.

The 253.7 nm analyzing lasers at the two detectors are in parallel planes and are at angles θ_1 and θ_2 with respect to the z-axis, respectively. These define the direction with respect to which the spin components are determined (see Fig. 3). Corresponding to the two possible nuclear spin components up and down, a ^{199}Hg atom in the electronic ground state has

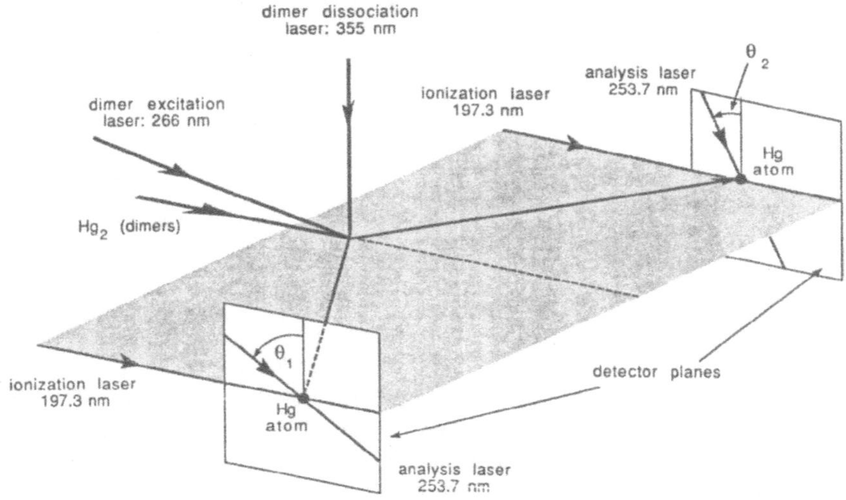

Fig. 3. Scheme of the experimental setup. The different lasers, the molecular beam, and two atoms originating from a dissociation process are shown in the detectors. The quantization axis with respect to which the analysis of the spins is performed is the propagation direction of the 253.7 nm lasers. See text for details.

the two possible Zeeman sublevels $m_F=\pm 1/2$. Assuming that the 253.7 nm laser light is left circular polarized, the selection rule gives Δm_F = -1, and the propagation direction defines the quantization axis. Therefore only those atoms in the ground state $m_F=+1/2$ can be excited to level 2 and subsequently ionized (see Fig. 2.). Theoretical calculations of the efficiency of this scheme show, that an efficiency > 99.5 % is possible [11]. This high efficiency can be explained due to the the large oscillator strengths of the transitions involved and the favorable ratio of non-radiative to radiative life time of the auto-ionizing level 3.

A possible leakage of the unwanted spin component (spin down in our example) is due to transitions from the m_F=-1/2 sublevel to the $6^3P_1^\circ(F=3/2)$ m_F=-3/2 level, which is detuned by 22 GHz. Non-perfect polarization of the light is of course another factor; it should be possible, however, to control the polarization well enough, so that the first is the dominant part of the leakage. Calculations show, that transitions via the $m_F=\pm 3/2$ hyperfine level are suppressed effectively due to the large detuning (ε_s < 0.1 %), so that excellent discrimination between the two spin states can be expected [11].

After the analysis of the spin state, the Hg atoms must be detected. High efficiency and low background are very important. The count rate of the experiment should not exceed approximately 1 Hg pair per second, since otherwise the probability of observing an atom in each detector from different dissociating dimers becomes non-negligible and correlations can not be correctly identified.

The final detection is performed via secondary electrons in channeltrons. The ion as well as the photo electron are used to produce secondary electrons at different metal electrodes. For our electrode configuration and acceleration voltages approximately 12 secondary electrons from a Hg^+ Ion and 5 from the associated electron are expected. A detection efficiency of almost unity should, therefore, be achieved for both the electron and the ion. This can be further increased employing an OR-detection scheme, i.e., the Hg atom is detected if either the electron or the ion are detected. Of course this leads to a high sensitivity to background events from ionization of residual Hg atoms or electrons produced by photo effect from scattered light. The former is highly suppressed by cooling the detection regions to ≈ 77 K. The latter is suppressed by (1) using nickel coated surfaces in the detection region (The Ni work function exceeds the 253 nm photoelectric threshold.); (2) carefully collimating the 197 nm radiation and (3) using electron optics to select only those electrons from the detection volume. If the detection efficiencies are as high as expected, an AND-detection becomes possible, which makes background signals less significant.

3.1 Quantum Mechanical Prediction

The QM prediction for the strong Bell inequality, S_{QM} Eq. 3, can be calculated with the help of the above mentioned parameters

$$S_{QM}(135°, 0°, 225°, 90°) = 1.159 > 1. \tag{4}$$

Since the Bell inequality is violated by the quantum mechanical prediction, the experiment fulfills all requirements for a test of the BCH-inequality. Furthermore locality can also be enforced. Using fast electro-optical modulators the polarization direction of linearly polarized light can be switched very fast and in a random way. With the use of beam splitting polarizers the propagation direction of the light and hence the quantization axis can therefore also be changed very fast. Thus, the direction of the quantization axis at one detector can be switched before a light signal can have "informed" the second detector about this change. It should be noted, that in order to rigorously enforce the locality condition the time a signal needs to travel the distance between the two detectors must be greater than the total time it takes to select a random number, switch to the appropriate polarization for the detection, ionize the Hg atoms, detect the electrons and finally read the result into the computer.

The verification of the parameters (g, η, $\varepsilon_{\pm}$) needed to evaluate the quantum mechanical prediction Eq. 3 is a crucial part of the experiment. If no violation of the inequalities is observed, it must be proven that the values of these parameters are such that the experimental setup was indeed capable of testing the BCH-inequality against quantum mechanics.

If the quantum mechanical predictions for the singles rates $R_{1+}(\theta)$, $R_{2+}(\theta)$ and for the coincidence rates $R_{++}(\theta_1,\theta_2)$, $R_{+-}(\theta_1,\theta_2)$ are assumed to be correct, we have shown that these parameters can be directly determined by measuring certain single and coincidence rates for special angles [13]. Of course it is desirable to independently measure these parameters without having to rely on quantum mechanics. This is especially the case for our experiment, since quantum mechanics is to be tested against LHV-theories. These parameters can, in fact, be determined by a few measurements that depend on simple general arguments and are not based on quantum mechanics.

Specifically, the absolute detection efficiencies for the electron and ion detection can be determined by detecting electrons and ions simultaneously. This capability is based on the fact that an ion and its corresponding photoelectron are strongly correlated. Having measured the count rates for ions N_i, electrons N_e and coincidence counts N_c, respectively, one can determine the three unknowns, the detection efficiencies for ions η_i and electrons η_e as well as the total number N_a of ionized Hg atoms by solving the following three equations:

$$N_i = \eta_i N_a, \quad N_e = \eta_e N_a, \quad N_c = \eta_e \eta_i N_a. \tag{5}$$

It should be noted that by using this procedure the detection efficiencies can be constantly monitored for both detectors independently during the course of the experiment.

The determination of the ionization efficiencies ε_+, ε_- and the conditional probability g is slightly more involved. First, the detection efficiencies can be measured as a function of the laser intensity. Naturally they only depend on the detector performance. The total number of ionized Hg atoms, however, does depend on the laser intensity. From a simple argument one sees, that N_a is given by $N_a=\varepsilon_+ N/2$, where N is the actual number of Hg atoms entering the detector. Since the population of the Hg atoms is equally split between the two Zeeman levels in the ground state, a factor of 1/2 was introduced in the above relation. Now, ε_+ can be determined by observing a saturation in the number of ionized atoms that are produced at high laser intensities. Next, an equation analogous to Eq. 5 also holds in the case when the 253.7 nm light is linear polarized; the only difference now is that in this case **all** the Hg atoms will be ionized and not just those in one Zeeman sublevel. Thus, the leakage ε_m is equal to zero in this case. Therefore, the number N_a' of the ion/electron pairs observed when using linear polarized light is given by $N_a'=\varepsilon_M$ N. We now have two equations with two unknowns ε_M and ε_+, respectively. Clearly, ε_m can now be calculated as well. Finally, the conditional probability g can be determined by using linear polarized light at both detectors and simply determining the count rate for one detector under the condition that an atom has been detected in the other detector.

A prototype of one detector has recently been finished. Soon, first measurements on the detection efficiency will be performed. After that it is merely a question of time until Bohm's gedankenexperiment, in a modified form using laser spectroscopy, will be realized.

The work is supported by the Robert A. Welch Foundation Grant No. A-1218, the Texas Advanced Research Program Grant 010366-17, and the National Science Foundation Grant PHY-9221038.

REFERENCES

1. A. Einstein, B. Podolsky, and N. Rosen, *Phys. Rev.* **47**, 777 (1935).
2. J. Bell, *Physics* **1**, 195 (1964); reprinted in J.S. Bell, *Speakable and Unspeakable in Quantum Mechanics* (Cambridge University Press, Cambridge, 1987).
3. J. Clauser and M. Horne, *Phys. Rev. D* **10**, 526 (1974).
4. J. Clauser and A. Shimony, *Rep. Prog. Phys* **41**, 1881 (1978).
5. A. Duncan and H. Kleinpoppen, in *Quantum Mechanics Versus Local Realism-The Einstein Podolsky-Rosen Experiment*, F. Selleri, ed. (Plenum, New York, 1988), Chap. 7, pp. 175–218.
6. J. Rarity and P. Tapster, *Phys. Rev. Lett.* **64**, 2495 (1990).
7. P. Pearle, *Phys. Rev. D* **2**, 1418 (1970).
8. A. Aspect, J. Dalibard, and G. Roger, *Phys. Rev. Lett.* **49**, 1804 (1982).
9. A. Zeilinger, *Phys. Lett. A* **118**, 1 (1986).
10. P. Kwiat, P. Eberhard, A. Steinberg, and R. Chiao, *Phys. Rev. A* **49**, 3209 (1994). S. Huelga, M. Ferrero, and E. Santos, *Europhys. Lett.* **27**, 181 (1994). M. Freyberger, P. Aravind, M. Horne, and A. Shimony, *Phys. Rev. A* **53**, 1232 (1996).
11. E. Fry, T. Walther, and S. Li, *Phys. Rev. A* **52**, 4381 (1995).
12. T. Walther and E. Fry, *Z. Naturforsch.* (1997), proceedings of the workshop in honor of E.C.G. Sudarshan.
13. E. Fry and T. Walther, in *Experimental Metaphysics - Quantum Mechanical Studies for Abner Shimony, Vol. I*, R. Cohen and J. Stachel, eds. (Kluwer Academic, Dordrecht, 1996).

WHAT IS ACHIEVED BY DECOHERENCE?

H. D. Zeh

Institut für Theoretische Physik, Universität Heidelberg
Philosophenweg 19
D-69120 Heidelberg, Germany

A short critical review of the concept of decoherence, its consequences, and its possible implications for the interpretation of quantum mechanics is given.

Key words: decoherence, classical limit, superselection, quantum events, quantum causality, quantum measurement.

1. WHAT IS *MEANT* BY DECOHERENCE?

There seems to be some confusion in the literature not only on what may actually be achieved by decoherence, but also on how this concept has to be defined. I will here "consistently" use it in terms of wave functions (not "histories"), since state vectors represent the established kinematical concept of quantum theory, while events (of which histories are often assembled) will turn out to be derivable in a certain sense in terms of wave functions just by means of decoherence.

Therefore, by decoherence I mean the practically irreversible and practically unavoidable (in general approximate) disappearance of certain phase relations from the states of local systems by interaction with their environment according to the Schrödinger equation. Since phase relations cannot absolutely disappear in a unitary evolution, this disappearance can only represent *delocalization*, which means that the phases are not "there" any more, neither in the system nor in the environment, although they still exist in the total state that describes both of them in accordance with quantum nonlocality.

This can also be described as the dynamical disappearance of non-diagonal elements from the subsystem density matrices in a certain basis.

M. Ferrero and A. van der Merwe (eds.),
New Developments on Fundamental Problems in Quantum Physics, 441–451.

However, the concept of a density matrix would then first have to be defined (or justified) in terms of the conceptually presumed wave functions. This does not present a problem "for all practical purposes" (FAPP), that is, when using the "rules" of quantum theory for calculating probabilities for outcomes of measurements, for example in the form of expectation values

$$\langle A_{\rm local}\rangle = Trace[A_{\rm local}\rho_{\rm total}] = Trace[A_{\rm local}\rho_{\rm local}] \qquad (1)$$

in a Hilbert space $\mathcal{H}_{\rm total} = \mathcal{H}_{\rm local} \bigotimes \mathcal{H}_{env}$, where $\rho_{\rm local} = Trace_{env}[\rho_{\rm total}]$, while $\rho_{\rm total} = |\psi_{total}\rangle\langle\psi_{total}|$ may represent a wave function. Obviously, this procedure FAPP makes essential use of the probability interpretation (albeit without using any assumptions about the conceptual nature of the elements of the corresponding statistical ensembles, or on how they may possibly arise dynamically). Conclusions derived by means of the density matrix should therefore not be used to justify the probability interpretation itself in order to avoid circular arguments.

Nonlocal phase relations are thus equivalent to entangled states of spatially separated systems (quantum nonlocality). They can be conveniently described by their Schmidt canonical form

$$\psi_{\rm total} = \sum_n \sqrt{p_n}\phi_n^{\rm local}\Phi_n^{env} \quad , \qquad (2)$$

which is generically unique, and directly corresponds to the diagonal representation of the density matrix for both subsystems, e.g.,

$$\rho_{\rm local} = \sum_n |\phi_n^{\rm local}\rangle p_n \langle\phi_n^{\rm local}| \quad . \qquad (3)$$

This density matrix has a form *as though* it represented an ensemble of wave functions ϕ_n^{local} with respective probabilities p_n. Equation (2) demonstrates, however, that it does not. This confusion between proper and "improper" mixtures has given rise to the most frequent misinterpretation of decoherence as leading to ensembles, and thus as deriving the collapse of the wave function as a stochastic process. Such an unjustified replacement of improper by proper mixtures is clearly part of the *quantum state diffusion model* (insofar as it goes beyond decoherence), and it is similarly used when "selecting" *consistent histories* from superpositions of histories (thus not from ensembles). Any subsequent restriction "FAPP only" will here solely refer to this "as though" — not to the arising entanglement with the environment or its dynamical consequences for the corresponding formal density matrix. It does neither mean "wrong" nor "useless".

For time-dependent states $\psi(t)$, their entanglement (2) must in general also depend on time. It may thus be entirely *caused* by interaction, for example according to von Neumann's (unitary part of) ideal measurements

$$(\sum_n c_n\phi_n^{\rm local})\Phi_0^{env} \to \sum_n c_n\phi_n^{\rm local}\Phi_n^{env} \quad . \qquad (4)$$

This measurement-like process describes *pure decoherence* of the system in neglecting any re-action (or "perturbation" by the environment). Decoherence has therefore also been called "continuous measurement" (continuous in the sense of a permanent and unavoidable coupling).

The dynamical evolution described by Eq. (4) contains an obvious *arrow of time*, based on the belief that entanglement must be *retarded*, that is, have causes in the past ("quantum causality"). Although entanglement is not a *statistical* correlation (not based on incomplete information), Eq. (4) is analogous to Boltzmann's assumption of molecular chaos, since it neglects any *re*coherence (any local effects of the arising quantum correlations in the future). In both cases, the existence of fluctuations demonstrates that this time arrow is not a law, but a fact in accordance with time-symmetric laws.

Genuine measurements (for example, of the passage of a particle through a slit of an interference device) are known to destroy interference for trivial reasons: the events following the observed passages may be counted separately, and may then simply be added. However, this is not required for decoherence, since (a) there need not be definite passages, and (b) the state of the environment need not correspond to a controllable "pointer state", from which information about the passage could be retrieved. For this reason one should rather speak of a measurement-*like* interaction with the environment.

Decoherence is thus a normal consequence of interacting quantum mechanical systems. It can hardly be denied to occur — but it cannot explain anything that could not have been explained before. Remarkable is only its quantitative (realistic) aspect that seems to have been overlooked for a long time. Entanglement is the norm — not the exception —, and it must have important consequences. In particular, macroscopic systems are never approximately isolated, and must thus not be described by wave packets obeying a Schrödinger equation (not even when a WKB approximation applies). Furthermore, since a macroscopic system together with its environment is even more macroscopic, quantum theory can provide a consistent description only when applied to the universe as a whole. That means, one has in principle to assume $\psi_{\text{total}} = \psi_{\text{universe}}$.

2. ELEMENTARY SYSTEMS UNDER DECOHERENCE

In order to illustrate the consequences of realistically taking into account the environment of quantum systems, this section lists some important examples of decoherence. Because of the rich literature on the subject I will here refrain from citing individual papers, and instead refer only to our recent review [1], which contains a bibliography considerably exceeding the length of this article. Further contributions — in particular some impressive work by the Madrid group(s) — can be found in these proceedings.

2.1 Trajectories of Macroscopic Variables

Interfering paths of mass points are equivalent to spatial waves. In a two-slit experiment with "bullets" (dust particles or even large molecules), not only their passage through the slits, but the whole path would unavoidably be measured under *all* realistic circumstances. No superposition of different positions of such objects can "be there" and lead to interference in the probability distributions for subsequent local measurements. In this respect, classical objects resemble alpha-particles in a cloud chamber. The resulting entanglement leads to a density matrix for their center of mass motion *as though* it represented an ever-increasing ensemble of narrow wave packets following slightly stochastic trajectories. This is dynamically described by a master equation

$$i\frac{\partial\rho(x,x',t)}{\partial t} = \frac{1}{2m}\left(\frac{\partial^2}{\partial x'^2} - \frac{\partial^2}{\partial x^2}\right)\rho - i\Lambda(x-x')^2\rho \quad , \tag{5}$$

which (including its coefficient Λ) can thus be derived from a universal Schrödinger equation. One does not have to *postulate* a phenomenological semigroup to characterize open systems.

Since this situation has been sufficiently discussed in the literature, let me here only emphasize that the required "measurements" must indeed be practically irreversible in order to avoid the possibility of "quantum erasure" (restoration of interference). Such a restoration would require, however, that every single scattered particle were completely recovered. (Experiments demonstrating quantum erasure in certain microscopic systems do not represent restoration of the *whole* initial superposition.) Every classical phenomenon, even in "reversible" classical mechanics, is based on such irreversible quantum processes, with a permanent production of entropy that is macroscopically negligible but large in terms of bits.

2.2 Molecular Configurations and Robust States

Chiral molecules, such as sugar, represent another simple example of systems under decoherence. They are described by wave functions, but in contrast to the analogous spin-3 state of the ammonia molecule, for example, not by energy eigenstates. The reason is that it is their chirality (not their parity) which is continuously "measured" by the scattering of air molecules (for sugar under normal conditions on a decoherence time scale of the order $10^{-9}sec$). The measurement of parity in sugar, or the preparation of (very short-lived) sugar molecules in their energy eigenstates, is therefore practically excluded, since it would require an even stronger coupling to the measurement device. (The molecule PH_3, discussed at this conference by Gonzalo, forms an amazing intermediate situation, with parity eigenstates expected to exist under exceptional environmental conditions.)

A further dynamical consequence then holds that each individual molecule in a bag of sugar must retain its chirality, while a parity state — if it had come into existence in a mysterious or expensive way — would almost immediately "collapse" into a local mixture of both chiralities with equal probabilities. Parity is thus not conserved for sugar molecules. The resulting mixture would also be diagonal in the parity representation if the diagonal elements were *exactly* equal. However, every *actually* resulting value of chirality would be permanent, that is, it would always be confirmed when measured again.

This dynamical "robustness" of the chiral *wave function* seems to characterize what we call a "real" (in the operational sense) or "classical" property — just like the spot on a photographic plate, or any other "pointer state" of a measurement device. It also seems to be essential for the physical realization of memory (such as in DNA, brains or computers — with the exception of quantum computers, which are extremely vulnerable to decoherence, as was discussed here by Ekert). Robustness is thereby compatible with a (regular) time dependence, as exemplified by the wave packet describing the center of mass motion of a bullet.

Chemists are used to describing the motion of nuclei in large molecules classically (for example by rigid configurations), while representing the electrons by wave functions. In general, they attribute this asymmetry to a Born-Oppenheimer approximation. This argument is definitely insufficient, since a straight-forward application of this approximation to molecules would lead to energy (and angular momentum) eigenstates, with vibrational and rotational spectra as known for the hydrogen (or ammonia) molecule. The same insufficient argument is now found in *quantum gravity*, where it is used to justify classical spacetime by employing a BO approximation with respect to the Planck mass (see Sect. 2.4).

Pseudo-classical behaviour can in both cases be explained by means of decoherence again: the positions of nuclei are permanently measured by scattering molecules or photons. But why only the nuclei (or ions), and why not even they in very small molecules? The answer is quantitative and based on a delicate balance between internal dynamics and interaction with the environment, whereby the density of states plays an essential role. (Much work on details remains to be done!) Depending on the quantitative situation, one will either obtain an approximately unitary evolution, a master equation (with a time asymmetry arising from quantum causality), or complete freezing of the motion (quantum Zeno effect). The situation becomes simple again only for a free nucleus, which is described by Eq. (5).

2.3 Charge Superselection

Gauss' law in the form $q = \int \mathbf{E} \cdot d\mathbf{S}/4\pi$ tells us that a local charge is correlated with its Coulomb field at any distance. It is a matter of taste whether this correlation is regarded as kinematics, or as *caused* in the form of the retarded field resulting from the charge in the past. Conceived quantum superpositions of different charge values,

$$\sum_q c_q \psi_q^{total} = \sum_q c_q \chi_q \Psi_q\{field\} = \sum_q c_q \chi_q \prod_r \Psi_q\{field(r)\} \quad , \qquad (6)$$

would therefore be nonlocal. Here, χ_q is the local charge state, $\Psi_q\{field\} = \prod_r \Psi_q\{field(r)\}$ a pseudo-classical field functional representing its Coulomb field, symbolically written as a direct product of quantum states on spheres with radius r. The local charge system itself (possibly including its field within a small sphere) would then be described by a density matrix of the form

$$\rho_{local} = \sum_q |\chi_q\rangle |c_q|^2 \langle\chi_q| \quad , \qquad (7)$$

since the field states outside r can be assumed to be mutually orthogonal for different charges q. The charge is therefore decohered by its own Coulomb field, and no superselection rule has to be *postulated*.

While this result is satisfactory for the theory, a more practical question is at what distance, and on what time scale, two compensating charges (or a charge in a superposition of different positions) are decohered (here possibly in a reversible manner) by the resulting dipole field. Theoretical and experimental contributions to an answer can be found in these proceedings (cf. Sols or Hasselbach). Notice that decoherence will thus allow one to distinguish between quantum field theory and action at a distance (where decoherence would occur only after the field has reached and changed the state of absorbing matter).

The gravitational field caused by any mass is analogous to the Coulomb field of a charge. Superpositions of different masses should therefore be decohered by the quantum state of spatial curvature. However, there is no elementary mass, and nobody has as yet reliably estimated at what mass difference decoherence by the correlated quantum state of curvature becomes essential. Quite obviously, some superposition of different mass-energy remains dynamically relevant in the form of (observed) time-dependent quantum states of local systems.

2.4 Classical Fields and Gravity

Not only the quantum states of charged particles are decohered by their fields, also the quantum field states are in general decohered by source particles on which they react. *Coherent states* (which represent classical fields) have in this situation been shown to be robust in a similar way as chiral molecules or wave packets for the positions of macroscopic objects. This is obviously the reason why no superpositions of macroscopically different "mean fields," or different vacua (as they may arise through spontaneous symmetry breaking), have ever been observed.

These arguments must as well apply to quantum gravity. One does not have to know its precise form (for example, after its expected unification with other forces) in order to conclude the existence of entangled superpositions of matter and geometry (as far as this distinction remains valid). Therefore, the classical appearance of spacetime geometry, with its lightcone structure as presumed in conventional quantum field theory, is no reason not to quantize gravity, since it should be explained by decoherence in the same way as the classical appearance of a bullet or an electromagnetic field. The resulting density matrix (functional) for the gravitational tensor field must be expected to behave as though it represented a statistical mixture of different classical curvature states (to which the observer is correlated — see Sect. 4). The beauty of Einstein's theory can hardly be ranked so much higher than that of Maxwell's in order to justify its exemption from a well established and general principle of physics (in particular, as this would lead to incompatibilities with the uncertainty relations for matter — known from the Bohr-Einstein debate).

The entropy characterizing a black hole (or the thermal field of an accelerating Unruh detector) is thus due to a similar "local perspective" of a strongly entangled subsystem as the entropy produced according to the master equation (5) for a mass point. An event horizon need not be different from any other pseudo-classical property, and even the disappearance of information behind a classical horizon would not describe a *real* collapse. For example, a succession of simultaneities carrying global states which contain an Unruh detector along an external world line may be chosen to remain all entirely outside the horizon (as described by the Schwarzschild time coordinate).

2.5 Quantum Jumps

Exponential decay of excited states is usually regarded as the standard manifestation of a fundamental quantum indeterminism. Stochastic decay into a single channel would in fact lead to an exact exponentially decreasing non-decay probability. However, the Schrödinger equation for a mass point in a potential barrier, for example, leads to an *approximately* exponential time-dependence *of the wave function*. It represents the *superposition* (not an ensemble) of different decay times. The corresponding interference is incompatible with exact exponential decay even in free space, and it has been confirmed in reflecting cavities as "coherent state vector revival".

On the other hand, if the decay fragments interact strongly with the surrounding matter, any interference between "decayed" and "not yet decayed" must practically irreversibly vanish on a very short decoherence time scale at every moment of time. The corresponding apparent quantum jumps at almost discrete times are observed as "clicks" of a Geiger counter (in analogy to pseudo-classical local "spots" on a photographic plate). They have also been seen with individual atoms coupled to laser fields as a sudden appearance and disappearance of "dark periods". For electrons tunneling from the tip of a metal needle, decoherence between different emission times may explain the short longitudinal coherence lengths reported at this conference by Hasselbach (while the dipole moments discussed in Sect. 2.3 are in these experiments represented by the *transversal* separation of the interfering paths). A decaying system strongly interacting with its environment is thus effectively described by a stochastic process with discrete decay times, and the resulting time dependence will be almost exactly exponential (as long as quantum causality governs decoherence).

It seems that this situation of continuously monitored decay has led to the myth of quantum theory as a stochastic theory that describes fundamental discontinuous quantum *events*. For example, Bohr formulated in 1928 that "the essence (of quantum theory) may be expressed in the so-called quantum postulate, which attributes *to any atomic process* an essential discontinuity, or rather individuality ..." (my italics). If this were correct, we would not even be able to describe lasers. Heisenberg and Pauli similarly emphasized that their preference for matrix mechanics originated in its (as it now seems misleading) superiority in describing discontinuities. According to the Schrödinger equation, the underlying entanglement processes occur smoothly — but recall now that all conclusions regarding ensembles are still based FAPP on the yet unspecified probability interpretation (1)!

3. CONSEQUENCES FOR THE *MOTIVATION* OF VARIOUS INTERPRETATIONS OF QUANTUM THEORY

How has the *fundamental* probability interpretation, used in Eq. (1), then to be understood? And where precisely does it have to be applied in view of the problem of where to position the Heisenberg cut? The greatly differing opinions of physicists on this point are surprising, and proof that the answer does not matter FAPP. Obviously, decoherence cannot give an answer either, although it may help to recognize some interpretations as being essentially motivated by prejudice or tradition.

Many seem to believe, in the tradition of Heisenberg, that Eq. (1)

describes probabilities for *classical* quantities to "enter reality". However, we have seen that these classical quantities can be described by robust wave packets (in a sense which comes close to Schrödinger's original attempts, but overcomes the problem of their dispersion by means of decoherence). It is then not at all surprising that classical concepts, if nonetheless assumed to apply, have to be restricted either in validity or in their observability in order to avoid contradictions. There is a rich vocabulary for de- or circumscribing the first possibility (uncertainty, complementarity, new logic, negative probabilities, consistency etc.). Those of these restrictions which are well defined can easily be *explained* in terms of wave packets (for example, by means of the Fourier theorem). The second possibility is represented by Bohm's "surrealistic trajectories". As a reflection on some contributions to this conference let me also emphasize that it appears useless and even quite misleading to describe a certain subset of experiments within classical concepts while simply neglecting the crucial rest.

A second class of interpretations suggests a description of reality in terms of yet unknown ("hidden") variables. However, the evidence related to Bell's inequalities clearly demonstrates (within limitations which must hold for *all* empirical proof — as known at least since David Hume, and as some physicists now seem to be rediscovering in detail) that these new variables would have to contain precisely the same nonlocality as entangled wave functions on configuration space.

Then why reject the wave function itself as describing "real" physical states? I think that this natural position is now strongly supported by the beautiful experiments with individual atoms in cavities (Schrödinger cat states, quantum engineering, phase space tomography etc.), which were discussed at this conference by several speakers. Thus, in a third class of interpretations of the fundamental probabilities one considers jumps between *wave functions* (usually described by a "stochastic Schrödinger equation"). However, why should the fundamental processes represented by (1) then occur precisely where they *appear* to occur and are readily described as decoherence?

4. CONSEQUENCES OF DECOHERENCE FOR THE INTERPRETATION OF QUANTUM MEASUREMENTS

The probabilities described by Eq. (1) are usually understood as referring to *measurements*. In many situations we may be satisfied just with entangled states, but individual measurements have definite outcomes, while von Neumann's or any similar unitary interaction leads to their superposition. This contrast defines what is usually called the *measurement problem*. John Bell insisted that we should not speak of measurements, but that would not resolve the dilemma. He obviously meant that we should treat measurements (and the occurrence of events?) as a normal physical process. That is precisely what many people are trying to do — but in terms of which fundamental kinematical concepts?

When describing macroscopic objects (such as measurement and registration devices) FAPP by the apparent ensembles of wave functions resulting through decoherence, we have in fact applied the probability interpretation (1) to them. Completion of a measurement for this purpose thus requires to include the reading of the pointer state. The corresponding chain of interactions would read

$$\begin{aligned} &\left(\sum_n c_n\phi_n\right)\psi_0^{app}\chi_0^{env}\Phi_0^{obs} \stackrel{\text{meas.}}{\longrightarrow} \left(\sum_n c_n\phi_n\psi_n^{app}\right)\chi_0^{env}\Phi_0^{obs} \\ &\stackrel{\text{decoh.}}{\longrightarrow} \left(\sum_n \phi_n\psi_n^{app}\chi_n^{env}\right)\Phi_0^{obs} \stackrel{\text{obs.}}{\longrightarrow} \phi_{n_0}\psi_{n_0}^{app}\chi_{n_0}^{env}\Phi_{n_0}^{obs} \end{aligned} \tag{8}$$

with probability $|c_{n_0}|^2$, where the last step is itself a long and complex chain. Only when it has been completed is the assumption of a collapse of the wave function really empirically indicated. Probabilities can then only be understood as representing frequencies $f(n)$ in the results $n_{01}, n_{02}, \ldots, n_{0N}$ of long series of N equivalent measurements, described by *one* final wave function containing an observer state $\Phi^{obs}_{n_{01},n_{02},\ldots,n_{0N}}$ (cf. Mittelstaedt's contribution to this volume).

If a real collapse occurred before the onset of decoherence, it would almost certainly have directly observable consequences. If it occurred later in the chain of interactions, decoherence could remain essential for the classical appearance of the world. This situation would then lead to a "partial Everett interpretation", in which certain components of the wave function, although they existed, would appear to be absent to an appropriate observer state. If decoherence itself practically triggered a collapse (similar to the GRW model), it would again be dangerously close to observation, since the environmental situation may be altered to some extent, while fundamental dynamical terms had to be fixed.

Thus recall von Neumann's motivation for the collapse: his aim was to re-establish a psycho-physical parallelism based on definite states $\Phi^{obs}_{n_0}$ of the observer (by whatever system an observer may be physically represented). However, there *is* a definite state Φ^{obs}_n in *each* robust component of a global superposition $\sum c_n\phi_n\psi_n^{app}\chi_n^{env}\Phi_n^{obs}$. Each component behaves then dynamically *as though* it represented the complete and only world (even though this behaviour is here derived from the assumption that it does *not*).

These branching robust components describe microscopic histories which do not *individually* determine their past. The essential reason for the dynamical independence (robustness) of the arising branches in their future (the absence of recoherence) is not simply the linearity of the Schrödinger equation, but paradoxically the very same retarded nature of quantum entanglement (quantum causality — Eq. (4), which seems to be responsible for the consistency (hence existence) of documents about *macroscopic* history (the "fixed past").

According to this picture it is the apparently observed quasi-classical world that exists only FAPP (*viz.*, with respect to each Φ_n^{obs} and his or her "friends" $\Phi_n^{obs'}$). The choice between the Everett interpretation and a vaguely located collapse thus remains presently a matter of taste.

Acknowledgement. I wish to thank Erich Joos and Claus Kiefer for their comments.

REFERENCES

1. D. Giulini, E. Joos, C. Kiefer, J. Kupsch, I.-O. Stamatescu, and H.D. Zeh, *Decoherence and the Appearance of a Classical World in Quantum Theory* (Springer, Berlin, 1996).

AUTHOR INDEX

Fundamental Theories of Physics

Series Editor: Alwyn van der Merwe, *University of Denver, USA*

1. M. Sachs: *General Relativity and Matter.* A Spinor Field Theory from Fermis to Light-Years. With a Foreword by C. Kilmister. 1982 ISBN 90-277-1381-2
2. G.H. Duffey: *A Development of Quantum Mechanics.* Based on Symmetry Considerations. 1985 ISBN 90-277-1587-4
3. S. Diner, D. Fargue, G. Lochak and F. Selleri (eds.): *The Wave-Particle Dualism.* A Tribute to Louis de Broglie on his 90th Birthday. 1984 ISBN 90-277-1664-1
4. E. Prugovečki: *Stochastic Quantum Mechanics and Quantum Spacetime.* A Consistent Unification of Relativity and Quantum Theory based on Stochastic Spaces. 1984; 2nd printing 1986 ISBN 90-277-1617-X
5. D. Hestenes and G. Sobczyk: *Clifford Algebra to Geometric Calculus.* A Unified Language for Mathematics and Physics. 1984 ISBN 90-277-1673-0; Pb (1987) 90-277-2561-6
6. P. Exner: *Open Quantum Systems and Feynman Integrals.* 1985 ISBN 90-277-1678-1
7. L. Mayants: *The Enigma of Probability and Physics.* 1984 ISBN 90-277-1674-9
8. E. Tocaci: *Relativistic Mechanics, Time and Inertia.* Translated from Romanian. Edited and with a Foreword by C.W. Kilmister. 1985 ISBN 90-277-1769-9
9. B. Bertotti, F. de Felice and A. Pascolini (eds.): *General Relativity and Gravitation.* Proceedings of the 10th International Conference (Padova, Italy, 1983). 1984 ISBN 90-277-1819-9
10. G. Tarozzi and A. van der Merwe (eds.): *Open Questions in Quantum Physics.* 1985 ISBN 90-277-1853-9
11. J.V. Narlikar and T. Padmanabhan: *Gravity, Gauge Theories and Quantum Cosmology.* 1986 ISBN 90-277-1948-9
12. G.S. Asanov: *Finsler Geometry, Relativity and Gauge Theories.* 1985 ISBN 90-277-1960-8
13. K. Namsrai: *Nonlocal Quantum Field Theory and Stochastic Quantum Mechanics.* 1986 ISBN 90-277-2001-0
14. C. Ray Smith and W.T. Grandy, Jr. (eds.): *Maximum-Entropy and Bayesian Methods in Inverse Problems.* Proceedings of the 1st and 2nd International Workshop (Laramie, Wyoming, USA). 1985 ISBN 90-277-2074-6
15. D. Hestenes: *New Foundations for Classical Mechanics.* 1986 ISBN 90-277-2090-8; Pb (1987) 90-277-2526-8
16. S.J. Prokhovnik: *Light in Einstein's Universe.* The Role of Energy in Cosmology and Relativity. 1985 ISBN 90-277-2093-2
17. Y.S. Kim and M.E. Noz: *Theory and Applications of the Poincaré Group.* 1986 ISBN 90-277-2141-6
18. M. Sachs: *Quantum Mechanics from General Relativity.* An Approximation for a Theory of Inertia. 1986 ISBN 90-277-2247-1
19. W.T. Grandy, Jr.: *Foundations of Statistical Mechanics.* Vol. I: *Equilibrium Theory.* 1987 ISBN 90-277-2489-X
20. H.-H von Borzeszkowski and H.-J. Treder: *The Meaning of Quantum Gravity.* 1988 ISBN 90-277-2518-7
21. C. Ray Smith and G.J. Erickson (eds.): *Maximum-Entropy and Bayesian Spectral Analysis and Estimation Problems.* Proceedings of the 3rd International Workshop (Laramie, Wyoming, USA, 1983). 1987 ISBN 90-277-2579-9

22. A.O. Barut and A. van der Merwe (eds.): *Selected Scientific Papers of Alfred Landé. [1888-1975]*. 1988 ISBN 90-277-2594-2
23. W.T. Grandy, Jr.: *Foundations of Statistical Mechanics.* Vol. II: *Nonequilibrium Phenomena.* 1988 ISBN 90-277-2649-3
24. E.I. Bitsakis and C.A. Nicolaides (eds.): *The Concept of Probability.* Proceedings of the Delphi Conference (Delphi, Greece, 1987). 1989 ISBN 90-277-2679-5
25. A. van der Merwe, F. Selleri and G. Tarozzi (eds.): *Microphysical Reality and Quantum Formalism, Vol. 1.* Proceedings of the International Conference (Urbino, Italy, 1985). 1988 ISBN 90-277-2683-3
26. A. van der Merwe, F. Selleri and G. Tarozzi (eds.): *Microphysical Reality and Quantum Formalism, Vol. 2.* Proceedings of the International Conference (Urbino, Italy, 1985). 1988 ISBN 90-277-2684-1
27. I.D. Novikov and V.P. Frolov: *Physics of Black Holes.* 1989 ISBN 90-277-2685-X
28. G. Tarozzi and A. van der Merwe (eds.): *The Nature of Quantum Paradoxes.* Italian Studies in the Foundations and Philosophy of Modern Physics. 1988 ISBN 90-277-2703-1
29. B.R. Iyer, N. Mukunda and C.V. Vishveshwara (eds.): *Gravitation, Gauge Theories and the Early Universe.* 1989 ISBN 90-277-2710-4
30. H. Mark and L. Wood (eds.): *Energy in Physics, War and Peace.* A Festschrift celebrating Edward Teller's 80th Birthday. 1988 ISBN 90-277-2775-9
31. G.J. Erickson and C.R. Smith (eds.): *Maximum-Entropy and Bayesian Methods in Science and Engineering.* Vol. I: *Foundations.* 1988 ISBN 90-277-2793-7
32. G.J. Erickson and C.R. Smith (eds.): *Maximum-Entropy and Bayesian Methods in Science and Engineering.* Vol. II: *Applications.* 1988 ISBN 90-277-2794-5
33. M.E. Noz and Y.S. Kim (eds.): *Special Relativity and Quantum Theory.* A Collection of Papers on the Poincaré Group. 1988 ISBN 90-277-2799-6
34. I.Yu. Kobzarev and Yu.I. Manin: *Elementary Particles. Mathematics, Physics and Philosophy.* 1989 ISBN 0-7923-0098-X
35. F. Selleri: *Quantum Paradoxes and Physical Reality.* 1990 ISBN 0-7923-0253-2
36. J. Skilling (ed.): *Maximum-Entropy and Bayesian Methods.* Proceedings of the 8th International Workshop (Cambridge, UK, 1988). 1989 ISBN 0-7923-0224-9
37. M. Kafatos (ed.): *Bell's Theorem, Quantum Theory and Conceptions of the Universe.* 1989 ISBN 0-7923-0496-9
38. Yu.A. Izyumov and V.N. Syromyatnikov: *Phase Transitions and Crystal Symmetry.* 1990 ISBN 0-7923-0542-6
39. P.F. Fougère (ed.): *Maximum-Entropy and Bayesian Methods.* Proceedings of the 9th International Workshop (Dartmouth, Massachusetts, USA, 1989). 1990 ISBN 0-7923-0928-6
40. L. de Broglie: *Heisenberg's Uncertainties and the Probabilistic Interpretation of Wave Mechanics.* With Critical Notes of the Author. 1990 ISBN 0-7923-0929-4
41. W.T. Grandy, Jr.: *Relativistic Quantum Mechanics of Leptons and Fields.* 1991 ISBN 0-7923-1049-7
42. Yu.L. Klimontovich: *Turbulent Motion and the Structure of Chaos.* A New Approach to the Statistical Theory of Open Systems. 1991 ISBN 0-7923-1114-0

Fundamental Theories of Physics

43. W.T. Grandy, Jr. and L.H. Schick (eds.): *Maximum-Entropy and Bayesian Methods.* Proceedings of the 10th International Workshop (Laramie, Wyoming, USA, 1990). 1991 ISBN 0-7923-1140-X
44. P.Pták and S. Pulmannová: *Orthomodular Structures as Quantum Logics.* Intrinsic Properties, State Space and Probabilistic Topics. 1991 ISBN 0-7923-1207-4
45. D. Hestenes and A. Weingartshofer (eds.): *The Electron.* New Theory and Experiment. 1991 ISBN 0-7923-1356-9
46. P.P.J.M. Schram: *Kinetic Theory of Gases and Plasmas.* 1991 ISBN 0-7923-1392-5
47. A. Micali, R. Boudet and J. Helmstetter (eds.): *Clifford Algebras and their Applications in Mathematical Physics.* 1992 ISBN 0-7923-1623-1
48. E. Prugovečki: *Quantum Geometry.* A Framework for Quantum General Relativity. 1992 ISBN 0-7923-1640-1
49. M.H. Mac Gregor: *The Enigmatic Electron.* 1992 ISBN 0-7923-1982-6
50. C.R. Smith, G.J. Erickson and P.O. Neudorfer (eds.): *Maximum Entropy and Bayesian Methods.* Proceedings of the 11th International Workshop (Seattle, 1991). 1993 ISBN 0-7923-2031-X
51. D.J. Hoekzema: *The Quantum Labyrinth.* 1993 ISBN 0-7923-2066-2
52. Z. Oziewicz, B. Jancewicz and A. Borowiec (eds.): *Spinors, Twistors, Clifford Algebras and Quantum Deformations.* Proceedings of the Second Max Born Symposium (Wrocław, Poland, 1992). 1993 ISBN 0-7923-2251-7
53. A. Mohammad-Djafari and G. Demoment (eds.): *Maximum Entropy and Bayesian Methods.* Proceedings of the 12th International Workshop (Paris, France, 1992). 1993 ISBN 0-7923-2280-0
54. M. Riesz: *Clifford Numbers and Spinors* with Riesz' Private Lectures to E. Folke Bolinder and a Historical Review by Pertti Lounesto. E.F. Bolinder and P. Lounesto (eds.). 1993 ISBN 0-7923-2299-1
55. F. Brackx, R. Delanghe and H. Serras (eds.): *Clifford Algebras and their Applications in Mathematical Physics.* Proceedings of the Third Conference (Deinze, 1993) 1993 ISBN 0-7923-2347-5
56. J.R. Fanchi: *Parametrized Relativistic Quantum Theory.* 1993 ISBN 0-7923-2376-9
57. A. Peres: *Quantum Theory: Concepts and Methods.* 1993 ISBN 0-7923-2549-4
58. P.L. Antonelli, R.S. Ingarden and M. Matsumoto: *The Theory of Sprays and Finsler Spaces with Applications in Physics and Biology.* 1993 ISBN 0-7923-2577-X
59. R. Miron and M. Anastasiei: *The Geometry of Lagrange Spaces: Theory and Applications.* 1994 ISBN 0-7923-2591-5
60. G. Adomian: *Solving Frontier Problems of Physics: The Decomposition Method.* 1994 ISBN 0-7923-2644-X
61 B.S. Kerner and V.V. Osipov: *Autosolitons.* A New Approach to Problems of Self-Organization and Turbulence. 1994 ISBN 0-7923-2816-7
62. G.R. Heidbreder (ed.): *Maximum Entropy and Bayesian Methods.* Proceedings of the 13th International Workshop (Santa Barbara, USA, 1993) 1996 ISBN 0-7923-2851-5
63. J. Peřina, Z. Hradil and B. Jurčo: *Quantum Optics and Fundamentals of Physics.* 1994 ISBN 0-7923-3000-5
64. M. Evans and J.-P. Vigier: *The Enigmatic Photon.* Volume 1: The Field $\boldsymbol{B}^{(3)}$. 1994 ISBN 0-7923-3049-8
65. C.K. Raju: *Time: Towards a Constistent Theory.* 1994 ISBN 0-7923-3103-6
66. A.K.T. Assis: *Weber's Electrodynamics.* 1994 ISBN 0-7923-3137-0

Fundamental Theories of Physics

67. Yu. L. Klimontovich: *Statistical Theory of Open Systems.* Volume 1: A Unified Approach to Kinetic Description of Processes in Active Systems. 1995
ISBN 0-7923-3199-0; Pb: ISBN 0-7923-3242-3
68. M. Evans and J.-P. Vigier: *The Enigmatic Photon.* Volume 2: Non-Abelian Electrodynamics. 1995 ISBN 0-7923-3288-1
69. G. Esposito: *Complex General Relativity.* 1995 ISBN 0-7923-3340-3
70. J. Skilling and S. Sibisi (eds.): *Maximum Entropy and Bayesian Methods.* Proceedings of the Fourteenth International Workshop on Maximum Entropy and Bayesian Methods. 1996 ISBN 0-7923-3452-3
71. C. Garola and A. Rossi (eds.): *The Foundations of Quantum Mechanics – Historical Analysis and Open Questions.* 1995 ISBN 0-7923-3480-9
72. A. Peres: *Quantum Theory: Concepts and Methods.* 1995 (see for hardback edition, Vol. 57) ISBN Pb 0-7923-3632-1
73. M. Ferrero and A. van der Merwe (eds.): *Fundamental Problems in Quantum Physics.* 1995 ISBN 0-7923-3670-4
74. F.E. Schroeck, Jr.: *Quantum Mechanics on Phase Space.* 1996 ISBN 0-7923-3794-8
75. L. de la Peña and A.M. Cetto: *The Quantum Dice.* An Introduction to Stochastic Electrodynamics. 1996 ISBN 0-7923-3818-9
76. P.L. Antonelli and R. Miron (eds.): *Lagrange and Finsler Geometry.* Applications to Physics and Biology. 1996 ISBN 0-7923-3873-1
77. M.W. Evans, J.-P. Vigier, S. Roy and S. Jeffers: *The Enigmatic Photon.* Volume 3: Theory and Practice of the $\boldsymbol{B}^{(3)}$ Field. 1996 ISBN 0-7923-4044-2
78. W.G.V. Rosser: *Interpretation of Classical Electromagnetism.* 1996
ISBN 0-7923-4187-2
79. K.M. Hanson and R.N. Silver (eds.): *Maximum Entropy and Bayesian Methods.* 1996
ISBN 0-7923-4311-5
80. S. Jeffers, S. Roy, J.-P. Vigier and G. Hunter (eds.): *The Present Status of the Quantum Theory of Light.* Proceedings of a Symposium in Honour of Jean-Pierre Vigier. 1997
ISBN 0-7923-4337-9
81. M. Ferrero and A. van der Merwe (eds.): *New Developments on Fundamental Problems in Quantum Physics.* 1997 ISBN 0-7923-4374-3
82. R. Miron: *The Geometry of Higher-Order Lagrange Spaces.* Applications to Mechanics and Physics. 1997 ISBN 0-7923-4393-X
83. T. Hakioğlu and A.S. Shumovsky (eds.): *Quantum Optics and the Spectroscopy of Solids.* Concepts and Advances. 1997 ISBN 0-7923-4414-6
84. A. Sitenko and V. Tartakovskii: *Theory of Nucleus.* Nuclear Structure and Nuclear Interaction. 1997 ISBN 0-7923-4423-5
85. G. Esposito, A.Yu. Kamenshchik and G. Pollifrone: *Euclidean Quantum Gravity on Manifolds with Boundary.* 1997 ISBN 0-7923-4472-3
86. R.S. Ingarden, A. Kossakowski and M. Ohya: *Information Dynamics and Open Systems.* Classical and Quantum Approach. 1997 ISBN 0-7923-4473-1
87. K. Nakamura: *Quantum versus Chaos.* Questions Emerging from Mesoscopic Cosmos. 1997 ISBN 0-7923-4557-6
88. B.R. Iyer and C.V. Vishveshwara (eds.): *Geometry, Fields and Cosmology.* Techniques and Applications. 1997 ISBN 0-7923-4725-0

KLUWER ACADEMIC PUBLISHERS – DORDRECHT / BOSTON / LONDON

GPSR Compliance

The European Union's (EU) General Product Safety Regulation (GPSR) is a set of rules that requires consumer products to be safe and our obligations to ensure this.

If you have any concerns about our products, you can contact us on ProductSafety@springernature.com

In case Publisher is established outside the EU, the EU authorized representative is:

Springer Nature Customer Service Center GmbH
Europaplatz 3
69115 Heidelberg, Germany

Batch number: 10809439

Printed by Printforce, the Netherlands